Every Coin Has Two Sides

Every Coin Has Two Sides: An Introduction to Causal Inference in Pharmaceutical Statistics introduces basic and advanced statistical inference methods and causal inference methods relevant to pharmaceutical statistics. This book distills seventy fundamental ideas and concepts—symbolized as gold coins, each with two sides—essential for mastering asymptotic statistics, causal inference, semiparametric statistics, and targeted learning. This book covers causal thinking that has increasingly become important in the planning, design, conduct, analysis, and interpretation of clinical studies. Progressing from basic statistical concepts to advanced targeted learning techniques, the book highlights the fusion of two cultures of statistical modelling. This book is suitable for graduate students in statistics, biostatistics, public health, and data science who are looking to pursue a career in the pharmaceutical industry, as well as for clinical statisticians and epidemiologists working in the pharmaceutical industry.

Key Features:

- Causal inference book for clinical statisticians in the pharmaceutical industry and graduate students
- Introduction to asymptotic statistics, causal inference, semiparametric statistics, and targeted learning
- Aligning with FDA and ICH guidance documents and covering different stages of clinical studies
- Seventy fundamental ideas and concepts—symbolized as gold coins, each with two sides

Yixin Fang is Director of Statistics and Senior Research Fellow at AbbVie Inc. He obtained his Ph.D. in Statistics from Columbia University and is an experienced statistician and data scientist who has a history of working in both the biopharmaceutical industry and academia. He is an elected Fellow of the American Statistical Association.

Chapman & Hall/CRC Biostatistics Series
Series Editors: Mark Chang, *Boston University, USA*

Biostatistics for Bioassay
Ann Yellowlees and Matthew Stephenson

Power and Sample Size in R
Catherine M. Crespi

R for Health Technology Assessment
Gianluca Baio, Howard Thom, and Petros Pechlivanoglou

Advanced Statistical Analytics for Health Data Science with SAS and R
Ding-Geng Chen and Jeffrey Wilson

Group Sequential and Adaptive Methods for Clinical Trials
Christopher Jennison and Bruce W. Turnbull

Machine Learning for Microbiome Statistics
Yinglin Xia and Jun Sun

Dose-exposure-response Modeling: Methods and Practical Implementation
Jixian Wang

Comparative Effectiveness and Personalized Medicine Research Using Real-World Data
Thomas P. A. Debray, Tri-Long Nguyen, and Robert W. Platt

Master Protocol Clinical Trials for Evidence Generation
Strategies, Designs, Operations, and Case Studies
Edited by Ruixiao Lu, Jingjing Ye, Chengxing (Cindy) Lu, and William Wang

Does This Treatment Cause That Outcome?
The Science of Estimating a Treatment Effect and Why It Matters
Stephen J. Ruberg

Empirical Likelihood Method in Survival Analysis
With R implementation, Second Edition
Mai Zhou

Bayesian Methods in Health Technology Assessment
Gianluca Baio

Every Coin Has Two Sides
An Introduction to Causal Inference in Pharmaceutical Statistics
Yixin Fang

Biomarkers in Drug Development
Statistical Methods and Applications
Guillaume Desachy and Gina D'Angelo

Medical Biostatistics
Fifth Edition
Abhaya Indrayan and Rajeev Kumar Malhotra

For more information about this series, please visit: HYPERLINK "http://www.routledge.com/Chapman--Hall-CRC-Biostatistics-Series/book-series/CHBIOSTATIS"www.routledge.com/Chapman--Hall-CRC-Biostatistics-Series/book-serie

Every Coin Has Two Sides

An Introduction to Causal Inference in Pharmaceutical Statistics

Yixin Fang

CRC Press
Taylor & Francis Group
Boca Raton London New York

CRC Press is an imprint of the
Taylor & Francis Group, an **informa** business
A CHAPMAN & HALL BOOK

Front cover image: grayjay/Shutterstock

First edition published 2027
by CRC Press
2385 NW Executive Center Drive, Suite 320, Boca Raton FL 33431

and by CRC Press
4 Park Square, Milton Park, Abingdon, Oxon, OX14 4RN

CRC Press is an imprint of Taylor & Francis Group, LLC

ISBN: 978-1-041-06300-1 (hbk)
ISBN: 978-1-041-06449-7 (pbk)
ISBN: 978-1-003-63546-8 (ebk)

DOI: 10.1201/9781003635468

Typeset in Latin Modern font
by KnowledgeWorks Global Ltd.

Publisher's note: This book has been prepared from camera-ready copy provided by the authors.

For my two kids, who are also interested in statistics

Contents

Preface

"Dao begets One, One begets Two, Two begets Three, Three begets all things." – Lao Zi

In Lao Zi's *Dao De Jing*, "two" symbolizes the duality of *yin* and *yang*. In the pharmaceutical industry, this notion of "two" is echoed in two-arm randomized controlled clinical trials (RCTs), widely regarded as the gold standard for evaluating the safety and efficacy of investigational new drugs. The two core techniques of an RCT, randomization and blinding, are often conceptually or verbally described by the simple act of "flipping a coin."

We can already observe three twos: the two arms of a trial, the two fundamental aspects of a drug—safety and efficacy, and the two core techniques used in an RCT—randomization and blinding. Naturally, the mental image we often use to describe the process of randomization—a coin flip—also reflects duality, with its two sides: heads and tails.

The phrase "every coin has two sides" reminds us that every situation can have two different perspectives or aspects, often contrasting or complementary to each other. Throughout my journey of learning and applying causal inference methods in pharmaceutical statistics, I have encountered many such "coins," revealing a deeper understanding of key concepts and methodology. In this book, I hope to share with you some "gold coins," representing some fundamental insights, that I believe could be valuable to you as well.

This book is organized into four parts: an introduction to asymptotic statistics, an introduction to causal inference, an introduction to semiparametric statistics, and an introduction to targeted learning. These form a table:

	Statistics	Causal Inference
Basic	I: Asymptotic Statistics	II: Causal Inference
Advanced	III: Semiparametric Statistics	IV: Targeted Learning

The table has two dimensions. The first dimension consists of two columns, one for statistics and the other for causal inference. The second dimension consists of two rows, one for foundational material and the other for more advanced topics. Readers are free to explore any of the four resulting series based on their interests, whether by row (basic or advanced) or by column (statistics or causal inference).

This book is intended for statisticians, data scientists, and epidemiologists working in the pharmaceutical industry, as well as for students aspiring to enter this field. It may also be suitable for graduate students in statistics or biostatistics programs. Recommended prerequisites include courses in Calculus, Linear Algebra, Introductory Probability, Introductory Statistics, and Theoretical Statistics.

Compared to other books on statistics and causal inference, this one offers a distinctive feature: it summarizes fundamental concepts and methods as "gold coins," each accompanied by self-contained mathematical explanations that can be understood without relying on external references. This feature is reflected in the title of the book. Like my previous book, *Causal Inference in Pharmaceutical Statistics* (2024), this book also focuses on causal inference in pharmaceutical statistics, but offers deeper coverage, broader scope, and clearer explanations.

The writing of this book was inspired by many of the monographs cited throughout its pages. Since books are no longer written with pencil and paper, I am deeply grateful for the modern tools that made this work possible: Overleaf for LaTeX typesetting, Google Scholar for literature search, and ChatGPT for grammar refinement.

I would like to take this opportunity to express my heartfelt gratitude to my family for their unwavering love and support; to my professors at the University of Science and Technology of China and Columbia University for their inspiring teaching and dedicated mentorship; and to my collaborators—Prof. Junhui Wang, Prof. Yuanjia Wang, Prof. Jinfeng Xu, Dr. Hana Lee, Dr. Mandy Jin, and many others—whose insights have greatly shaped my thinking. I am deeply grateful to Dr. Jie Chen, Dr. Mark Levenson, and Dr. Ivan Chan for their generous support in my career development; to Prof. Mark van der Laan and Dr. Susan Gruber for their exceptional training in targeted learning; and to my colleagues and leaders at AbbVie—including Hongwei Wang, the late Meijing Wu, Yabing Mai, Dai Feng, Zuoyi Zhang, Tianyu Zhan, Jane Zhang, and many others, and especially our functional vice presidents, Yihua Gu and Xun Chen—whose encouragement has been invaluable. I am profoundly grateful to my manager and mentor, Dr. Weili He, for her thoughtful guidance in my transition from academia to industry and for her continued support throughout my career.

Lastly, I am sincerely grateful to David Grubbs at Chapman & Hall/CRC Press for his kind assistance and support throughout the publishing process.

Part I

An Introduction to Asymptotic Statistics

1

The Number Two

1.1 Two, Dos, 2, II ...

You might argue that 2 (two) is just an ordinary number, the natural number following 1 and preceding 3. Then why is 2 so important and special that I even highlight it as the title of the first chapter?

In fact, 2 is a special number. Two is the only even prime number. Two represents the concept of duality such as up/down, left/right, positive/negative, life/death, male/female, on/off, true/false, and yes/no. It is the base of the binary system in computer science. It is associated with themes of balance, harmony, love, and faith. In statistics, the number two is almost everywhere, such as null and alternative hypotheses, bias–variance trade-off, parametrics and nonparametrics, frequentist and Bayesian, and Type I and Type II errors, along with many famous statistical results with two great names such as Gauss–Markov theorem, Neyman–Pearson lemma, Cramér–Rao lower bound, and Box–Cox transformation. In addition, every coin has two sides, and every statistician has two arms.

This book will cover essential topics, theoretical and applied, that clinical statisticians need to understand for effectively planning, designing, conducting, analyzing, and interpreting clinical studies—whether interventional or non-interventional.

Each topic in this book is linked to the number two. Because of their value, let us think of them as "gold coins"—each with two sides. It's time to dig deep and uncover these gold coins of knowledge.

1.2 The Most Useful Equation in Statistics

The most elegant equation in mathematics is Euler's identity:[1]

$$e^{\pi i} + 1 = 0.$$

As the greatest mathematician of the 18th century, Leonhard Euler (1707–1783) magically combined five fundamental mathematical constants: 0, 1, π, e (Euler's number), and i (the imaginary unit) into an elegant equation.

DOI: 10.1201/9781003635468-2

Mathematicians often use the term "elegant" to describe a mathematical solution, proof, or equation due to its simplicity, clarity, and insightful nature. Statisticians often use the term "useful" to describe a statistical model, concept, or method due to its wide impact; as George Box said, "all models are wrong, but some are useful."

Then, what is the most useful equation in statistics?

Gold Coin # 1. *The probability density function of a normal distribution, denoted as* $\mathcal{N}(\mu, \sigma^2)$, *is*

$$f(x) = \frac{1}{\sqrt{2\pi}\sigma} e^{-\frac{1}{2}\left(\frac{x-\mu}{\sigma}\right)^2},$$

which involves three 2s (one in the fraction, one beneath the square root, and one in the exponent), two parameters (mean μ *and standard deviation* σ*), two fundamental mathematical constants (*π *and* e*), and the idea of standardization,* $\frac{x-\mu}{\sigma}$.

The *normal distribution* is also called the *Gaussian distribution*, named after the greatest mathematician of the 19th century, Carl Friedrich Gauss (1777–1855).* Throughout this book, we will see why the expression of its density function is considered the most useful equation in statistics. The following are some of the reasons:

- The central limit theorem states that the sampling distribution of the sample mean will be normally distributed asymptotically;
- Most of the confidence interval estimates and p-values to be discussed in this book are based on the central limit theorem;
- The efficiency theory, which is the foundation for the causal inference methods to be discussed, is based on the asymptotic normality;
- The normal distribution is the building block for three other important distributions, the chi-squared, Student's t, and Fisher's F distributions.

*I learned *Stigler's Law of Eponymy* from a footnote in *The Lady Tasting Tea* (Salsburg, 2001; page 26). The law, proposed by statistician Stephen Stigler in 1980, states that no scientific discovery is named after its original discoverer. Ironically, the law exemplifies its own principle, as Stigler credited the idea to sociologist Robert K. Merton. A classic example of this law in statistics is the *normal distribution*. Although the distribution is widely known as the *Gaussian distribution*, it was first derived by Abraham de Moivre in 1733 as an approximation to the binomial distribution. Gauss later rediscovered and popularized it in the early 19th century through his work on measurement errors and the method of least squares. This law applies to many named distributions and discoveries discussed in this book, though they will not be identified individually each time.

1.3 Mean and Standard Deviation

A parametric distribution is one that can be completely described using a finite set of parameters. The normal distribution is fully characterized by two parameters: the *mean* μ and the *standard deviation* σ (or equivalently, the mean μ and the variance σ^2), making it a parametric distribution.

Let $X \sim \mathcal{N}(\mu, \sigma^2)$ denote that the random variable X is distributed with the distribution $\mathcal{N}(\mu, \sigma^2)$. Then, for any $a, b \in [-\infty, \infty]$, if $a < b$, the probability that X falls in (a, b) can be expressed as

$$\mathbb{P}(a < X < b) = \int_a^b \frac{1}{\sqrt{2\pi}\sigma} e^{-\frac{1}{2}\left(\frac{x-\mu}{\sigma}\right)^2} \, dx.$$

A special normal distribution is called the standard normal distribution, denoted as $\mathcal{N}(0, 1)$. In $\mathcal{N}(0, 1)$, the two parameters are 0 (the additive identity) and 1 (the multiplicative identity), which are the two most fundamental numbers in mathematics. The process of standardization can be expressed as

$$\text{If } X \sim \mathcal{N}(\mu, \sigma^2), \quad \text{then } Z = \frac{X - \mu}{\sigma} \sim \mathcal{N}(0, 1).$$

It can be shown that the probability density function of Z is

$$\phi(z) = \frac{1}{\sqrt{2\pi}} e^{-\frac{1}{2}z^2}.$$

In fact, it can be verified that $\phi(z)$ defines a valid probability density function by confirming that (see Exercise 1.1)

$$\int_{-\infty}^{\infty} \frac{1}{\sqrt{2\pi}} e^{-\frac{1}{2}z^2} \, dz = 1.$$

Furthermore, for any $a < b$,

$$\begin{aligned}
&\mathbb{P}(a < Z < b) \\
=&\mathbb{P}\left(a < \frac{X - \mu}{\sigma} < b\right) \\
=&\mathbb{P}\left(a\sigma + \mu < X < b\sigma + \mu\right) \\
=&\int_{a\sigma+\mu}^{b\sigma+\mu} \frac{1}{\sqrt{2\pi}\sigma} e^{-\frac{1}{2}\left(\frac{x-\mu}{\sigma}\right)^2} \, dx \\
=&\int_{a\sigma+\mu}^{b\sigma+\mu} \frac{1}{\sqrt{2\pi}} e^{-\frac{1}{2}\left(\frac{x-\mu}{\sigma}\right)^2} \, d\left(\frac{x-\mu}{\sigma}\right), \\
=&\int_a^b \frac{1}{\sqrt{2\pi}} e^{-\frac{1}{2}z^2} \, dz,
\end{aligned}$$

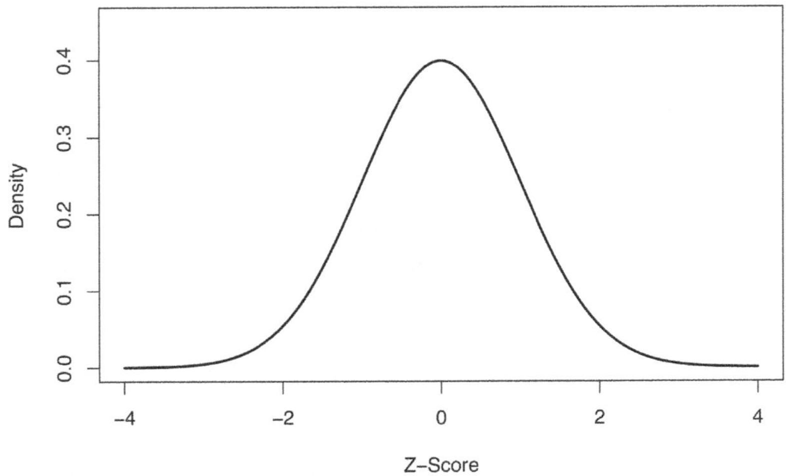

FIGURE 1.1
The density of $\mathcal{N}(0,1)$

where the penultimate equality highlights the role of the Jacobian determinant (here $1/\sigma$), and the final equality follows by substituting $z = (x - \mu)/\sigma$. $\square$

Due to standardization, it is unnecessary to create infinitely many normal tables for different values of μ and σ; a single table for the standard normal $\mathcal{N}(0,1)$ is sufficient.

Thus, the density of any normal distribution $\mathcal{N}(\mu, \sigma^2)$ can be understood once the density of the standard normal distribution $\mathcal{N}(0,1)$, which is a bell-shaped curve symmetric around 0 as shown in Figure 1.1, has been comprehended. In particular, after the percentiles of $\mathcal{N}(0,1)$ have been tabulated in the normal table, the percentiles of any $\mathcal{N}(\mu, \sigma^2)$ can be obtained from this table through standardization. For any $a < b$,

$$\mathbb{P}(a < X < b) = \mathbb{P}\left(\frac{a-\mu}{\sigma} < Z < \frac{b-\mu}{\sigma}\right) = \int_{\frac{a-\mu}{\sigma}}^{\frac{b-\mu}{\sigma}} \phi(z)\, dz.$$

For example, since $\mathbb{P}(|Z| > c)$ is approximately $68\%, 95\%$, and 99.7%, for $c = 1, 2$, and 3, respectively, it follows that $\mathbb{P}(|X - \mu| > c\sigma)$ takes the same approximate values for $c = 1, 2$, and 3, for any μ and σ. This is known as the 68–95–99.7 rule.

More accurately, $\mathbb{P}(|X - \mu| > 2\sigma) = 95.44\%$ and $\mathbb{P}(|X - \mu| > 1.96\sigma) = 95\%$. Nevertheless, the rounded value 2 in $\mathbb{P}(|X - \mu| < 2\sigma) \doteq 95\%$ is easier to remember and interpret than the more accurate value 1.96.

Transitioning from the case of a normally distributed variable to a general random variable, we assume that the reader has completed an introductory course in probability theory and is therefore familiar with the definitions of expectation (a.k.a. mean) and variance, which are two of the most commonly used summary measures of a variable.[2†]

For a random variable Y with density function $p(y)$, its expectation and variance are defined as follows:

$$\mathbb{E}(Y) = \int_{-\infty}^{\infty} y\, p(y)\, dy,$$

$$\mathbb{V}(Y) = \int_{-\infty}^{\infty} [y - \mathbb{E}(Y)]^2\, p(y)\, dy,$$

respectively. The variance can be expressed as (see Exercise 1.2)

$$\mathbb{V}(Y) = \mathbb{E}(Y^2) - [\mathbb{E}(Y)]^2.$$

The standard deviation of Y is defined as

$$\sigma(Y) = \sqrt{\mathbb{V}(Y)}.$$

It can be shown that the expectation and variance of $Z \sim \mathcal{N}(0,1)$ are 0 and 1, respectively (see Exercise 1.3). Since $X = \sigma Z + \mu$, the mean and variance of $X \sim \mathcal{N}(\mu, \sigma^2)$ are

$$\mathbb{E}(X) = \sigma\mathbb{E}(Z) + \mu = \mu,$$

$$\mathbb{V}(X) = \sigma^2\mathbb{V}(Z) = \sigma^2.$$

Gold Coin # 2. *If $X \sim \mathcal{N}(\mu, \sigma^2)$, then its expectation (a.k.a. mean) is μ, which indicates the center of the distribution, and its variance is σ^2 (that is, its standard deviation is σ), which quantifies the spread of the distribution around the expectation. By standardizing X, the standardized variable $Z = \frac{X-\mu}{\sigma}$ follows the standard normal distribution $\mathcal{N}(0,1)$.*

---

†The textbook *A Course in Probability Theory* by Chung (2001)[2] is only one of many excellent books on introductory probability theory. It is cited here merely as an example; it may not necessarily be the best reference. It happens to be the book that I studied when I was a graduate student. This remark applies to most of the references cited throughout this book. For instance, when citing a source for the bootstrap method, I did not reference Efron's original paper,[3] *Bootstrap Methods: Another Look at the Jackknife*, published in *The Annals of Statistics* in 1979, but instead his 1994 book with Tibshirani,[4] *An Introduction to the Bootstrap*, since that is the textbook from which I learned the method. The underlying reason is that it is often difficult to identify and cite the most original paper for a given concept or a given method—again, a manifestation of Stigler's Law of Eponymy.

1.4 Why Does the Universe Choose the Number 2?

In *A Brief History of Time*,[5] Stephen Hawking defines *entropy*—within the framework of the second law of thermodynamics—as a measure of disorder or randomness in a system. He writes, "The increase of disorder or entropy is what distinguishes the past from the future, giving a direction to time."

Let us explore the concept of entropy in statistics. When flipping a fair coin, there is a half–half chance of getting heads or tails. This complete unpredictability leads to maximum entropy, as each outcome is equally likely. In contrast, if the coin is unfair—meaning one side is more likely to land than the other—the uncertainty of the outcome decreases, resulting in lower entropy.

To understand why a fair coin has a higher entropy than an unfair one, consider a discrete variable X with possible outcomes $\{x^1, \ldots, x^m\}$ and the corresponding probabilities $\{p_1, \ldots, p_m\}$. The entropy $H(X)$ is defined as:[6]

$$H(X) = -\sum_{j=1}^{m} p_j \log p_j.$$

The logarithm in the definition of entropy is usually taken with base 2, in which case the entropy is measured in bits. However, in this book, the natural logarithm (base e) is used, allowing the entropy to be measured in nats.

For example, for a coin with $\mathbb{P}(\text{heads}) = p$ and $\mathbb{P}(\text{tails}) = 1-p$, $0 < p < 1$, the entropy is given by

$$H(p) = -p \log p - (1-p)\log(1-p).$$

It can be shown that the entropy $H(p)$ attains its maximum value at $p = 1/2$, corresponding to a fair coin. To demonstrate this, differentiate $H(p)$ with respect to p and equate it to zero, resulting in the following equation,

$$-\log p - 1 + \log(1-p) + 1 = 0,$$

whose solution is $p = 1/2$. This is illustrated in Figure 1.2.

Similarly, for a continuous variable X with density $p(x)$, the entropy $H(X)$ is defined as:

$$H(X) = -\int_{-\infty}^{\infty} p(x) \log p(x)\, dx,$$

where the integration is performed over the support of X, given by

$$\text{supp}(X) = \{x : p(x) > 0\}.$$

In particular, the entropy of the normal distribution $\mathcal{N}(\mu, \sigma^2)$ is given by

$$H(X) = -\int_{-\infty}^{\infty} \left[\frac{1}{\sqrt{2\pi}\sigma} e^{-\frac{1}{2}\left(\frac{x-\mu}{\sigma}\right)^2}\right] \log\left[\frac{1}{\sqrt{2\pi}\sigma} e^{-\frac{1}{2}\left(\frac{x-\mu}{\sigma}\right)^2}\right] dx.$$

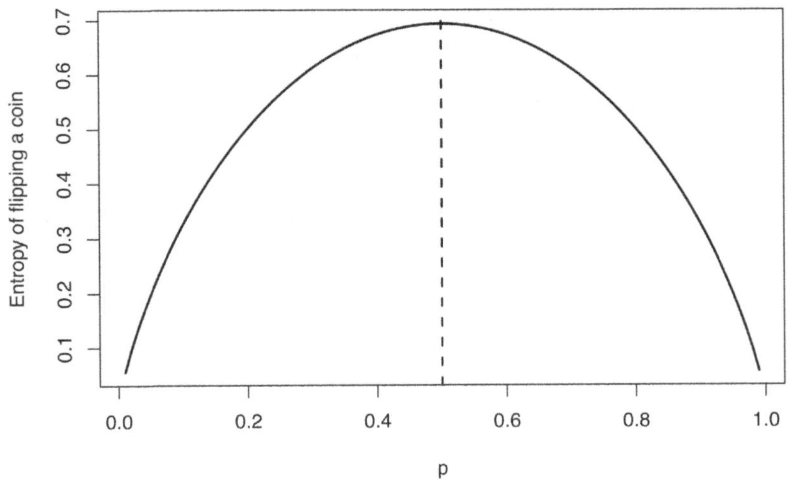

FIGURE 1.2
The plot of function $H(p)$ over $(0,1)$

This expression can be simplified as follows:

$$\begin{aligned} H(X) &= \int_{-\infty}^{\infty} \left[\log(\sqrt{2\pi}\sigma) + \frac{(x-\mu)^2}{2\sigma^2} \right] f(x)\, dx \\ &= \log(\sqrt{2\pi}\sigma) \int_{-\infty}^{\infty} f(x) dx + \frac{1}{2\sigma^2} \int_{-\infty}^{\infty} (x-\mu)^2 f(x)\, dx \\ &= \log(\sqrt{2\pi}\sigma) + \frac{1}{2\sigma^2} \mathbb{V}(X) \\ &= \log(\sqrt{2\pi}\sigma) + 1/2. \end{aligned}$$

The final expression shows that the entropy—and thus the uncertainty—of a normal distribution increases with the standard deviation σ.

Recall that one of the reasons that the normal distribution plays such a central role in statistics is its connection to the central limit theorem, which is maybe the most fundamental result in statistical theory. As we recognize the profound significance of this theorem, which states that the distribution of a normalized sample mean converges to the standard normal distribution, we are wondering: Why does the limit distribution in the central limit theorem take the form of a normal distribution?

This question motivates the title of this section: Why does the universe choose the number 2? The answer may lie in the *maximum entropy principle*,[7] which suggests that the normal distribution emerges as the most "uninformative" distribution under certain constraints—fixed mean and variance.

Gold Coin # 3. *The maximum entropy principle states that, among all continuous distributions with a given mean and variance, the normal distribution has the highest entropy.*

To establish the maximum entropy principle, the method of Lagrange multipliers is applied to maximize the entropy subject to the constraints $\mathbb{E}(X) = \mu$ and $\mathbb{V}(X) = \sigma^2$. The corresponding Lagrangian is defined as

$$\mathcal{L} = -\int_{-\infty}^{\infty} p(x) \log p(x)\, dx + \lambda_1 \left[\int_{-\infty}^{\infty} p(x)\, dx - 1\right] + \lambda_2 \left[\int_{-\infty}^{\infty} xp(x)\, dx - \mu\right] + \lambda_3 \left[\int_{-\infty}^{\infty} (x-\mu)^2 p(x)\, dx - \sigma^2\right].$$

To maximize the Lagrangian, the functional derivative of $\mathcal{L}$ with respect to $p(x)$ is taken and set equal to zero (see Exercise 1.4). After the constraints are applied, the resulting solution to the Lagrangian equation is found to be

$$p(x) = \frac{1}{\sqrt{2\pi}\sigma} e^{-\frac{1}{2}\left(\frac{x-\mu}{\sigma}\right)^2},$$

which corresponds to the normal distribution $\mathcal{N}(\mu, \sigma^2)$. □

The exponent of 2 in the normal distribution plays a critical role in determining the shape of the curve, specifically how quickly the density decays as the value of x deviates from the mean μ. To further explore the question "why does the universe choose the number 2," the generalized normal distribution is considered,[8] defined as:

$$p(x) = \frac{\alpha}{2\beta\Gamma(1/\alpha)} \exp\left\{-\left|\frac{x-\mu}{\beta}\right|^{\alpha}\right\},$$

where $\Gamma(\cdot)$ denotes the gamma function and the parameter α controls the shape of the curve. When $\alpha = 2$, the Gaussian distribution is obtained, while setting $\alpha = 1$ produces the Laplace distribution, which is characterized by heavier tails than the Gaussian distribution. The mean and variance of the generalized normal distribution are[9]

$$\mathbb{E}(X) = \mu \quad \text{and} \quad \mathbb{V}(X) = \frac{\beta^2\Gamma(3/\alpha)}{\Gamma(1/\alpha)}.$$

Applying the same method used to calculate the entropy of the normal distribution, the entropy of the generalized normal distribution is calculated as follows[9] (see Exercise 1.5):

$$H(X) = \frac{1}{\alpha} - \log\left[\frac{\alpha}{2\beta\Gamma(1/\alpha)}\right].$$

Without loss of generality, let β be chosen such that

$$\mathbb{V}(X) = \frac{\beta^2\Gamma(3/\alpha)}{\Gamma(1/\alpha)} = 1,$$

which leads to

$$\beta = \sqrt{\frac{\Gamma(1/\alpha)}{\Gamma(3/\alpha)}}.$$

Consequently, the entropy of the generalized normal distribution with unit variance is given by

$$H_1(X;\alpha) = \frac{1}{\alpha} - \log\left[\frac{\alpha\Gamma^{1/2}(3/\alpha)}{2\Gamma^{3/2}(1/\alpha)}\right].$$

As illustrated in Figure 1.3, the maximum of the entropy function $H_1(X;\alpha)$ is attained at $\alpha = 2$, precisely corresponding to the normal distribution. Thus, the maximum entropy principle is verified, at least partially.

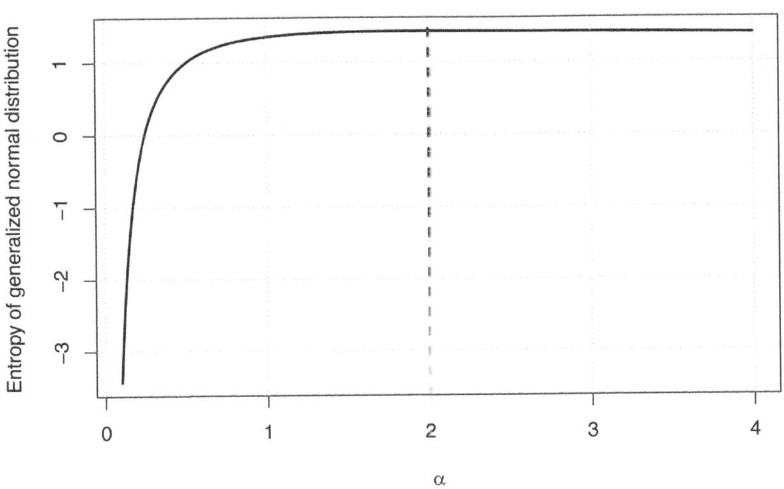

FIGURE 1.3
The plot of entropy of generalized normal distribution with unit variance

1.5 Exercises

Exercise 1.1

Show that

$$\int_{-\infty}^{\infty} \frac{1}{\sqrt{2\pi}} e^{-\frac{1}{2}z^2}\, dz = 1.$$

Exercise 1.2

Show that

$$\mathbb{V}(Y) = \mathbb{E}(Y^2) - [\mathbb{E}(Y)]^2.$$

Exercise 1.3

If $Z \sim \mathcal{N}(0,1)$, show that

$$\mathbb{E}(Z) = 0 \quad \text{and} \quad \mathbb{V}(Z) = 1.$$

Exercise 1.4

Complete the proof of the maximum entropy principle subject to the constraints $\int_{-\infty}^{\infty} xp(x)\, dx = \mu$ and $\int_{-\infty}^{\infty} (x-\mu)^2 p(x)\, dx = \sigma^2$.

Exercise 1.5

Calculate the entropy of the generalized normal distribution with parameters μ, α, and β.

2

Fundamentals of Asymptotic Statistics

2.1 I.I.D.

In statistics, it is common to begin with the assumption that

$$X_1, \ldots, X_n \text{ are independent and identically distributed (i.i.d.) with } X \sim P,$$

where P is the common probability distribution. The distribution P will be examined in both the continuous and discrete cases.

Gold Coin # 4. *Assumption: $X_1, \ldots, X_n$ are independent and identically distributed (i.i.d.) with $X \sim P$. This assumption has two components: (1) independent and (2) identically distributed.*

This assumption is needed for all of the theoretical results presented in this book. Although certain results can indeed be extended to settings where this assumption does not hold, such extensions are beyond the scope of this book. Given its fundamental role, the assumption will be examined in detail in the following three subsections.

2.1.1 Sample and Population

Two primary scenarios are commonly considered for the generation of random variables $X_1, \ldots, X_n$.

Scenario 1: A random experiment is performed repeatedly n times, with the outcome recorded as X_i at each repetition, $i = 1, \ldots, n$. For example, if a fair coin is flipped n times, the resulting sequence of outcomes can be denoted by independent and identically distributed random variables $X_1, \ldots, X_n$, where $\mathbb{P}(X_i = \text{heads}) = \mathbb{P}(X_i = \text{tails}) = 1/2$, $i = 1, \ldots, n$.

Scenario 2: A sample of size n is selected from a finite population of N subjects. The way in which sampling is conducted plays an important role.

DOI: 10.1201/9781003635468-3

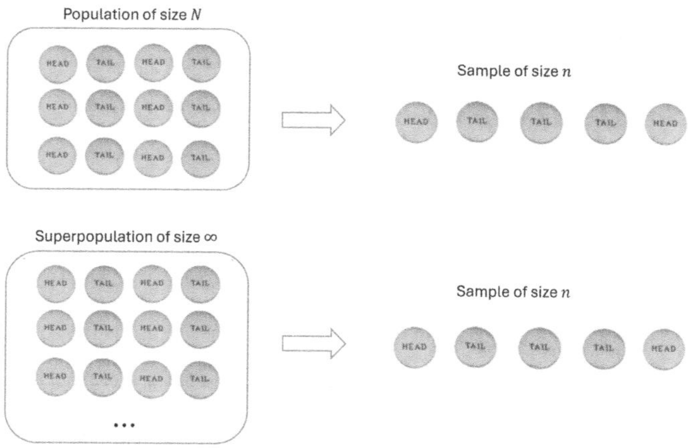

FIGURE 2.1
Simple random sample from a population or superpopulation

For example, in simple random sampling, each subject in the population is given an equal probability of being chosen.

Figure 2.1 presents two examples in which a simple random sample is drawn, one sample from a finite population and the other sample from an infinitely large population.

Suppose a finite population consists of N one-sided coins (purely as a thought experiment, since in reality every coin has two sides), with half showing heads (coded as 1) and the other half showing tails (coded as 0), where N is an even number. From this population, a simple random sample of size n is selected, and the observed values are denoted by $X_1, \ldots, X_n$. These variables are identically distributed in the sense that

$$\mathbb{P}(X_i = 1) = \mathbb{P}(X_i = 0) = 1/2, \quad i = 1, \ldots, n.$$

However, independence is not satisfied. For any pair, X_i and X_j, the covariance is given by

$$\begin{aligned}\mathrm{Cov}(X_i, X_j) &= \mathbb{E}(X_i X_j) - \mathbb{E}(X_i)\mathbb{E}(X_j) \\ &= \frac{N/2(N/2-1)}{N(N-1)} - \left(\frac{1}{2}\right)^2 \\ &= -\frac{1}{4(N-1)}.\end{aligned}$$

When N is large, this correlation becomes negligible, making it reasonable to treat $X_1, \ldots, X_n$ as approximately independent and identically distributed.

For a more rigorous justification of this approximation, continue the above thought experiment and imagine an infinitely large population composed equally of heads and tails. From this imagined population, a simple random sample of size n is drawn, and the observed values are denoted by $X_1, \ldots, X_n$. In this case, the variables $X_1, \ldots, X_n$ are precisely independent and identically distributed. Such an infinitely large population will be referred to as a *superpopulation*[10] in the next subsection.

2.1.2 Population and Superpopulation

In a clinical study, the *target population* is specified by a set of inclusion and exclusion criteria.

Assume that there are J inclusion criteria and K exclusion criteria. Let $\mathcal{I}_j$ denote the set of all subjects who satisfy the jth inclusion criterion, and let $\mathcal{E}_k$ denote the set of all subjects who satisfy the kth exclusion criterion. Then, the population of interest, $\mathcal{P}$, is defined as

$$\mathcal{P} = \left(\cup_{j=1}^{J} \mathcal{I}_j\right) \cap \left(\cup_{k=1}^{K} \mathcal{E}_k\right)^c,$$

where the subscript c denotes the complement.

In a clinical study, the process of enrolling participants typically involves the following three steps:

1. Selection of study sites and specification of the study timeline;
2. Obtaining informed consent from potential participants;
3. Screening of participants based on inclusion/exclusion criteria.

It is important to note that the sample of participants sampled through this process is selected by *convenience sampling* rather than *random sampling*. Consequently, the sample may not be representative of the target population $\mathcal{P}$, potentially leading to selection bias.

The task at hand is to conceptually "eliminate" the aforementioned selection bias. To this end, the concept of a *superpopulation* is introduced as a theoretical framework. The list of study sites is imagined to be extended indefinitely to include all possible sites globally universally, and the study timeline is similarly extended to include any time in the future. In this hypothetical construction, the superpopulation consists of all individuals who are similar to the enrolled subjects—those who exist or may exist at any place and at any time, be it real or imagined.

This superpopulation is denoted by $\mathcal{P}^*$, from which the enrolled sample is considered a simple random sample.

2.1.3 Continuous and Discrete Variables

After the "$X_1, \ldots, X_n$ are i.i.d." part of the assumption that "$X_1, \ldots, X_n$ are i.i.d. with $X \sim P$" has been discussed, attention now turns to the remaining part: "$X \sim P$."

Continuous Variable

Let Ω denote the sample space of the random variable X, defined as the set of all possible values that X may assume. In the case of a continuous variable, Ω is represented by a range, typically expressed as an interval such as $(-\infty, \infty)$, $[0, \infty)$, or $[a, b]$. A continuous random variable X can be characterized by its *probability density function* $p(x)$, satisfying the following conditions:

$$p(x) \geq 0 \quad \text{and} \quad \int_{\Omega} p(x)\, dx = 1.$$

Two examples of continuous distributions are the Gaussian distribution and the Laplace distribution. The density function of the Gaussian distribution is specified by two parameters and is given by

$$p(x \mid \mu, \sigma) = \frac{1}{\sqrt{2\pi}\sigma} \exp\left\{ -\frac{1}{2}\left(\frac{x-\mu}{\sigma}\right)^2 \right\},$$

while the density function of the Laplace distribution is also specified by two parameters and is given by

$$p(x \mid \mu, b) = \frac{1}{2b} \exp\left\{ -\frac{|x-\mu|}{b} \right\}.$$

Discrete Variable

A discrete random variable X is restricted to a finite or countably infinite set of possible values; that is, the sample space Ω is composed of distinct, countable outcomes. A discrete random variable can be characterized by its *probability mass function* $p(x)$, satisfying the following conditions:

$$p(x) \geq 0 \quad \text{and} \quad \sum_{x \in \Omega} p(x) = 1.$$

Two examples of discrete distributions are the Bernoulli distribution and the Poisson distribution. The Bernoulli distribution is used to model the probability of a binary outcome, such as heads/tails (as in the coin-flipping example), success/failure, yes/no, or 1/0. Without loss of generality, a Bernoulli distribution coded as 1/0 can be defined by

$$p(1) = p \quad \text{and} \quad p(0) = 1 - p,$$

where the distribution is specified by a single parameter p. The Poisson distribution, which is also specified by a single value λ, is used to model the probability of a discrete variable that assumes non-negative integer values,

$$p(k) = \frac{e^{-\lambda}\lambda^k}{k!}, \quad k = 0, 1, 2, \ldots$$

Unified Notation*

With a slight abuse of notation, let $p(x)$ denote either the probability density function (PDF) for a continuous random variable or the probability mass function (PMF) for a discrete random variable. To further unify the treatment of both types of variables, the cumulative distribution function (CDF) is defined as

$$P(x) = \mathbb{P}(X \leq x).$$

For a continuous random variable, the CDF is given by

$$P(x) = \int_{-\infty}^{x} p(x')\, dx',$$

whereas for a discrete random variable, the CDF is given by

$$P(x) = \sum_{x' \leq x} p(x').$$

It is often more convenient to work with the CDF than with the PDF and PMF. Note that the following identity holds:

$$\begin{aligned} 1 &= \int_{-\infty}^{\infty} dP(x) \\ &= \int_{\Omega} p(x)\, dx \quad \text{(for continuous random variables)} \\ &= \sum_{x \in \Omega} p(x). \quad \text{(for discrete random variables)} \end{aligned}$$

The expected value of a random variable X can be expressed as

$$\begin{aligned} \mathbb{E}(X) &= \int_{-\infty}^{\infty} x\, dP(x) \\ &= \int_{\Omega} x p(x)\, dx \quad \text{(for continuous random variables)} \\ &= \sum_{x \in \Omega} x p(x). \quad \text{(for discrete random variables)} \end{aligned}$$

*You may be wondering why some sections of this book are rather lengthy, with expressions that may at times seem tedious or somewhat repetitive. This is intentional: the unification of two complementary or even competing concepts is a central theme of this book. Therefore, I consistently provide detailed explanations whenever such unifications are discussed.

More generally, for any function $g(X)$, the expected value is given by

$$
\begin{aligned}
\mathbb{E}[g(X)] &= \int_{-\infty}^{\infty} g(x)\, dP(x) \\
&= \int_{\Omega} g(x)p(x)\, dx \quad \text{(for continuous random variables)} \\
&= \sum_{x\in\Omega} g(x)p(x). \quad \text{(for discrete random variables)}
\end{aligned}
$$

In particular, the probability that X lies in the set A can be written as

$$
\begin{aligned}
\mathbb{P}(X \in A) = \mathbb{E}[I_A(X)] &= \int_{-\infty}^{\infty} I_A(x)\, dP(x) \\
&= \int_{A} p(x)\, dx \quad \text{(for continuous random variables)} \\
&= \sum_{x\in A} p(x), \quad \text{(for discrete random variables)}
\end{aligned}
$$

where $I_A(x)$ is the indicator function, equal to 1 if $x \in A$ and 0 otherwise.

Furthermore, the definition of CDF can be extended from a univariate random variable X to a random vector

$$
\mathbf{X} = (X^{(1)}, \ldots, X^{(d)})^\top.
$$

In this case, the CDF is defined as

$$
P(\mathbf{x}) = P(x^{(1)}, \ldots, x^{(d)}) = \mathbb{P}\Big(X^{(1)} \le x^{(1)}, \ldots, X^{(d)} \le x^{(d)}\Big),
$$

where $\mathbf{x} = (x^{(1)}, \ldots, x^{(d)})^\top$. Accordingly,

$$
\begin{aligned}
\int dP(\mathbf{x}) &= 1, \\
\mathbb{E}[\mathbf{X}] &= \int \mathbf{x}\, dP(\mathbf{x}), \\
\mathbb{E}[g(\mathbf{X})] &= \int g(\mathbf{x})\, dP(\mathbf{x}), \\
\mathbb{P}(\mathbf{X} \in A) &= \int_A dP(\mathbf{x}).
\end{aligned}
$$

Finally, if the random vector $\mathbf{X}$ contains both continuous and discrete components, the corresponding PDF and PMF can be unified through the use of a *base measure.*

Let μ denote the base measure, defined as the product of the Lebesgue measure for the continuous variables and the counting measure for the discrete variables. Let $p(\mathbf{x})$ denote the density of $\mathbf{X}$ with respect to μ. Then the integral of $p(\mathbf{x})$ over the sample space, the expectations of $\mathbf{X}$ and $g(\mathbf{X})$, and the probability of $\mathbf{X} \in A$ can be written as

$$\begin{aligned}\int p(\mathbf{x})\, \mu(d\mathbf{x}) &= 1,\\ \mathbb{E}[\mathbf{X}] &= \int \mathbf{x}\, p(\mathbf{x})\, \mu(d\mathbf{x}),\\ \mathbb{E}[g(\mathbf{X})] &= \int g(\mathbf{x})\, p(\mathbf{x})\, \mu(d\mathbf{x}),\\ \mathbb{P}(\mathbf{X} \in A) &= \int_A p(\mathbf{x})\, \mu(d\mathbf{x}).\end{aligned}$$

2.2 Convergence

2.2.1 Convergence in Probability and in Distribution

Let $\{T_n\}_{n=1}^{\infty}$ be a sequence of random variables. The sequence $\{T_n\}$ is said to be o_p of a function $g(n)$ if, for every $\epsilon > 0$,

$$\mathbb{P}\left(\left|\frac{T_n}{g(n)}\right| > \epsilon\right) \to 0, \quad \text{as } n \to \infty.$$

This result is denoted by $T_n = o_p\left(g(n)\right)$.

In particular, by choosing $g(n) = 1$, the sequence $\{T_n\}$ is said to be $o_p(1)$ if, for every $\epsilon > 0$,

$$\mathbb{P}\left(|T_n| > \epsilon\right) \to 0, \quad \text{as } n \to \infty.$$

If there exists a constant θ such that $T_n = \theta + o_p(1)$, or equivalently $T_n - \theta = o_p(1)$, then T_n is said to *converge in probability* to θ, denoted as

$$T_n \xrightarrow{p} \theta, \quad \text{as } n \to \infty.$$

In addition to o_p (pronounced "small oh in probability"), there is O_p (pronounced "big oh in probability"), which is not discussed here. Instead, the focus will be on a special case of $O_p(1)$, that is, convergence in distribution.

A sequence of random variables $\{T_n\}_{n=1}^{\infty}$ is said to *converge in distribution* to a random variable T_∞, denoted by

$$T_n \xrightarrow{d} T_\infty, \quad \text{as } n \to \infty,$$

if, for every x at which the CDF of T_∞ is continuous,

$$\mathbb{P}(T_n \le x) \to \mathbb{P}(T_\infty \le x), \quad \text{as } n \to \infty.$$

With the definitions of convergence in probability and convergence in distribution now established, two fundamental theorems in probability and statistics can be introduced: the Law of Large Numbers (LLN) and the Central Limit Theorem (CLT).

Gold Coin # 5. *Assume that $X_1, \ldots, X_n$ are i.i.d. with $X \sim P$.*

- *(LLN) If $\mathbb{E}(X) = \mu$ is finite, then*

$$\overline{X}_n = \frac{1}{n}\sum_{i=1}^{n} X_i \xrightarrow{p} \mu, \quad \text{as } n \to \infty.$$

- *(CLT) If $\mathbb{E}(X) = \mu$ and $\mathbb{V}(X) = \sigma^2$ are finite, then*

$$\sqrt{n}\left(\overline{X}_n - \mu\right) \xrightarrow{d} \mathcal{N}(0, \sigma^2), \quad \text{as } n \to \infty.$$

The reader is referred to any classical textbook on probability theory[2,11] for rigorous proofs of the LLN and CLT. In the following two subsections, a proof of a weaker version of the LLN will be employed to illustrate the concept of convergence in probability, and a heuristic argument for the CLT will be provided in the context of $X_1, \ldots, X_n$ representing outcomes from n independent trials of flipping a fair coin. These are not intended as formal proofs, but rather as means to foster an appreciation for the mathematical elegance of these foundational results.

2.2.2 Law of Large Numbers

Suppose that $\mathbb{E}(X) = \mu$ and $\mathbb{V}(X) = \sigma^2$ are finite. (Note that these conditions are the same conditions required for the CLT, and are stronger than those required for the LLN. This is the reason why it is called a weaker version of the LLN.) Then,

$$\overline{X}_n = \frac{1}{n}\sum_{i=1}^{n} X_i \xrightarrow{p} \mu, \quad \text{as } n \to \infty.$$

By the definition of convergence in probability, it suffices to show that

$$\mathbb{P}\left(\left|\overline{X}_n - \mu\right| \geq \epsilon\right) \to 0, \quad \text{as } n \to \infty.$$

To establish this, Chebyshev's inequality (see Exercise 2.1) is applied:

$$\mathbb{P}(|\overline{X}_n - \mu| \geq \epsilon) \leq \frac{\mathbb{V}(\overline{X}_n)}{\epsilon^2}.$$

Note that the upper bound is equal to

$$\frac{1}{n^2\epsilon^2}\mathbb{V}\left(\sum_{i=1}^{n} X_i\right) = \frac{n\sigma^2}{n^2\epsilon^2} = \frac{\sigma^2}{n\epsilon^2},$$

which converges to zero as $n \to \infty$. Thus, the weaker LLN is proved. □

2.2.3 Central Limit Theorem

At first glance, the CLT may seem surprising: the distribution of

$$\sqrt{n}\left(\overline{X}_n - \mu\right)$$

is shown to converge in distribution to the normal distribution $\mathcal{N}(0, \sigma^2)$, regardless of how "abnormal" the original distribution P may be.

For example, consider n i.i.d. trials of flipping a fair coin. The outcome of the ith trial is denoted by X_i, taking the value 1 for heads and -1 for tails. The random variable X_i is known as a Rademacher random variable, which takes on the values 1 and -1 with equal probability and satisfies $\mathbb{E}(X_i) = 0$ and $\mathbb{V}(X_i) = 1$.

The de Moivre–Laplace Central Limit Theorem, named after Abraham de Moivre (1667–1754) and Pierre-Simon Laplace (1749–1827), historically the first version of the CLT, states that if $X_1, \ldots, X_n$ are i.i.d. Rademacher random variables, then

$$\frac{1}{\sqrt{n}}\sum_{i=1}^{n} X_i \xrightarrow{d} \mathcal{N}(0,1), \quad \text{as } n \to \infty.$$

Many methods exist for proving the general CLT; for instance, the moment generating function (MGF) is considered a powerful tool. However, rather than presenting a formal proof of this well-known result, a heuristic proof of the De Moivre–Laplace Theorem will be provided in order to highlight its elegance and intuitive appeal.

For this goal, let $S_n = \sum_{i=1}^{n} X_i$. When $S_{2n} = 2k$, it follows that among the $2n$ trials, $n+k$ are heads (value 1) and $n-k$ are tails (value -1). Therefore,

$$\mathbb{P}(S_{2n} = 2k) = \binom{2n}{n+k}\left(\frac{1}{2}\right)^{2n} = \frac{(2n)!}{2^{2n}(n+k)!(n-k)!}.$$

Using Stirling's approximation (see Exercise 2.2), it can be shown that

$$\mathbb{P}(S_{2n} = 2k) \doteq \frac{1}{\sqrt{\pi n}}e^{-x^2/2}, \quad \text{as } \frac{2k}{\sqrt{2n}} \to x.$$

From the exponential term $e^{-x^2/2}$ on the right-hand side, the limit distribution is inferred to be the standard normal distribution $\mathcal{N}(0,1)$. To verify

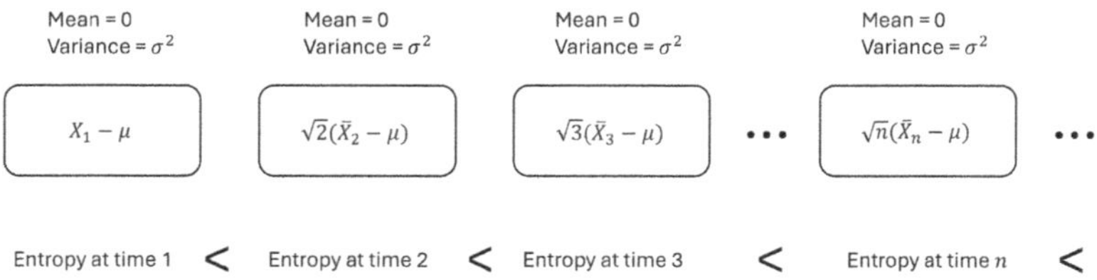

FIGURE 2.2
The maximum entropy principle behind the CLT

this, the connection between discrete differentiation and continuous differentiation is employed. Specifically,

$$f(x) \doteq \frac{\mathbb{P}\left(\frac{S_{2n}}{\sqrt{2n}} \le \frac{2k}{\sqrt{2n}}\right) - \mathbb{P}\left(\frac{S_{2n}}{\sqrt{2n}} \le \frac{2(k-1)}{\sqrt{2n}}\right)}{\frac{2k}{\sqrt{2n}} - \frac{2(k-1)}{\sqrt{2n}}}$$

$$= \frac{\mathbb{P}(S_{2n} = 2k)}{\frac{2}{\sqrt{2n}}} \doteq \sqrt{\frac{n}{2}} \cdot \frac{1}{\sqrt{\pi n}} e^{-x^2/2} = \frac{1}{\sqrt{2\pi}} e^{-x^2/2}.$$

Hence, a heuristic proof of the theorem is completed. □

One might naturally wonder why the limiting distribution in the CLT must be the normal distribution. A similar question was raised in Chapter 1: "Why does the university choose the number 2?" That is, why does the exponential term in the limiting distribution take the specific form $e^{-x^2/2}$, rather than alternatives such as $e^{-|x|}$, $e^{-|x|^3}$, or e^{-x^4}? What makes $e^{-x^2/2}$ so fundamental in this context?

Figure 2.2 is intended to offer an intuitive explanation of the maximum entropy principle behind the CLT. At time 1, the centered variable $X_1 - \mu$ is obtained. At time 2, the scaled average of two variables is obtained, namely $\sqrt{2}(\overline{X}_2 - \mu)$. At time 3, the scaled average of three variables is obtained, namely $\sqrt{3}(\overline{X}_3 - \mu)$. This process continues in the same fashion.

Each of the random variables $X_1 - \mu$, $\sqrt{2}(\overline{X}_2 - \mu)$, $\sqrt{3}(\overline{X}_3 - \mu)$, $\ldots$ is centered at zero and has variance σ^2. It can be reasonably believed that, over time, the entropy of these averages increases, since the average of a larger number of independent variables becomes more "random" than the average of fewer variables, under the constraint of fixed mean and variance.

A natural question arises: What is the limiting distribution associated with this increasing entropy? According to the maximum entropy principle, the limiting distribution is expected to be the distribution that maximizes entropy among all distributions with mean zero and variance σ^2, namely the normal distribution $\mathcal{N}(0, \sigma^2)$.

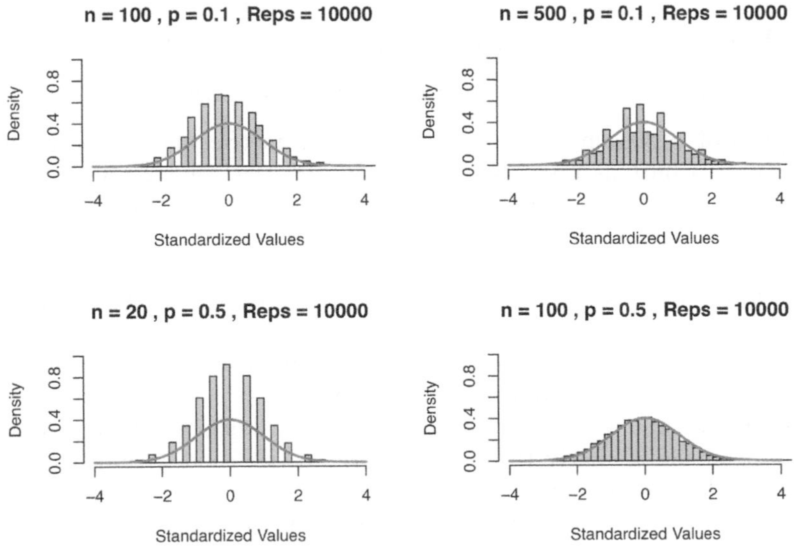

FIGURE 2.3
Simulation results to visualize the CLT

This reasoning suggests that the limit of the scaled average of an increasing number of variables is the normal distribution $\mathcal{N}(0, \sigma^2)$. □

To visualize the CLT (see Exercise 2.3), the statistical software R is used to generate n random variables $X_1, \ldots, X_n$ that are i.i.d. with $X \sim \text{Bernoulli}(p)$. The standardized statistic is then computed as

$$Z = \frac{\sqrt{n}\left(\overline{X}_n - p\right)}{\sqrt{p(1-p)}}.$$

Four parameter settings are considered: $(n, p) = (100, 0.1)$, $(500, 0.1)$, $(20, 0.5)$, and $(100, 0.5)$. For each setting, the data generating process is repeated 10,000 times, and the resulting standardized values are used to construct histograms. The resulting histograms are shown in Figure 2.3.

In the top panel of the figure, corresponding to $p = 0.1$, it can be observed that convergence to the standard normal distribution occurs relatively slowly. In contrast, the bottom panel of the figure corresponding to $p = 0.5$ shows a faster convergence. This difference in convergence speed is influenced by the entropy of the original Bernoulli distribution: Since the entropy of Bernoulli(0.5) is greater than that of Bernoulli(0.1), faster convergence is observed in the former case.

2.3 Probability and Empirical Distributions

2.3.1 Parametric and Nonparametric Distributions

Two types of probability distribution are: parametric and nonparametric. The four distributions presented in Section 2.1—each named after a great mathematician—are considered examples of parametric distributions, as they are characterized by a finite number of parameters.

A nonparametric distribution is defined as one that does not assume a specific form governed by a finite set of parameters. Under the assumption that $X_1, \ldots, X_n$ are i.i.d. with $X \sim P$, a realistic perspective necessitates the acknowledgment that P may be completely unknown, and thus ideally no parametric assumptions should be imposed on P.

It should be noted that certain parametric distributions can be saturated, thereby behaving similarly to nonparametric distributions. Among the four parametric distributions discussed earlier, only the Bernoulli distribution is saturated, as it possesses a sufficient number of parameters to exactly represent the distribution. Specifically, it contains one parameter for a two-point sample space, satisfying the condition that the number of parameters equals one less than the cardinality of the sample space.

More generally, for a discrete variable with m possible outcomes (denoted $x^1, \ldots, x^m$), a distribution with $m-1$ parameters,

$$p_j = \mathbb{P}(X = x^j), \quad j = 1, \ldots, m-1,$$

is considered saturated.

2.3.2 Empirical Distribution

Suppose that $X_1, \ldots, X_n$ are i.i.d. with $X \sim P_0$. The distribution P_0 is regarded as the true, but unknown, probability distribution (defined in terms of its CDF). In a realistic setting, no assumptions should be made regarding the form of P_0; it is treated as entirely unknown. Given the data $\{X_1, \ldots, X_n\}$, the question then arises: How can P_0 be estimated?

The empirical distribution is defined as

$$\widehat{P}_n(x) = \frac{1}{n} \sum_{i=1}^{n} I(X_i \leq x),$$

where $I(\cdot)$ denotes the indicator function.

By the LLN, the consistency of the empirical distribution can be shown:

$$\widehat{P}_n(x) = \frac{1}{n} \sum_{i=1}^{n} I(X_i \leq x) \xrightarrow{p} \mathbb{E}\left[I(X \leq x)\right] = P_0(x).$$

Furthermore, the CLT implies its asymptotic normality:

$$\sqrt{n}\left[\widehat{P}_n(x) - P_0(x)\right] \xrightarrow{d} \mathcal{N}\big(0, P_0(x)[1 - P_0(x)]\big),$$

noting that $\mathbb{V}[I(X \leq x)] = P_0(x)[1 - P_0(x)]$. The consistency and asymptotic normality will be discussed in detail in Section 2.5.

Thus, for any fixed x, the empirical CDF $\widehat{P}_n(x)$ converges in probability to the true CDF $P_0(x)$, and its properly scaled deviation converges in distribution to a Gaussian random variable as $n \to \infty$. In fact, uniform convergence in probability of $\widehat{P}_n(x)$ to $P_0(x)$ can be shown, and the standardized empirical CDF can be demonstrated to converge in distribution to a Gaussian process.[12]

Rather than delving into these advanced asymptotic results, attention will now be turned to an important conceptual interpretation: the empirical CDF can be viewed as the distribution function of an implicit random variable, denoted X^*.

After one realization of the data, each X_i takes a fixed value, say $X_i = x_i$, for $i = 1, \ldots, n$. If all values $x_1, \ldots, x_n$ are distinct, then a discrete random variable X^* can be defined with the following probability mass function:

$$\mathbb{P}(X^* = x_i) = \frac{1}{n}, \quad i = 1, \ldots, n.$$

It can be shown that the CDF of X^* is equal to $\widehat{P}_n$. In fact,

$$P_{X^*}(x) = \mathbb{P}(X^* \leq x) = \frac{1}{n}\sum_{i=1}^{n} I(x_i \leq x).$$

A conceptual construction of this random variable X^* involves imagining an urn containing n balls labeled with the observed values $x_1, \ldots, x_n$; X^* is then obtained by randomly drawing a ball and assigning its label to the variable.

This reasoning remains valid even if some of the observed values are tied. Let $y_1, \ldots, y_m$ denote the m distinct values among $x_1, \ldots, x_n$, and define

$$\widehat{p}_j = \frac{1}{n}\sum_{i=1}^{n} I(x_i = y_j), \quad j = 1, \ldots, m.$$

Then, X^* is defined as a discrete random variable with the following probability mass function:

$$\mathbb{P}(X^* = y_j) = \widehat{p}_j, \quad j = 1, \ldots, m.$$

Again, it can be shown that the CDF of X^* is equal to $\widehat{P}_n$, since

$$\begin{aligned}
P_{X^*}(x) = \mathbb{P}(X^* \leq x) &= \sum_{j=1}^{m} \widehat{p}_j I(y_j \leq x) \\
&= \sum_{j=1}^{m} \left[\frac{1}{n} \sum_{i=1}^{n} I(x_i = y_j) \right] I(y_j \leq x) \\
&= \frac{1}{n} \sum_{i=1}^{n} \sum_{j=1}^{m} I(x_i = y_j) I(y_j \leq x) \\
&= \frac{1}{n} \sum_{i=1}^{n} I(x_i \leq x).
\end{aligned}$$

In this case as well, the same urn model applies: X^* is obtained by randomly selecting a ball from an urn containing the observed values $\{x_1, \ldots, x_n\}$ and assigning its label to X^*.

In summary, the following two data generating procedures are equivalent:

- Draw $X^* \sim \widehat{P}_n$, where $\widehat{P}_n$ is the empirical distribution based on $X_1, \ldots, X_n$;
- Randomly draw a ball from the urn of n balls labeled by $X_1, \ldots, X_n$ and assign its label to X^*.

2.3.3 Bootstrapping

If either of the two equivalent data generating procedures described above is repeated n times, a simple random sample $X_1^*, \ldots, X_n^*$ is obtained. This sample is referred to as a *bootstrap sample*.[4] That is, a bootstrap sample can be generated by resampling from $\{X_1, \ldots, X_n\}$ with replacement, according to the following procedure:

- Randomly draw a ball from the urn of n balls labeled by $X_1, \ldots, X_n$, assign its label to X_1^*, and return the ball to the urn;
- Randomly draw a ball from the urn of n balls labeled by $X_1, \ldots, X_n$, assign its label to X_2^*, and return the ball to the urn;
- ...
- Randomly draw a ball from the urn of n balls labeled by $X_1, \ldots, X_n$, assign its label to X_n^*, and return the ball to the urn.

After these n steps, the random variables $X_1^*, \ldots, X_n^*$ are obtained, which are i.i.d. with $X^* \sim \widehat{P}_n$.

2.3.4 Bijective Relationship

Two distinct worlds are considered: one in which the random variable X is distributed according to P_0, and another in which the random variable X^* follows the empirical distribution $\widehat{P}_n$. A set of "bijective" relationships can be established between these two worlds.

Gold Coin # 6. *The "bijective" relationship between the true distribution P_0 and the empirical distribution $\widehat{P}_n$:*

- $X \sim P_0$ *vs.* $X^* \sim \widehat{P}_n$
- $\mathbb{E}_{P_0}(X)$ *vs.* $\mathbb{E}_{\widehat{P}_n}(X^*)$
- $\mathbb{E}_{P_0}[g(X)]$ *vs.* $\mathbb{E}_{\widehat{P}_n}[g(X^*)]$ *for any function* g
- $\theta(P_0)$ *vs.* $\theta(\widehat{P}_n)$ *for any parameter* θ
- $X_1, \ldots, X_n$ *are i.i.d. with* P_0 *vs.* $X_1^*, \ldots, X_n^*$ *are i.i.d. with* $\widehat{P}_n$
- $T(X_1, \ldots, X_n)$ *vs.* $T(X_1^*, \ldots, X_n^*)$ *for any statistic* T
- $\cdots$

The relationships listed above are now explained in greater detail. In the world where $X \sim P_0$, the expectation of a function of X is defined as

$$\mathbb{E}_{P_0}(X) = \int x \, dP(x),$$
$$\mathbb{E}_{P_0}[g(X)] = \int g(x) \, dP(x).$$

In contrast, under the empirical distribution where $X^* \sim \widehat{P}_n$, the corresponding expectations are computed as

$$\mathbb{E}_{\widehat{P}_n}(X) = \int x \, d\widehat{P}_n(x) = \frac{1}{n}\sum_{i=1}^{n} X_i,$$
$$\mathbb{E}_{\widehat{P}_n}[g(X^*)] = \int g(x) \, d\widehat{P}_n(x) = \frac{1}{n}\sum_{i=1}^{n} g(X_i).$$

These expressions show that population expectations are mirrored by their sample analogs under the empirical distribution.

Based on the above bijective relationship, the sampling distribution and bootstrap distribution of a given statistic can be defined and related. A statistic T is defined as a function of data. In the world where the data consist of $X_1, \ldots, X_n$, the statistic is written as $T(X_1, \ldots, X_n)$. In the empirical world based on $X_1^*, \ldots, X_n^*$, the same functional form is applied to produce $T(X_1^*, \ldots, X_n^*)$. There is a connection between the distribution of $T(X_1, \ldots, X_n)$ and that of $T(X_1^*, \ldots, X_n^*)$.

The distribution of $T(X_1, \ldots, X_n)$ is known as the *sampling distribution*, whereas the distribution of $T(X_1^*, \ldots, X_n^*)$ is known as the *bootstrap distribution*. In particular, the variance of the sampling distribution corresponds to the variance of the bootstrap distribution. A detailed discussion of the bootstrap method will be provided in the next chapter.

2.4 Estimand and Estimator

Assume that $X_1, \ldots, X_n$ are i.i.d. with $X \sim P_0$.

An *estimand* is defined as the target of estimation, that is, some quantity to be estimated. Often, an estimand is some feature of the population that is unknown but of inferential interest. The estimand may be the full distribution P_0 itself or a specific functional of it, denoted by

$$\theta_0 = \theta(P_0).$$

An *estimator* is defined as a function of the observed data $\mathcal{D} = \{X_1, \ldots, X_n\}$, denoted by

$$\widehat{\theta}_n = \widehat{\theta}(\mathcal{D}).$$

The process by which an estimand $\theta(P_0)$ is inferred using an estimator $\widehat{\theta}(\mathcal{D})$ is referred to as *estimation*. Once a specific realization of the dataset, denoted by $\{x_1, \ldots, x_n\}$, has been observed, the corresponding value $\widehat{\theta}(x_1, \ldots, x_n)$ is known as the *estimate*.

Thus, a family of related concepts can be identified, called the "ESTIMA-club" in this book: *estimation*, *estimand*, *estimator*, and *estimate*—as a verb and as a noun.

2.4.1 Plug-in Estimator

In the preceding section, it was shown that when the probability distribution P_0 itself is regarded as the estimand, the empirical distribution $\widehat{P}_n$ serves as an estimator with desirable asymptotic properties.

Consider the settings where the estimand is not the full distribution but a functional aspect of it, denoted by

$$\theta_0 = \theta(P_0).$$

The question arises: How can an estimator for θ_0 be constructed?

By leveraging the bijective relationship between the world governed by $X \sim P_0$ and that governed by $X^* \sim \widehat{P}_n$, it is natural to consider the following *plug-in estimator*:

$$\widehat{\theta}_n = \theta(\widehat{P}_n).$$

To better understand this approach, two examples are considered.

Example 1: Population Mean

Assume that the estimand of interest is the population mean of X,

$$\theta_0 = \theta(P_0) = \mathbb{E}_{P_0}(X) = \int x \, dP_0(x).$$

The corresponding plug-in estimator is

$$\widehat{\theta}_n = \theta(\widehat{P}_n) = \mathbb{E}_{\widehat{P}_n}(X^*) = \int x \, d\widehat{P}_n(x) = \frac{1}{n}\sum_{i=1}^{n} X_i,$$

which coincides with the sample mean. Hence, the sample mean is interpreted as the plug-in estimator of the population mean. □

Remark:

By convention, $\mathbb{E}(X)$ without a subscript indicates that the expectation is taken with respect to the true underlying distribution P_0. The same convention applies to the variance and covariance operators, denoted by $\mathbb{V}(X)$ and $\mathrm{Cov}(X)$, respectively.

In some contexts (e.g., Example 1), to emphasize the dependence on the true distribution P_0, the notations $\mathbb{E}_{P_0}(X)$ and $\mathbb{V}_{P_0}(X)$ are used. In other situations, to highlight the dependence on a parametric distribution P_θ, the notations $\mathbb{E}_\theta(X)$ and $\mathbb{V}_\theta(X)$ are employed, where expectations and variances are taken under $X \sim P_\theta$. In particular, $\mathbb{E}_0(X)$ and $\mathbb{V}_0(X)$ refer to the expectation and variance under the true distribution P_0.

Therefore, when no confusion arises, $\mathbb{E}(X)$ and $\mathbb{E}_0(X)$ are used interchangeably, both denoting the expectation with respect to the true underlying distribution P_0.

Example 2: Population Variance

Assume that the estimand is the population variance of X,

$$\theta_0 = \theta(P_0) = \mathbb{E}_{P_0}(X^2) - [\mathbb{E}_{P_0}(X)]^2 = \int x^2\, dP_0(x) - \left(\int x\, dP_0(x)\right)^2.$$

The plug-in estimator is given by

$$\begin{aligned}\widehat{\theta}_n = \theta(\widehat{P}_n) &= \mathbb{E}_{\widehat{P}_n}(X^{*2}) - \left[\mathbb{E}_{\widehat{P}_n}(X^*)\right]^2 \\ &= \int x^2\, d\widehat{P}_n(x) - \left[\int x\, d\widehat{P}_n(x)\right]^2 \\ &= \frac{1}{n}\sum_{i=1}^n X_i^2 - \left(\frac{1}{n}\sum_{i=1}^n X_i\right)^2 \\ &= \frac{1}{n}\sum_{i=1}^n \left(X_i - \overline{X}_n\right)^2,\end{aligned}$$

which corresponds to the sample variance. Thus, the sample variance is obtained as the plug-in estimator of the population variance. □

However, it should be noted that the above plug-in estimator for variance $\mathbb{V}_{P_0}(X)$ is biased (see Exercise 2.4). In fact,

$$\mathbb{E}_{P_0}\left[\frac{1}{n}\sum_{i=1}^n \left(X_i - \overline{X}_n\right)^2\right] = \frac{n-1}{n}\mathbb{V}_{P_0}(X).$$

To correct for this bias, the sample variance is often defined using the following unbiased estimator:

$$S^2 = \frac{1}{n-1}\sum_{i=1}^n \left(X_i - \overline{X}_n\right)^2,$$

which satisfies

$$\mathbb{E}_{P_0}(S^2) = \mathbb{V}_{P_0}(X).$$

2.4.2 Bias and Variance

The *mean square error* (MSE) is widely utilized as a metric to evalute the performance of an estimator in the estimation of an estimand. When the estimand $\theta(P_0)$ is estimated using the estimator $\widehat{\theta}(\mathcal{D})$, the MSE is defined as

$$\text{MSE} = \mathbb{E}\left[\widehat{\theta}(\mathcal{D}) - \theta(P_0)\right]^2.$$

A decomposition of the MSE can be performed as follows:

$$\begin{aligned}
\text{MSE} &= \mathbb{E}\left\{\widehat{\theta}(\mathcal{D}) - \mathbb{E}[\widehat{\theta}(\mathcal{D})] + \mathbb{E}[\widehat{\theta}(\mathcal{D})] - \theta(P_0)\right\}^2 \\
&= \mathbb{E}\left\{\widehat{\theta}(\mathcal{D}) - \mathbb{E}[\widehat{\theta}(\mathcal{D})]\right\}^2 + \left|\mathbb{E}[\widehat{\theta}(\mathcal{D})] - \theta(P_0)\right|^2 \\
&\quad + 2\mathbb{E}\left\{\left(\widehat{\theta}(\mathcal{D}) - \mathbb{E}[\widehat{\theta}(\mathcal{D})]\right)\left(\mathbb{E}[\widehat{\theta}(\mathcal{D})] - \theta(P_0)\right)\right\} \\
&= \mathbb{V}\left[\widehat{\theta}(\mathcal{D})\right] + \left|\mathbb{E}[\widehat{\theta}(\mathcal{D})] - \theta(P_0)\right|^2 \\
&= \text{Variance} + \text{Bias}^2,
\end{aligned}$$

where the first term is interpreted as the variance of the estimator, and the second term corresponds to the square of the bias. The bias is defined as

$$\text{Bias} = \mathbb{E}[\widehat{\theta}(\mathcal{D})] - \theta(P_0).$$

The cross term in the above expansion is equal to zero, as shown below:

$$\begin{aligned}
&\mathbb{E}\left\{\left(\widehat{\theta}(\mathcal{D}) - \mathbb{E}[\widehat{\theta}(\mathcal{D})]\right)\left(\mathbb{E}[\widehat{\theta}(\mathcal{D})] - \theta(P_0)\right)\right\} \\
&= \left(\mathbb{E}[\widehat{\theta}(\mathcal{D})] - \theta(P_0)\right) \cdot \mathbb{E}\left\{\widehat{\theta}(\mathcal{D}) - \mathbb{E}[\widehat{\theta}(\mathcal{D})]\right\} \\
&= \left(\mathbb{E}[\widehat{\theta}(\mathcal{D})] - \theta(P_0)\right) \cdot 0 \\
&= 0.
\end{aligned}$$

Gold Coin # 7. *A decomposition of the mean square error (MSE) of an estimator in the estimation of an estimand is given by*

$$MSE = Variance + Bias^2.$$

An estimator is considered effective when both its variance and its bias are small.

2.4.3 Asymptotic and Non-Asymptotic Properties

Let $\widehat{\theta}_n$ be an estimator for the estimand $\theta_0 = \theta(P_0)$. It is desired that the estimator possesses certain favorable properties. These properties are typically classified into two categories: *asymptotic* and *non-asymptotic* properties.

The asymptotic properties are used to characterize the behavior of the estimator in the limit as the sample size n tends to infinity. In the next section, two desirable asymptotic properties will be discussed: *consistency* and *asymptotic normality*.

Although the main focus of this book is placed on the study of asymptotic properties, for the sake of completeness, a brief discussion of non-asymptotic properties is presented here.

Non-asymptotic properties are defined as the behaviors of an estimator in finite samples, that is, for a fixed sample size n. As discussed earlier, the accuracy of an estimator is commonly quantified using the MSE,

$$\text{MSE}\left(\widehat{\theta}_n\right) = \mathbb{V}_{P_0}\left(\widehat{\theta}_n\right) + \left|\mathbb{E}_{P_0}\left(\widehat{\theta}_n\right) - \theta_0\right|^2,$$

which is the sum of the variance and the squared bias. The MSE is often employed to compare the non-asymptotic performance of different estimators.

For instance, given two estimators, $\widehat{\theta}_n^{(1)}$ and $\widehat{\theta}_n^{(2)}$, the estimator associated with the smaller MSE is typically preferred. From the decomposition of the MSE, it can be seen that a trade-off often exists between variance and bias; some estimators are characterized by a smaller bias but a larger variance, while others exhibit the opposite pattern. The optimal estimator is often identified by appropriately balancing these two components.

Another classical criterion is based on the concept of the uniformly minimum variance unbiased estimator (UMVUE),[13] which refers to the unbiased estimator with the smallest variance among all unbiased estimators. According to this criterion, optimality is determined within the class of all unbiased estimators. An estimator $\widehat{\theta}_n$ is said to be unbiased if

$$\mathbb{E}_{P_0}\left(\widehat{\theta}_n\right) = \theta_0.$$

The appeal of the UMVUE lies in its simultaneous satisfaction of two desirable properties: unbiasedness and minimum variance. A comprehensive theoretical foundation has been developed around this concept, including the definitions of sufficiency and completeness, the Rao–Blackwell theorem, and the Lehmann–Scheffé theorem. However, UMVUEs are known to exist only for specific parametric families, and their practical applicability in modern statistics is therefore limited. For this reason, emphasis will later be placed on the concept of *efficiency*,[12] which is an asymptotic notion analogous to the UMVUE but more broadly applicable in contemporary statistical theory.

This section concludes with a classical example of UMVUE (see Exercise 2.5). Suppose $X_1, \ldots, X_n$ are i.i.d. with $X \sim \mathcal{N}(\mu, \sigma^2)$. If the estimand of interest is μ, then

$$\widehat{\mu}_n = \frac{1}{n}\sum_{i=1}^{n} X_i$$

serves as an unbiased estimator. According to the theory of completeness and sufficiency,[13] $\widehat{\mu}_n$ is identified as the unique UMVUE.

2.5 Consistency and Asymptotic Normality

Assume that $X_1, \ldots, X_n$ are i.i.d. with $X \sim P_0$. Let $\theta_0 = \theta(P_0)$ denote the estimand of interest, and let $\widehat{\theta}_n = \theta(\widehat{P}_n)$ denote an estimator of this estimand. It is desired that $\widehat{\theta}_n$ possess certain desirable asymptotic properties. The two key properties typically required of an estimator are:

1. **Consistency:**

$$\widehat{\theta}_n \xrightarrow{p} \theta_0, \quad \text{as } n \to \infty,$$

2. **Asymptotic Normality:**

$$\sqrt{n}(\widehat{\theta}_n - \theta_0) \xrightarrow{d} \mathcal{N}(0, \Sigma_{\theta_0}), \quad \text{as } n \to \infty,$$

where Σ_{θ_0} represents the asymptotic variance (if θ_0 is scalar) or the asymptotic covariance matrix (if θ_0 is vector-valued).

Example 1 (continued): Population Mean

If the estimand is the population mean, i.e., $\theta_0 = \mathbb{E}_{P_0}(X)$, then the corresponding plug-in estimator is given by

$$\widehat{\theta}_n = \mathbb{E}_{\widehat{P}_n}(X^*) = \frac{1}{n}\sum_{i=1}^{n} X_i.$$

By the LLN, the following convergence in probability is obtained:

$$\widehat{\theta}_n = \frac{1}{n}\sum_{i=1}^{n} X_i \xrightarrow{p} \mathbb{E}_{P_0}(X) = \theta_0,$$

demonstrating the consistency of the estimator. Furthermore, by the CLT, it is established that

$$\sqrt{n}(\widehat{\theta}_n - \theta_0) = \sqrt{n}\left(\frac{1}{n}\sum_{i=1}^{n} X_i - \mathbb{E}_{P_0}(X)\right) \xrightarrow{d} \mathcal{N}(0, \mathbb{V}_{P_0}(X)),$$

demonstrating the asymptotic normality of the estimator. □

Example 2 (continued): Population Variance

If the estimand is the population variance, i.e., $\theta_0 = \mathbb{V}_{P_0}(X)$, then the plug-in estimator is expressed as

$$\widehat{\theta}_n = \mathbb{V}_{\widehat{P}_n}(X^*) = \frac{1}{n}\sum_{i=1}^{n} X_i^2 - \left(\frac{1}{n}\sum_{i=1}^{n} X_i\right)^2.$$

To establish both the consistency and asymptotic normality of $\widehat{\theta}_n$, in addition to the LLN and CLT, the Continuous Mapping Theorem and the Slutsky's Lemma are also utilized.[12]

The Continuous Mapping Theorem. For a continuous function g, if $U_n \xrightarrow{p} U$, then $g(U_n) \xrightarrow{p} g(U)$; Similarly, if $U_n \xrightarrow{d} U$, then $g(U_n) \xrightarrow{d} g(U)$.

The Slutsky's Lemma. If $U_n \xrightarrow{d} U$ and $V_n \xrightarrow{p} c$ for some constant c, then (1) $U_n + V_n \xrightarrow{d} U + c$; and (2) $V_n U_n \xrightarrow{d} cU$.

To prove the consistency, by the LLN, it holds that

$$\frac{1}{n}\sum_{i=1}^{n} X_i \xrightarrow{p} \mathbb{E}_{P_0}(X) \quad \text{and} \quad \frac{1}{n}\sum_{i=1}^{n} X_i^2 \xrightarrow{p} \mathbb{E}_{P_0}(X^2).$$

Hence, by the continuous mapping theorem,

$$\widehat{\theta}_n = \frac{1}{n}\sum_{i=1}^{n} X_i^2 - \left(\frac{1}{n}\sum_{i=1}^{n} X_i\right)^2 \xrightarrow{p} \mathbb{E}_{P_0}(X^2) - \mathbb{E}^2_{P_0}(X) = \mathbb{V}_{P_0}(X) = \theta_0,$$

establishing the consistency of the estimator.

To prove the asymptotic normality, the estimator is rewritten as:

$$\widehat{\theta}_n = \frac{1}{n}\sum_{i=1}^{n}(X_i - \mu_0)^2 - (\overline{X}_n - \mu_0)^2,$$

where $\mu_0 = \mathbb{E}_{P_0}(X)$. Therefore,

$$\sqrt{n}(\widehat{\theta}_n - \theta_0) = \sqrt{n}\left(\frac{1}{n}\sum_{i=1}^{n}(X_i - \mu_0)^2 - \theta_0\right) - \sqrt{n}(\overline{X}_n - \mu_0)^2.$$

Applying the CLT to the first term yields:

$$\sqrt{n}\left(\frac{1}{n}\sum_{i=1}^{n}(X_i - \mu_0)^2 - \theta_0\right) \xrightarrow{d} \mathcal{N}(0, \sigma^2_{\theta_0}),$$

where

$$\sigma^2_{\theta_0} = \mathbb{V}_{P_0}\left[(X - \mu_0)^2\right],$$

under the assumption that $\mathbb{E}_{P_0}(X^4) < \infty$. The second term is rewritten as:

$$\sqrt{n}(\overline{X}_n - \mu_0) \cdot (\overline{X}_n - \mu_0),$$

which converges in probability to zero, since $\sqrt{n}(\overline{X}_n - \mu_0) \xrightarrow{d} \mathcal{N}(0, \mathbb{V}_{P_0}(X))$ and $(\overline{X}_n - \mu_0) \xrightarrow{p} 0$. By the Slutsky's lemma, it follows that

$$\sqrt{n}(\widehat{\theta}_n - \theta_0) \xrightarrow{d} \mathcal{N}(0, \sigma^2_{\theta_0}),$$

establishing the asymptotic normality of the estimator. □

Gold Coin # 8. *Let θ_0 be the estimand of interest, and let $\widehat{\theta}_n$ be an estimator. Two asymptotic properties are typically desired:*

(1) Consistency,

$$\widehat{\theta}_n \xrightarrow{p} \theta_0, \quad \text{as } n \to \infty;$$

(2) Asymptotic Normality,

$$\sqrt{n}\left(\widehat{\theta}_n - \theta_0\right) \xrightarrow{d} \mathcal{N}(0, \Sigma_{\theta_0}), \quad \text{as } n \to \infty.$$

2.6 Exercises

Exercise 2.1

Prove the Chebyshev's inequality:

$$\mathbb{P}(|X - \mu| \geq \epsilon) \leq \frac{\mathbb{V}(X)}{\epsilon^2}.$$

Exercise 2.2

Show that

$$\mathbb{P}(S_{2n} = 2k) \doteq \frac{1}{\sqrt{\pi n}} e^{-x^2/2}, \quad \text{as } \frac{2k}{\sqrt{2n}} \to x.$$

Exercise 2.3

Write R code to generate the plots in Figure 2.3.

Exercise 2.4

Show that

$$\mathbb{E}\left[\frac{1}{n}\sum_{i=1}^{n}\left(X_i - \overline{X}_n\right)^2\right] = \frac{n-1}{n}\mathbb{V}(X).$$

Exercise 2.5

Suppose $X_1, \ldots, X_n$ are i.i.d. with $X \sim \mathcal{N}(\mu, 1)$. Consider the following two unbiased estimators:

$$\widehat{\mu}_1 = \frac{1}{n}\sum_{i=1}^{n} X_i,$$
$$\widehat{\mu}_2 = \text{median}\{X_1, \ldots, X_n\}.$$

Use R for a simulation showing that $\widehat{\mu}_1$ has a smaller variance than $\widehat{\mu}_2$.

3

Statistical Inference

3.1 Two Great Statisticians

Ronald A. Fisher and Jerzy Neyman were two great statisticians in the history of modern statistics. The inclusion of a brief biography of them was initially considered for this chapter. However, this idea was set aside because of the extensive existing literature that already documents their lives and contributions in considerable detail. Instead, the first paragraph of the preface to *Fisher, Neyman, and the Creation of Classical Statistics* by Erich L. Lehmann—Neyman's first student at Berkeley—is cited to provide historical context and highlight their impact:[14]

> "Classical statistical theory—hypothesis testing, estimation, and the design of experiments and sample surveys—is mainly the creation of two men: Ronald A. Fisher (1890–1962) and Jerzy Neyman (1894–1981). Their contributions sometimes complemented each other, sometimes occurred in parallel, and, particularly at later stages, often were in strong opposition. The two men would not be pleased to see their names linked in this way, since throughout most of their working lives they detested each other. Nevertheless, they worked on the same problems, and through their combined efforts created a new discipline."

The famous debate between Fisher and Neyman is regarded as a gold coin. In this chapter, rather than dwelling on their dispute, attention is directed toward the powerful foundations they helped establish for statistical inference. Their contributions laid the groundwork for two essential tools: hypothesis testing and confidence interval estimation.

Gold Coin # 9. *The Fisher–Neyman debate between the two pioneering statisticians led to the development of several foundational concepts and methodologies in modern statistics. Among these, hypothesis testing and confidence interval estimation remain two of the most widely used tools in statistical inference.*

DOI: 10.1201/9781003635468-4

TABLE 3.1
Hypothetical data of the lady-tasting-tea example

Cup	True Preparation	Lady's Claim	Correct?
1	Milk-Tea	Milk-Tea	Yes
2	Milk-Tea	Tea-Milk	No
3	Tea-Milk	Tea-Milk	Yes
4	Tea-Milk	Tea-Milk	Yes
5	Tea-Milk	Milk-Tea	No
6	Milk-Tea	Milk-Tea	Yes
7	Tea-Milk	Tea-Milk	Yes
8	Milk-Tea	Milk-Tea	Yes

3.2 Testing

3.2.1 Significance Testing

The design of experiments, the principle of randomization, the concept of *p*-value, and the framework of significance testing are all foundational elements of modern statistics. Yet, where should our discussion begin?

"It was a summer afternoon in Cambridge, England, in the late 1920s. A group of university dons, their wives, and some guests were sitting around an outdoor table for afternoon tea. One of the women was insisting that the tea tasted different depending on whether the tea was poured into the milk or whether the milk was poured into the tea."

This scene, used by David Salsburg to open his book *The Lady Tasting Tea*,[15] has become emblematic of how statistical thinking transformed science in the 20th century.

In response to the claim, an experiment was designed by R. A. Fisher to test the null hypothesis that no such distinction could be made. In the experiment, the lady was presented with eight cups of tea, randomly arranged. Four of the cups were prepared by pouring milk into tea, and the remaining four by pouring tea into milk. The lady was then asked to identify the four cups prepared by pouring milk into tea.

Under the null hypothesis, it is assumed that the lady has no ability to distinguish between the two preparation methods, and that any success in classification is attributable to chance. A hypothetical dataset from this experiment is shown in Table 3.1.

Let T denote the number of correct identifications among the four cups that were actually prepared by pouring milk into tea. The test statistic T can take the values 0 (none correct), 1, 2, 3, or 4 (all correct). Based on the data presented in Table 3.1, the observed value of the test statistic is $T^{\text{obs}} = 3$.

As proposed by Fisher, the *p*-value is defined as the probability of observing a test statistic as extreme as, or more extreme than, the observed value

$T^{\text{obs}} = 3$, under the assumption that the null hypothesis is true. The null hypothesis posits that the lady does not possess the ability to distinguish the two methods of tea preparation.

To compute the p-value, it is assumed that under the null hypothesis the lady randomly selects four cups from the eight, identifying them as having been prepared by pouring milk into tea. The total number of such combinations is given by

$$\binom{8}{4} = 70.$$

Among these, the number of combinations yielding $T = 4$ is

$$\binom{4}{4} \times \binom{4}{0} = 1,$$

and the number of combinations yielding $T = 3$ is

$$\binom{4}{3} \times \binom{4}{1} = 16.$$

Therefore, the p-value associated with $T^{\text{obs}} = 3$ is given by

$$\mathbb{P}(T \geq 3 \mid \text{null hypothesis true}) = \frac{1+16}{70} = \frac{17}{70} \doteq 24.3\%.$$

Since the resulting p-value exceeds the conventional threshold of 5%, the result is not considered statistically significant. Thus, no significant evidence against the null hypothesis would be claimed, and the conclusion would be that the lady's ability to distinguish the teas is not supported by the data.

As reported by Salsburg in *The Lady Tasting Tea*,[15] Fairfield Smith, a colleague of Fisher, stated that in the actual experiment, the lady correctly identified all four cups. The p-value corresponding to $T^{\text{obs}} = 4$ is

$$\frac{1}{70} \doteq 1.4\% < 5\%,$$

which would be considered statistically significant. In such a case, Fisher would have been willing to reject the null hypothesis and conclude that the lady does possess the claimed ability.

This type of statistical reasoning is known as *significance testing*, and the test described above is known as *Fisher's exact test.*

3.2.2 Permutation and P-value

Let us create a new story as follows. Many guests at the aforementioned tea party expressed curiosity about whether they shared the same ability as the

TABLE 3.2
Simulated data for the bigger tea tasting experiment

Participants	Fisher's Note	Participant's Note	Correct? (1/0)
1	Tea-Milk	Milk-Tea	0
2	Tea-Milk	Tea-Milk	1
3	Milk-Tea	Tea-Milk	0
4	Milk-Tea	Milk-Tea	1
5	Tea-Milk	Tea-Milk	1
6	Milk-Tea	Milk-Tea	1
7	Milk-Tea	Milk-Tea	1
8	Milk-Tea	Milk-Tea	1
9	Milk-Tea	Tea-Milk	0
10	Tea-Milk	Tea-Milk	1
11	Tea-Milk	Tea-Milk	1
12	Tea-Milk	Tea-Milk	1
13	Milk-Tea	Milk-Tea	1
14	Tea-Milk	Tea-Milk	1
15	Milk-Tea	Tea-Milk	0
16	Tea-Milk	Milk-Tea	0
17	Milk-Tea	Tea-Milk	0
18	Milk-Tea	Milk-Tea	1
19	Tea-Milk	Milk-Tea	0
20	Milk-Tea	Milk-Tea	1

lady. This curiosity prompted Fisher to design a larger-scale experiment. A total of 20 interested participants were asked to line up. In a location out of view of the participants, tea was prepared according to the outcome of fair coin tosses: if heads appeared, milk was poured into tea; if tails appeared, tea was poured into milk. The preparation for each cup was recorded by Fisher.

In sequence, each participant was handed a cup of tea, tasted it, recorded their guess on a note, and submitted it to Fisher. The notes submitted by the participants were then compared to Fisher's records, and the results were summarized in Table 3.2.

The last column of the table displays the binary outcomes of each comparison: 1 for a correct guess and 0 for an incorrect one. Let $X_1, \ldots, X_n$ denote these outcomes, with $n = 20$. Each X_i is a Bernoulli random variable indicating correctness.

In fact, the data shown in Table 3.2 were generated using R software with a fixed random seed (see Exercise 3.1). Fisher's approach to significance testing can now be followed. The null hypothesis under consideration posits that no participant possesses any special ability to distinguish between the two methods of tea preparation. The test statistic is defined as the total number

TABLE 3.3
An example of permutation (1st of 1000 random imputations)

Participants	Permuted Fisher's Note	Participant's Note	1/0
1	Milk-Tea	Milk-Tea	1
2	Milk-Tea	Tea-Milk	0
3	Milk-Tea	Tea-Milk	0
4	Tea-Milk	Milk-Tea	0
5	Milk-Tea	Tea-Milk	0
6	Tea-Milk	Milk-Tea	0
7	Tea-Milk	Milk-Tea	0
8	Milk-Tea	Milk-Tea	1
9	Tea-Milk	Tea-Milk	1
10	Milk-Tea	Tea-Milk	0
11	Tea-Milk	Tea-Milk	1
12	Tea-Milk	Tea-Milk	1
13	Milk-Tea	Milk-Tea	1
14	Tea-Milk	Tea-Milk	1
15	Tea-Milk	Tea-Milk	1
16	Milk-Tea	Milk-Tea	1
17	Tea-Milk	Tea-Milk	1
18	Milk-Tea	Milk-Tea	1
19	Milk-Tea	Milk-Tea	1
20	Milk-Tea	Milk-Tea	1

of correct guesses:

$$T = \sum_{i=1}^{20} X_i.$$

In this instance, the observed test statistic is $T^{\text{obs}} = 13$.

Under the null hypothesis, Fisher's notes and participants' guesses are assumed to be statistically independent. Thus, the distribution of T under the null hypothesis can be approximated by permuting Fisher's notes and comparing them to the participants' fixed guesses. There are $20! = 2{,}432{,}902{,}008{,}176{,}640{,}000$ such permutations, each yielding a value of T. The distribution of these values represents the sampling distribution of T under the null hypothesis.

However, evaluating all 20! permutations is computationally infeasible. Instead, a feasible number of random permutations, say 1000, can be performed to approximate the null distribution (see Exercise 3.2). Table 3.3 illustrates one such permutation. In this example, the resulting test statistic is again 13 (by accident), denoted $T^{(1)}$.

The p-value is then approximated by the proportion of permutations where the permuted test statistic is at least as extreme as the observed:

$$p\text{-value} \doteq \frac{\sum_{j=1}^{1000} I(T^{(j)} \geq T^{\text{obs}})}{1000} = 0.181 > 0.05.$$

Since the approximate p-value exceeds the conventional threshold of 0.05, the result is not considered statistically significant. Therefore, it cannot be concluded that the participants possess the ability to distinguish whether milk is poured into tea or vice versa.

3.2.3 Hypothesis Testing

Neyman was in England during the late 1920s and early 1930s, a period considered crucial for the development of his collaboration with Egon Pearson (1895–1980), son of Karl Pearson (1857–1936). Through this collaboration, foundational contributions to modern hypothesis testing were made, including the formulation of the groundbreaking Neyman–Pearson Lemma.

Imagine that Neyman had also been present at the tea party attended by Fisher—clearly, this is merely a wishful imagining of the two great statisticians having enjoyed an afternoon together. The data displayed in Table 3.2 will now be examined as they might have been analyzed within Neyman's formal framework of hypothesis testing.

Assume that $X_1, \ldots, X_n$ are i.i.d. with $X \sim \text{Bernoulli}(p)$, where $\mathbb{P}(X = 1) = 1 - \mathbb{P}(X = 0) = p$. The null hypothesis asserting that no individual possesses the ability to distinguish between tea poured into milk and milk poured into tea corresponds to $H_0 \colon p = 0.5$.

Unlike Fisher, who would have specified only the null hypothesis, Neyman's method requires that an alternative hypothesis be stated concurrently. For instance,

$$H_0 \colon p = 0.5 \quad \text{vs.} \quad H_a \colon p = 0.8.$$

This point alternative can alternatively be replaced with a two-sided hypothesis $H_a \colon p \neq 0.5$ or a one-sided alternative $H_a \colon p > 0.5$. These forms will be considered later when confidence intervals are discussed.

With an alternative hypothesis specified, the Neyman–Pearson lemma can be invoked to derive the most powerful test (if it exists). In this particular case, it turns out that the most powerful test statistic coincides with that used by Fisher, namely $T = \sum_{i=1}^{n} X_i$. The sampling distribution of T follows a binomial distribution $\text{Bin}(n, p)$, with

$$\mathbb{P}_p(T = k) = \binom{n}{k} p^k (1-p)^{n-k}, \quad k = 0, 1, \ldots, n.$$

In particular, under the null hypothesis $p = 0.5$, the distribution of T becomes $\text{Bin}(n, 0.5)$. The p-value can therefore be calculated as follows:

$$\begin{aligned}
p\text{-value} &= \mathbb{P}_{T\sim\text{Bin}(n,0.5)}(T \geq T^{\text{obs}}) \\
&= \sum_{k=T^{\text{obs}}}^{n} \binom{n}{k} 0.5^k (1-0.5)^{n-k} \\
&= \sum_{k=13}^{20} \binom{20}{k} 0.5^{20} \\
&= 0.132 > 0.05.
\end{aligned}$$

Hence, the null hypothesis cannot be rejected at the 5% significance level.

At this stage, Neyman's approach to hypothesis testing appears similar to Fisher's significance testing. This shared foundation serves as a unifying perspective for both methods.

Gold Coin # 10. *There are two main approaches to hypothesis testing—Fisher's approach and Neyman's approach. For simplicity, we will refer to both approaches collectively as hypothesis testing throughout this book.*

For completeness, further contributions achievable through Neyman's hypothesis testing framework are outlined in what follows.

First, as noted earlier, the most powerful test for a given alternative can be developed using the Neyman–Pearson lemma.

Second, when an alternative hypothesis is specified, a critical value for a given significance level α (e.g., $\alpha = 0.05$) can be determined. In the current example, the critical value is 15, since

$$\mathbb{P}(T \geq 14 \mid H_0) = \sum_{k=14}^{20} \binom{20}{k} 0.5^{20} = 0.057,$$

$$\mathbb{P}(T \geq 15 \mid H_0) = \sum_{k=15}^{20} \binom{20}{k} 0.5^{20} = 0.021,$$

and rejection of the null hypothesis is warranted when $T^{\text{obs}} \geq 15$.

Third, the *power* of the test, defined as

$$\text{Power} = \mathbb{P}(\text{Reject } H_0 \mid H_a),$$

can be computed. Under $H_a: p = 0.8$, the power is

$$\mathbb{P}(T \geq 15 \mid H_a) = \sum_{k=15}^{20} \binom{20}{k} 0.8^k (1-0.8)^{20-k} = 0.804.$$

Fourth, both Type I and Type II errors can be evaluated:

$$\text{Type I error} = \mathbb{P}(\text{Reject } H_0 \mid H_0) = \mathbb{P}(T \geq 15 \mid p = 0.5) = 0.021,$$
$$\text{Type II error} = \mathbb{P}(\text{Accept } H_0 \mid H_a) = \mathbb{P}(T \leq 14 \mid p = 0.8) = 0.196.$$

Fifth, many other pairs of null and alternative hypotheses can be prespecified, such as:

$$\begin{aligned}
H_0 &: p = p_{\text{null}} \quad \text{vs.} \quad H_a : p = p_a,\\
H_0 &: p = p_{\text{null}} \quad \text{vs.} \quad H_a : p \neq p_{\text{null}},\\
H_0 &: p = p_{\text{null}} \quad \text{vs.} \quad H_a : p > p_{\text{null}},\\
H_0 &: p = p_{\text{null}} \quad \text{vs.} \quad H_a : p < p_{\text{null}},\\
H_0 &: p \leq p_{\text{null}} \quad \text{vs.} \quad H_a : p > p_{\text{null}},\\
H_0 &: p \geq p_{\text{null}} \quad \text{vs.} \quad H_a : p < p_{\text{null}}.
\end{aligned}$$

These and related topics have been extensively developed and the reader is encouraged to consult the classical texts[16,17] for further reading.

3.3 Confidence Interval

3.3.1 Construction and Interpretation

Given that both hypothesis testing and confidence interval estimation were developed by Neyman, it is not surprising that the two methods are closely interconnected.

Lower One-sided Confidence Interval

The approach that would have been taken by Neyman for the tea tasting experiment is revisited for the following pair of hypotheses:

$$H_0 : p = p_{\text{null}} \quad \text{vs.} \quad H_a : p > p_{\text{null}}.$$

Upon the observed test statistic, denoted T^{obs}, the p-value is computed as

$$p\text{-value} = \mathbb{P}\left(T \geq T^{\text{obs}} \mid p = p_{\text{null}}\right).$$

Given a prespecified significance level α (e.g., 0.01, 0.025, the conventional 0.05, etc.), a decision can be made regarding rejection of the null hypothesis, based on whether the p-value is less than α.

In the tea tasting example, if

$$p\text{-value} = \sum_{k=T^{\text{obs}}}^{n} \binom{n}{k} p_{\text{null}}^{k} (1 - p_{\text{null}})^{n-k} < \alpha,$$

then the null hypothesis $p = p_{\text{null}}$ is rejected in favor of the alternative $p > p_{\text{null}}$. For a given observed value T^{obs}, it is desired to determine which values of p_{null} would result in rejection of the null hypothesis and which would not.

It will be shown that a threshold exists such that, when the prespecified p_{null} is below this threshold, the null hypothesis is rejected, and when it is greater than or equal to the threshold, it is not rejected.

This threshold is denoted by $\widehat{p}_{\text{lower}}$ and is defined as the solution to

$$\mathbb{P}\left(T \geq T^{\text{obs}} \mid p = \widehat{p}_{\text{lower}}\right) = \sum_{k=T^{\text{obs}}}^{n} \binom{n}{k} \widehat{p}_{\text{lower}}^{k} (1 - \widehat{p}_{\text{lower}})^{n-k} = \alpha,$$

which is well-defined for $T^{\text{obs}} > 0$, because

$$g(p) = \sum_{k=T^{\text{obs}}}^{n} \binom{n}{k} p^k (1-p)^{n-k}$$

is a continuous and increasing function of p, with $g(0) = 0$ and $g(1) = 1$. When $T^{\text{obs}} = 0$, $\widehat{p}_{\text{lower}}$ is defined as 0.

The increasing nature of $g(p)$ is justified by noting that, for any $p_1 < p_2$, it holds that

$$g(p_1) = \mathbb{P}\left(T \geq T^{\text{obs}} \mid p = p_1\right) < \mathbb{P}\left(T \geq T^{\text{obs}} \mid p = p_2\right) = g(p_2).$$

It can now be verified whether $\widehat{p}_{\text{lower}}$ serves as the threshold of interest. In fact, if $p_{\text{null}} < \widehat{p}_{\text{lower}}$, then $g(p_{\text{null}}) < g(\widehat{p}_{\text{lower}}) = \alpha$, and the null hypothesis is rejected. Conversely, if $p_{\text{null}} \geq \widehat{p}_{\text{lower}}$, then $g(p_{\text{null}}) \geq g(\widehat{p}_{\text{lower}}) = \alpha$, and the null hypothesis is not rejected.

Accordingly, a one-sided $(1-\alpha)100\%$ confidence interval for the unknown true parameter p_0 has been constructed as

$$(1-\alpha)100\% \text{ CI} = [\widehat{p}_{\text{lower}}, 1].$$

It is important to distinguish between p_0 and p_{null}: p_0 represents the true value of the parameter p (i.e., the estimand of interest), whereas p_{null} denotes a tentative value specified under the null hypothesis $H_0 : p = p_{\text{null}}$.

Upper One-sided Confidence Interval

A similar procedure is now applied to the following pair of hypotheses:

$$H_0 : p = p_{\text{null}} \quad \text{vs.} \quad H_a : p < p_{\text{null}}.$$

Upon the observed test statistic, denoted T^{obs}, the p-value is computed as

$$p\text{-value} = \mathbb{P}\left(T \leq T^{\text{obs}} \mid p = p_{\text{null}}\right).$$

Given a prespecified significance level α, a decision regarding rejection of the null hypothesis is based on whether the p-value is less than α.

In the context of the tea tasting example, the p-value is given by

$$p\text{-value} = \sum_{k=0}^{T^{\text{obs}}} \binom{n}{k} p_{\text{null}}^k (1 - p_{\text{null}})^{n-k}.$$

If this value is less than α, the null hypothesis is rejected in favor of the alternative.

As in the previous case, a threshold can be defined such that, when the prespecified value p_{null} exceeds this threshold, the null hypothesis is rejected, and when it is less than or equal to the threshold, the null hypothesis is not rejected.

This threshold is denoted by $\widehat{p}_{\text{upper}}$ and is defined as the solution to

$$\mathbb{P}\left(T \leq T^{\text{obs}} \mid p = \widehat{p}_{\text{upper}}\right) = \sum_{k=0}^{T^{\text{obs}}} \binom{n}{k} \widehat{p}_{\text{upper}}^k (1 - \widehat{p}_{\text{upper}})^{n-k} = \alpha,$$

which is well-defined for $T^{\text{obs}} < n$, since the function

$$h(p) = \sum_{k=0}^{T^{\text{obs}}} \binom{n}{k} p^k (1 - p)^{n-k}$$

is continuous and strictly decreasing in p, with $h(0) = 1$ and $h(1) = 0$. When $T^{\text{obs}} = n$, the threshold is defined to be $\widehat{p}_{\text{upper}} = 1$.

To verify that $\widehat{p}_{\text{upper}}$ serves as the appropriate threshold, note that if $p_{\text{null}} > \widehat{p}_{\text{upper}}$, then $h(p_{\text{null}}) < h(\widehat{p}_{\text{upper}}) = \alpha$, and the null hypothesis is rejected. Conversely, if $p_{\text{null}} \leq \widehat{p}_{\text{upper}}$, then $h(p_{\text{null}}) \geq \alpha$, and the null hypothesis is not rejected.

Accordingly, an upper one-sided confidence interval for the unknown true parameter p_0 is constructed with confidence level $1 - \alpha$ as

$$(1 - \alpha)100\% \text{ CI} = [0, \widehat{p}_{\text{upper}}].$$

Two-sided Confidence Interval

The following pair of null and alternative hypotheses is now considered:

$$H_0 : p = p_{\text{null}} \quad \text{vs.} \quad H_a : p \neq p_{\text{null}}.$$

By combining the procedures used to construct one-sided confidence intervals, a two-sided confidence interval for the unknown parameter p_0 can be constructed with confidence level $1 - \alpha$:

$$(1 - \alpha)100\% \text{ CI} = [\widehat{p}_{\text{lower}}, \widehat{p}_{\text{upper}}],$$

where $\widehat{p}_{\text{lower}}$ and $\widehat{p}_{\text{upper}}$ are the solutions to

$$\mathbb{P}\left(T \geq T^{\text{obs}} \mid p = \widehat{p}_{\text{lower}}\right) = \alpha/2,$$
$$\mathbb{P}\left(T \leq T^{\text{obs}} \mid p = \widehat{p}_{\text{upper}}\right) = \alpha/2,$$

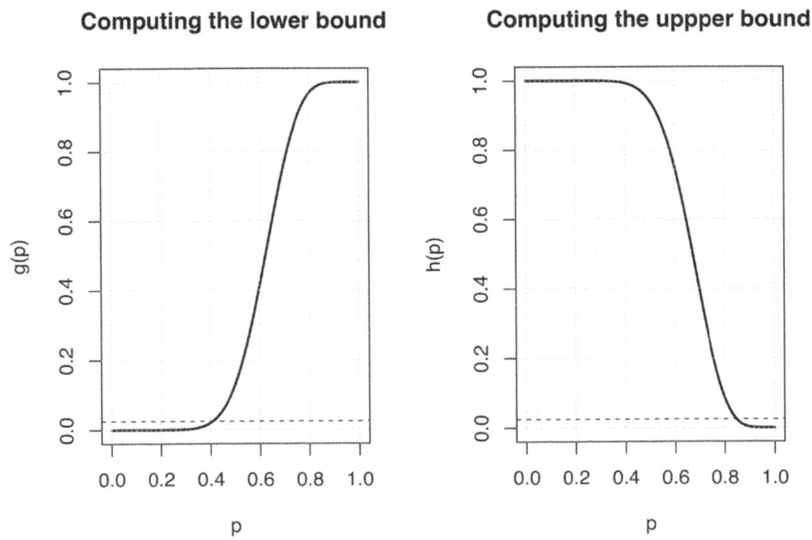

FIGURE 3.1
Demonstration of computing the lower and upper bounds

respectively. This interval is referred to as the *exact binomial confidence interval*, also known as the *Clopper–Pearson confidence interval.*

In the tea tasting example, the test statistic follows a binomial distribution, $T \sim \text{Bin}(n, p_0)$, with $n = 20$ and p_0 denoting the parameter of interest. Given the observed value $T^{\text{obs}} = 13$, a 95% confidence interval of p_0 is obtained (see Exercise 3.3), as illustrated in Figure 3.1:

$$95\% \text{ two-sided CI} = [0.408, 0.849].$$

Interpretation

Constructing confidence intervals can be challenging; interpreting them, however, is often even more difficult. Indeed, "it is easier done than said."

The tea tasting example from Table 3.2, which led to the 95% confidence interval $[0.408, 0.849]$, is now continued. A common—but incorrect—interpretation of this interval is that there is a 95% probability that the true parameter p_0 lies within $[0.408, 0.849]$. This interpretation is incorrect because the computed interval either contains the true value of p_0 (in which case the probability is 1), or it does not (in which case the probability is 0).

A textbook explanation of the statement "95% two-sided CI $=$ $[0.408, 0.849]$" is the following: "We are 95% confident that the true value p_0 lies between 0.408 and 0.849." However, this is simply a restatement of the mathematical expression and may not provide meaningful insight. Another textbook explanation is that "If you repeated the same sampling process infinitely many times and constructed a 95% confidence interval each time, then about 95% of those intervals would contain the true parameter." However, it is a statement about the procedure, not about your specific interval.

The difficulty in interpretation motivates the earlier discussion on the connection between confidence intervals and hypothesis testing. In the construction of confidence intervals, it has been shown that the confidence level is related to the significance level α—specifically, the confidence level is equal to $1-\alpha$ under this connection. For instance, a significance level of 5% corresponds to a 95% confidence level.

Given the observed value T^{obs}, for any $p' \in [0.408, 0.849]$, the null hypothesis $H_0 : p = p'$ in the pair

$$H_0 : p = p' \quad \text{vs.} \quad H_a : p \neq p'$$

would not be rejected at the 5% significance level. In other words, the set of values of p for which the null hypothesis would not be rejected—given T^{obs} and a 5% significance level—forms the 95% confidence interval for the true parameter p_0.

Gold Coin # 11. *Both hypothesis testing and confidence interval estimation were developed by Neyman; therefore, it is not surprising that the two methods are closely connected. The idea of the significance level α in hypothesis testing can be used to interpret the confidence level* $(1-\alpha)100\%$ *associated with a confidence interval.*

3.3.2 Point Estimator and Standard Error

The goal is to estimate the estimand of interest, denoted by θ_0. When an estimator $\widehat{\theta}_n$ is constructed for this purpose, it is referred to as a *point estimator.*[13] The uncertainty associated with a point estimator can be quantified using a confidence interval. An alternative approach for quantifying uncertainty is through the standard error of the point estimator.

Assume that $X_1, \ldots, X_n$ are i.i.d. with $X \sim P_0$. An estimand is defined as a function of the true distribution P_0, i.e., $\theta_0 = \theta(P_0)$. The standard error of the estimator $\widehat{\theta}_n$ is defined as

$$\text{SE}\left(\widehat{\theta}_n\right) = \left[\mathbb{V}_{P_0}(\widehat{\theta}_n)\right]^{1/2}.$$

The tea tasting example presented in Table 3.2 is now continued. Suppose that $X_1, \ldots, X_n$ are i.i.d. with $X \sim \text{Bernoulli}(p_0)$. The plug-in estimator for

$p_0 = \mathbb{E}_{P_0}(X) = \mathbb{P}_{P_0}(X = 1)$ is given by

$$\widehat{p}_n = \overline{X}_n = \frac{1}{n}\sum_{i=1}^{n} X_i = \frac{1}{n}\sum_{i=1}^{n} I(X_i = 1).$$

Consequently, the standard error of $\widehat{p}_n$ is expressed as

$$\mathrm{SE}\,(\widehat{p}_n) = [\mathbb{V}_{P_0}(\widehat{p}_n)]^{1/2} = \left[\frac{p_0(1-p_0)}{n}\right]^{1/2}.$$

An estimator for this standard error is

$$\widehat{\mathrm{SE}}\,(\widehat{p}_n) = \left[\frac{\widehat{p}_n(1-\widehat{p}_n)}{n}\right]^{1/2}.$$

Using the data from Table 3.2, the point estimate $\widehat{p}_n = 13/20 = 0.65$ is obtained, and the estimated standard error is calculated as

$$\widehat{\mathrm{SE}}(\widehat{p}_n) = \sqrt{\frac{0.65(1-0.65)}{20}} = 0.107.$$

By the law of large numbers, it follows that $\widehat{p}_n \xrightarrow{p} p_0$ as $n \to \infty$. Furthermore, by the continuous mapping theorem,

$$\sqrt{\widehat{p}_n(1-\widehat{p}_n)} \xrightarrow{p} \sqrt{p_0(1-p_0)}, \quad \text{as } n \to \infty.$$

In other words, both the point estimator $\widehat{p}_n$ for the estimand and the estimator $\widehat{\mathrm{SE}}(\widehat{p}_n)$ for the standard error are consistent.

3.3.3 Bootstrap Method

In the previous example, the standard error of $\widehat{p}_n$ was available in closed form. However, when an explicit formula for the standard error of an estimator is difficult to derive (see Exercise 3.4), the *bootstrap method* can be employed as an alternative to estimate it.[4]

As introduced in Chapter 2, the *empirical distribution* $\widehat{P}_n$ is defined as the probability distribution that assigns the probability mass $1/n$ to each observation X_i for $i = 1, \ldots, n$.

Let $\mathcal{D} = \{X_1, \ldots, X_n\}$ denote the observed dataset. An estimator of the estimand of interest is defined as a function of the data,

$$\widehat{\theta}_n = \widehat{\theta}(\mathcal{D}) = \widehat{\theta}(X_1, \ldots, X_n).$$

The distribution of $\widehat{\theta}_n$ under repeated sampling is referred to as its *sampling distribution*. To understand it, one may imagine that many datasets

$\mathcal{D}^{(j)} = \{X_1^{(j)}, \ldots, X_n^{(j)}\}$ are independently drawn from the population distribution P_0, and the corresponding values of the estimator are computed:

$$X_1^{(j)}, \ldots, X_n^{(j)} \overset{\text{i.i.d.}}{\sim} P_0 \quad \Rightarrow \quad \widehat{\theta}_n^{(j)} = \widehat{\theta}(X_1^{(j)}, \ldots, X_n^{(j)}), \quad j = 1, \ldots, M,$$

where M is a large number (e.g., $M = 10^6$). The empirical distribution of the resulting values $\{\widehat{\theta}_n^{(j)}\}_{j=1}^M$ approximates the true sampling distribution of $\widehat{\theta}_n$.

Following this idea, the bootstrap method approximates the sampling distribution by replacing the true distribution P_0 with the empirical distribution $\widehat{P}_n$. Specifically, multiple (say, $M = 2000$) bootstrap datasets $\mathcal{D}^{*(j)} = \{X_1^{*(j)}, \ldots, X_n^{*(j)}\}$ are generated by sampling from $\widehat{P}_n$, and corresponding estimator values are computed:

$$X_1^{*(j)}, \ldots, X_n^{*(j)} \overset{\text{i.i.d.}}{\sim} \widehat{P}_n \quad \Rightarrow \quad \widehat{\theta}_n^{*(j)} = \widehat{\theta}(X_1^{*(j)}, \ldots, X_n^{*(j)}), \quad j = 1, \ldots, M.$$

The distribution of the values $\{\widehat{\theta}_n^{*(j)}\}_{j=1}^M$ is referred to as the *bootstrap distribution* of $\widehat{\theta}_n$.

Each dataset $\{X_1^*, \ldots, X_n^*\}$ is called a *bootstrap sample*, generated by sampling with replacement from the observed data $\{X_1, \ldots, X_n\}$. An example of such a bootstrap sample in the context of the tea tasting experiment is shown in Table 3.4 (see Exercise 3.5). Based on that bootstrap sample, the point estimate $\widehat{p}_n^* = 14/20$ is obtained, compared to the original estimate $\widehat{p}_n = 13/20$.

TABLE 3.4
One bootstrap sample of the tea tasting data

Bootstrap Sample	Fisher's Note	Participant's Note	Outcome (1/0)
5	Tea-Milk	Tea-Milk	1
12	Tea-Milk	Tea-Milk	1
10	Tea-Milk	Tea-Milk	1
9	Milk-Tea	Tea-Milk	0
16	Tea-Milk	Milk-Tea	0
13	Milk-Tea	Milk-Tea	1
3	Milk-Tea	Tea-Milk	0
14	Tea-Milk	Tea-Milk	1
18	Milk-Tea	Milk-Tea	1
6	Milk-Tea	Milk-Tea	1
5	Tea-Milk	Tea-Milk	1
16	Tea-Milk	Milk-Tea	0
17	Milk-Tea	Tea-Milk	0
18	Milk-Tea	Milk-Tea	1
5	Tea-Milk	Tea-Milk	1
15	Milk-Tea	Tea-Milk	0
2	Tea-Milk	Tea-Milk	1
10	Tea-Milk	Tea-Milk	1
6	Milk-Tea	Milk-Tea	1
20	Milk-Tea	Milk-Tea	1

By repeating this process many times, a collection of bootstrap replicates $\{\widehat{p}_n^*\}$ is generated, from which a histogram can be constructed to approximate the bootstrap distribution of $\widehat{p}_n$.

There are two primary uses for the bootstrap distribution.[4] The first use is to estimate the standard error of the estimator $\widehat{\theta}_n$. To understand this, note that the standard error of $\widehat{\theta}_n$ is defined as the standard deviation of its sampling distribution. Given the conceptual "bijective" correspondence between the bootstrap distribution and the sampling distribution, the standard deviation of the bootstrap distribution can be employed to approximate that of the sampling distribution.

Accordingly, the standard error of $\widehat{\theta}_n$ can be estimated using the sample standard deviation of the bootstrap replicates $\{\widehat{\theta}_n^{*(j)}\}_{j=1}^M$:

$$\widehat{\mathrm{SE}}_{\mathrm{boot}} = \left[\frac{1}{M-1}\sum_{j=1}^{M}\left(\widehat{\theta}_n^{*(j)} - \frac{1}{M}\sum_{j=1}^{M}\widehat{\theta}_n^{*(j)}\right)^2\right]^{1/2}.$$

A second use of the bootstrap distribution is the construction of a $(1-\alpha)100\%$ confidence interval for the unknown parameter θ_0. Given the bijective correspondence between the bootstrap and sampling distributions, a similar relationship holds between their respective quantiles. Denote the lower and upper $\alpha/2$ quantiles of the bootstrap distribution by $\widehat{L}_{\alpha/2}$ and $\widehat{U}_{\alpha/2}$, respectively. Then, the interval

$$[\widehat{L}_{\alpha/2}, \widehat{U}_{\alpha/2}]$$

serves as a $(1-\alpha)100\%$ confidence interval for θ_0.

Based on the dataset in Table 3.2 and $M = 2000$ bootstrap replications, the bootstrap estimate of standard error is computed as $\widehat{\mathrm{SE}}_{\mathrm{boot}} = 0.106$, and the bootstrap percentile confidence interval is given by

$$[\widehat{L}_{2.5\%}, \widehat{U}_{2.5\%}] = [0.45, 0.85].$$

The histogram of 2000 bootstrap estimates is displayed in Figure 3.2.

3.4 A Unified Approach

In Chapter 1, it was stated that one of the most useful formulas in statistics is the density function of the normal distribution. In this section, a unified approach to statistical inference is presented. This approach is built upon the density function of the standard normal distribution $\mathcal{N}(0,1)$, given by

$$\phi(x) = \frac{1}{\sqrt{2\pi}}e^{-x^2/2},$$

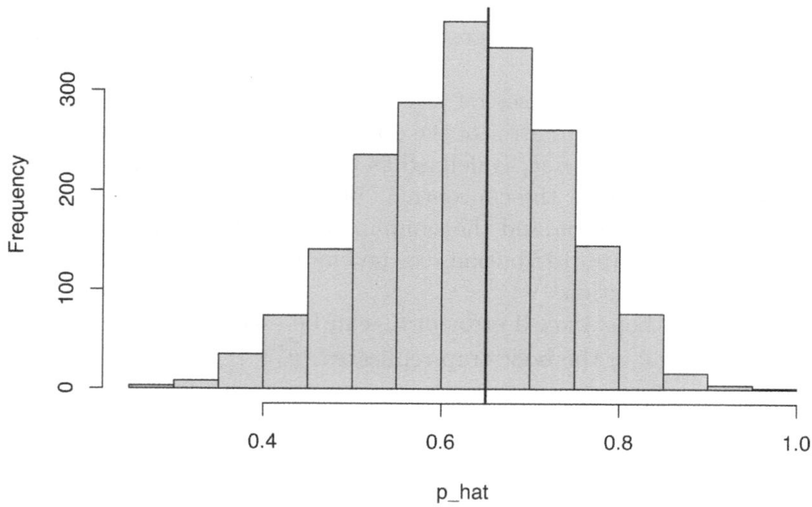

FIGURE 3.2
Histogram based on 2000 bootstrap samples

along with its corresponding cumulative distribution function,

$$\Phi(x) = \int_{-\infty}^{x} \frac{1}{\sqrt{2\pi}} e^{-t^2/2}\, dt.$$

3.4.1 Statistical Inference Based on Normality

Hypothesis Testing

Suppose $X \sim \mathcal{N}(\mu, 1)$, where the true but unknown parameter μ is denoted by μ_0. The following null and alternative hypotheses are to be tested:

$$H_0 : \mu = \mu_{\text{null}} \quad \text{vs.} \quad H_a : \mu \neq \mu_{\text{null}},$$

where μ_{null} represents a prespecified value under the null hypothesis. The statistic X is employed as the test statistic—its optimality under the Neyman–Pearson lemma is acknowledged but not detailed here. After a single trial, an observed value X^{obs} is obtained and the two-sided p-value is

$$\begin{aligned} p\text{-value} &= \mathbb{P}_{X\sim\mathcal{N}(\mu_{\text{null}},1)}\left(|X - \mu_{\text{null}}| \geq |X^{\text{obs}} - \mu_{\text{null}}|\right) \\ &= 2\mathbb{P}_{X\sim\mathcal{N}(\mu_{\text{null}},1)}\left(X - \mu_{\text{null}} \geq |X^{\text{obs}} - \mu_{\text{null}}|\right) \\ &= 2\mathbb{P}\left(\mathcal{N}(0,1) \geq |X^{\text{obs}} - \mu_{\text{null}}|\right) \\ &= 2\left[1 - \Phi\left(|X^{\text{obs}} - \mu_{\text{null}}|\right)\right]. \end{aligned}$$

In particular, if $\mu_{\text{null}} = 0$, then the p-value simplifies to $2\left[1 - \Phi\left(|X^{\text{obs}}|\right)\right]$.

Given a significance level α (e.g., 5%), the null hypothesis is rejected if the p-value is less than α; otherwise, it is not rejected.

Confidence Interval

Given the connection between confidence intervals and hypothesis testing, the set of values of μ_{null} for which the null hypothesis is not rejected at level α forms a $(1-\alpha)100\%$ confidence interval for μ_0. To ensure that $H_0 : \mu = \mu_{\text{null}}$ is not rejected, the following inequality must hold:

$$2\left[1 - \Phi\left(\left|X^{\text{obs}} - \mu_{\text{null}}\right|\right)\right] \geq \alpha.$$

Let $z_{\alpha/2}$ denote the $(1-\alpha/2)$ quantile of the standard normal distribution. The above condition is equivalent to

$$\left|X^{\text{obs}} - \mu_{\text{null}}\right| \leq z_{\alpha/2},$$

which in turn is equivalent to

$$X^{\text{obs}} - z_{\alpha/2} \leq \mu_{\text{null}} \leq X^{\text{obs}} + z_{\alpha/2}.$$

Hence, a $(1-\alpha)100\%$ confidence interval for μ_0 is given by

$$\left[X^{\text{obs}} - z_{\alpha/2}, X^{\text{obs}} + z_{\alpha/2}\right].$$

With this foundation in place, the unified approach to statistical inference can now be described in the next subsection.

3.4.2 Statistical Inference Based on Asymptotic Normality

Assume that $X_1, \ldots, X_n$ are i.i.d. with $X \sim P_0$. An estimand is defined as a functional of the true distribution P_0, that is, $\theta_0 = \theta(P_0)$.

Let $\mathcal{D} = \{X_1, \ldots, X_n\}$ denote the set of n observations. An estimator for the estimand of interest is represented as a function of $\mathcal{D}$, written as $\widehat{\theta}_n = \widehat{\theta}(\mathcal{D})$.

It is assumed that the following three asymptotic properties can be established: (1) consistency of $\widehat{\theta}_n$ for the estimand of interest,

$$\widehat{\theta}_n \xrightarrow{p} \theta_0, \quad \text{as } n \to \infty,$$

(2) asymptotic normality of $\widehat{\theta}_n$,

$$\sqrt{n}(\widehat{\theta}_n - \theta_0) \xrightarrow{d} \mathcal{N}(0, \sigma_0^2), \quad \text{as } n \to \infty,$$

and (3) consistency of $\widehat{\sigma}_n^2$ as an estimator for the asymptotic variance σ_0^2,

$$\widehat{\sigma}_n^2 \xrightarrow{p} \sigma_0^2, \quad \text{as } n \to \infty.$$

Based on these three asymptotic results, the procedures described in the previous subsection can be applied to conduct hypothesis testing and construct confidence intervals.

Hypothesis Testing

Assuming that the asymptotic normality result has been established, the following hypotheses are considered:

$$H_0 : \theta = \theta_{\text{null}} \quad \text{vs.} \quad H_a : \theta \neq \theta_{\text{null}},$$

where θ_{null} denotes a prespecified value. The test statistic is

$$T = \frac{\widehat{\theta}_n - \theta_{\text{null}}}{\widehat{\sigma}_n/\sqrt{n}} \mathrel{\dot{\sim}} \mathcal{N}(0,1), \quad \text{under } H_0,$$

where $\dot{\sim}$ indicates approximate distribution. The observed value of the above test statistic is denoted by T^{obs}.

Following Fisher's approach, the p-value is approximated by

$$p\text{-value} \doteq 2\mathbb{P}\left(\mathcal{N}(0,1) \geq |T^{\text{obs}}|\right) = 2\left[1 - \Phi\left(|T^{\text{obs}}|\right)\right].$$

For a given significance level α, according to Neyman's approach, the null hypothesis is rejected if the p-value is less than α; otherwise, it is not rejected.

Confidence Interval

To construct a $(1-\alpha)100\%$ confidence interval for θ_0, the set of all values of θ_{null} for which the null hypothesis $H_0 : \theta = \theta_{\text{null}}$ would not be rejected at level α is considered. The condition for non-rejection is given by:

$$2\left[1 - \Phi\left(\frac{|\widehat{\theta}_n - \theta_{\text{null}}|}{\widehat{\sigma}_n/\sqrt{n}}\right)\right] \geq \alpha,$$

which is equivalent to:

$$\frac{|\widehat{\theta}_n - \theta_{\text{null}}|}{\widehat{\sigma}_n/\sqrt{n}} \leq z_{\alpha/2},$$

and further equivalent to:

$$\widehat{\theta}_n - z_{\alpha/2} \cdot \frac{\widehat{\sigma}_n}{\sqrt{n}} \leq \theta_{\text{null}} \leq \widehat{\theta}_n + z_{\alpha/2} \cdot \frac{\widehat{\sigma}_n}{\sqrt{n}}.$$

Thus, an approximate $(1-\alpha)100\%$ confidence interval for θ_0 is given by:

$$\left[\widehat{\theta}_n - z_{\alpha/2} \cdot \frac{\widehat{\sigma}_n}{\sqrt{n}},\ \widehat{\theta}_n + z_{\alpha/2} \cdot \frac{\widehat{\sigma}_n}{\sqrt{n}}\right],$$

or equivalently,

$$\widehat{\theta}_n \pm z_{\alpha/2} \cdot \frac{\widehat{\sigma}_n}{\sqrt{n}}.$$

Gold Coin # 12. *Based on the asymptotic normality of* $\widehat{\theta}_n$, *the following expressions can be used:*

$$p\text{-value} \doteq 2\Phi\left(-\frac{|\widehat{\theta}_n - \theta_{null}|}{\widehat{\sigma}_n/\sqrt{n}}\right),$$

as an approximate p-value for testing $H_0 : \theta = \theta_0$ *versus* $H_a : \theta \neq \theta_0$, *and*

$$\widehat{\theta}_n \pm z_{\alpha/2} \cdot \frac{\widehat{\sigma}_n}{\sqrt{n}}$$

as an approximate $(1 - \alpha)100\%$ *confidence interval for* θ_0.

The unified approach described above is now applied to the data presented in Table 3.2, where $\widehat{p}_n = 0.65$ and $\widehat{\sigma}_n^2 = 0.65(1 - 0.65) = 0.2275$. To test the hypotheses $H_0 : p = 0.5$ vs. $H_a : p \neq 0.5$, the observed test statistic is computed as:

$$T^{\text{obs}} = \frac{0.65 - 0.5}{\sqrt{0.2275/20}} = 1.406.$$

The p-value is therefore approximately 0.159.

To obtain a 95% confidence interval for p_0, note that $z_{2.5\%} = 1.96$ and $1.96 \cdot \sqrt{0.2275/20} = 0.209$. Thus, the 95% confidence interval is given by:

$$0.65 \pm 0.209 = [0.441, 0.859].$$

3.4.3 The Moderate Number 2

An unsettled issue concerns the determination of α as the significance level in hypothesis testing and the selection of $1 - \alpha$ as the confidence level in the construction of confidence intervals. In the preceding example, a significance level of $\alpha = 5\%$ and a corresponding confidence level of $1 - \alpha = 95\%$ were considered. These particular values were chosen solely for illustrative purposes.

It is known that the use of the 5% significance level was introduced as a convention by Fisher, and its usage has become widespread in practice. However, the prevalence of this convention has led to ongoing discussions and controversies about its suitability.[18]

While the 5% threshold is not written in stone, its continued popularity invites thoughtful consideration. Within Confucian philosophy, the doctrine of *zhong yong*—emphasizing moderation, balance, and the avoidance of extremes—can offer a useful perspective. In political discourse, for example, this principle is applied by avoiding positions that are excessively right-leaning

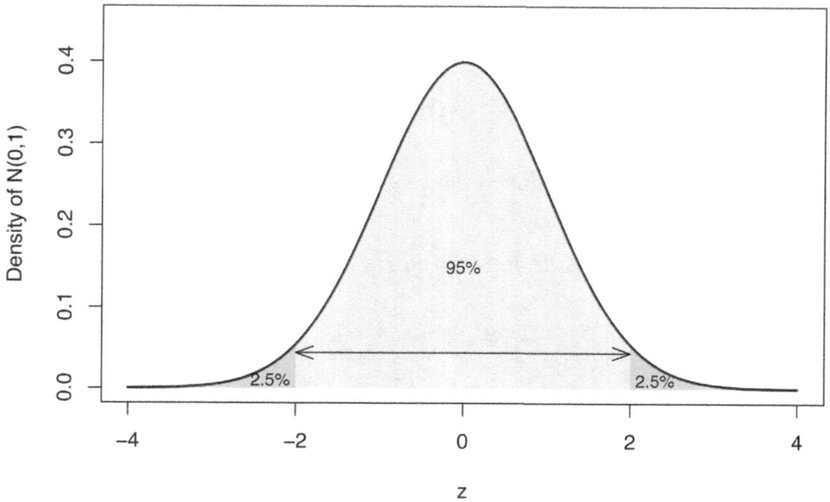

FIGURE 3.3
The 95% probability within 2

or excessively left-leaning. Similarly, a form of such a balance may be applied to statistical inference.

This notion of moderation can be related to the well-known empirical rule, also known as the 68–95–99.7 rule, which describes the distribution of data under a normal curve:

- Approximately 68% of the data lie within 1 standard deviation of the mean (i.e., between -1 and 1),
- Approximately 95% of the data lie within 2 standard deviations of the mean (i.e., between -2 and 2),
- Approximately 99.7% of the data lie within 3 standard deviations of the mean (i.e., between -3 and 3).

By applying the 68–95–99.7 rule to determine the significance level, three candidates are obtained: 32%, 5%, and 0.3%. A significance level of 32% would result in an unacceptably high rate of false positives, while a level of 0.03% would lead to an excessive number of false negatives. Therefore, in accordance with the doctrine of *zhong yong*, which emphasizes moderation and balance, a significance level of 5% may be regarded a reasonable compromise. This choice is illustrated in Figure 3.3.

Similarly, when the 68–95–99.7 rule is used to guide the selection of the confidence level, the values 68%, 95%, and 99.7% are identified. A confidence level of 68% would result in intervals that are too narrow and potentially unreliable, while a level approaching 99.7% may yield intervals that are excessively wide and of limited practical use. Once again, guided by the principle of *zhong yong*, the 95% confidence level can be viewed as a balanced and practical choice. This level is also depicted in Figure 3.3.

In summary, coverage within 1 standard deviation may be considered too limited, while coverage within 3 standard deviations may be considered excessive. In contrast, coverage within 2 standard deviations offers a moderate and intuitively satisfying balance. Thus, the value 2—corresponding to the 95% level—can be seen as a natural choice reflecting moderation.

3.5 Exercises

Exercise 3.1

Generate Table 3.2 in R.

Exercise 3.2

Generate Table 3.3 in R.

Exercise 3.3

Using the data in Table 3.2, verify that the 95% exact binomial confidence interval is $[0.408, 0.849]$.

Exercise 3.4

Assume $X_1, \ldots, X_n$ are i.i.d. with $X \sim p(x)$, where $p_0(x)$ is the probability density function of X. Let θ_0 be the median of $p(x)$ defined by the following:

$$\int_{-\infty}^{\theta_0} p_0(x)\,dx = \frac{1}{2}.$$

Let $\widehat{\theta}_n$ be the median of $X_1, \ldots, X_n$. The asymptotic normality of $\widehat{\theta}_n$ has been approved as

$$\sqrt{n}(\widehat{\theta}_n - \theta_0) \xrightarrow{d} \mathcal{N}\left(0, \frac{1}{4np_0^2(\theta_0)}\right).$$

Design two methods to estimate the standard error of $\widehat{\theta}_n$.

Exercise 3.5

Generate Table 3.4 in R.

4

Maximum Likelihood Estimation (MLE)

One of the most significant developments in 20th century statistics was the formulation of the maximum likelihood estimation (MLE) method. This method was initially introduced by Fisher and later advanced by other statisticians, including Neyman. The concept of MLE led to the development and adoption of numerous fundamental ideas and terms that are now routinely used by statisticians, such as "parameter," "estimation," "statistic," "consistency," "information," and "efficiency."[19]

4.1 Parametric MLE and Nonparametric MLE

4.1.1 Parametric MLE

Assume that $X_1, \ldots, X_n$ are i.i.d. with $X \sim p(x;\theta)$, where $p(x;\theta)$ denotes a parametric probability density function for a continuous random variable X or a parametric probability mass function for a discrete random variable X.

The likelihood function is defined as the joint probability of the observed data $X_1, \ldots, X_n$ given the parameter θ:

$$L(\theta; X_1, \ldots, X_n) = \prod_{i=1}^{n} p(X_i;\theta).$$

The likelihood function can be interpreted in two ways. When defining the likelihood, it is viewed as a function of the data $\{X_1, \ldots, X_n\}$. When the likelihood is used to estimate the parameter, it is treated as a function of θ.

The log-likelihood function is often preferred due to its computational simplicity, as the logarithm transforms the product into a sum:

$$\ell(\theta; X_1, \ldots, X_n) = \log L(\theta; X_1, \ldots, X_n) = \sum_{i=1}^{n} \log p(X_i;\theta).$$

DOI: 10.1201/9781003635468-5

The maximum likelihood estimator (MLE) of the parameter θ is defined as the value that maximizes the likelihood (or, equivalently, the log-likelihood):

$$\widehat{\theta}_n = \arg\max_{\theta} L(\theta; X_1, \ldots, X_n) = \arg\max_{\theta} \ell(\theta; X_1, \ldots, X_n).$$

The development of MLEs is demonstrated through two examples.

Example 1: Normal Distribution

Assume that $X_1, \ldots, X_n$ are i.i.d. with $X \sim p(x; \theta)$, where $\theta = (\mu, \sigma^2)^\top$ and

$$p(x; \theta) = \frac{1}{\sqrt{2\pi\sigma^2}} \exp\left\{-\frac{(x-\mu)^2}{2\sigma^2}\right\}.$$

The likelihood function is given by

$$\begin{aligned} L(\mu, \sigma^2; X_1, \ldots, X_n) &= \prod_{i=1}^{n} \frac{1}{\sqrt{2\pi\sigma^2}} \exp\left\{-\frac{(X_i-\mu)^2}{2\sigma^2}\right\} \\ &= (2\pi\sigma^2)^{-n/2} \exp\left\{-\frac{\sum_{i=1}^{n}(X_i-\mu)^2}{2\sigma^2}\right\}. \end{aligned}$$

The log-likelihood function is expressed as

$$\ell(\mu, \sigma^2; X_1, \ldots, X_n) = -\frac{n}{2}\log(2\pi\sigma^2) - \frac{1}{2\sigma^2}\sum_{i=1}^{n}(X_i-\mu)^2.$$

To find the MLEs, the first derivatives of the log-likelihood function with respect to μ and σ^2 are computed:

$$\begin{aligned} \frac{\partial \ell(\mu, \sigma^2)}{\partial \mu} &= \frac{1}{\sigma^2}\sum_{i=1}^{n}(X_i-\mu), \\ \frac{\partial \ell(\mu, \sigma^2)}{\partial \sigma^2} &= -\frac{n}{2\sigma^2} + \frac{1}{2\sigma^4}\sum_{i=1}^{n}(X_i-\mu)^2. \end{aligned}$$

As μ increases from $-\infty$ to ∞ and σ^2 increases from 0 to ∞, the partial derivatives are initially positive, reach zero, and then become negative, indicating that the log-likelihood attains a maximum at the solutions of the following equations:

$$\begin{aligned} \frac{1}{\sigma^2}\sum_{i=1}^{n}(X_i-\mu) &= 0, \\ -\frac{n}{2\sigma^2} + \frac{1}{2\sigma^4}\sum_{i=1}^{n}(X_i-\mu)^2 &= 0. \end{aligned}$$

Solving the above equations yields the MLEs:

$$\widehat{\mu}_n = \overline{X}_n,$$

$$\widehat{\sigma}_n^2 = \frac{1}{n}\sum_{i=1}^{n}(X_i - \overline{X}_n)^2.$$

The second derivatives can also be derived. The second derivative with respect to μ is

$$\frac{\partial^2 \ell(\mu, \sigma^2)}{\partial \mu^2} = -\frac{n}{\sigma^2},$$

the second derivative with respect to σ^2 is

$$\frac{\partial^2 \ell(\mu, \sigma^2)}{\partial (\sigma^2)^2} = \frac{n}{2\sigma^4} - \frac{1}{\sigma^6}\sum_{i=1}^{n}(X_i - \mu)^2,$$

and the mixed partial derivative with respect to μ and σ^2 is

$$\frac{\partial^2 \ell(\mu, \sigma^2)}{\partial \mu \partial \sigma^2} = \frac{\partial^2 \ell(\mu, \sigma^2)}{\partial \sigma^2 \partial \mu} = -\frac{1}{\sigma^4}\sum_{i=1}^{n}(X_i - \mu).$$

The Hessian matrix is therefore given by

$$H = \begin{pmatrix} -\frac{n}{\sigma^2} & -\frac{1}{\sigma^4}\sum_{i=1}^{n}(X_i - \mu) \\ -\frac{1}{\sigma^4}\sum_{i=1}^{n}(X_i - \mu) & \frac{n}{2\sigma^4} - \frac{1}{\sigma^6}\sum_{i=1}^{n}(X_i - \mu)^2 \end{pmatrix}.$$

At the MLEs $(\widehat{\mu}_n, \widehat{\sigma}_n^2)^\top$, this reduces to

$$H = \begin{pmatrix} -\frac{n}{\widehat{\sigma}_n^2} & 0 \\ 0 & -\frac{n}{2\widehat{\sigma}_n^4} \end{pmatrix} < 0.$$

Since the Hessian matrix is negative definite at the MLEs, the log-likelihood function is maximized at these values. □

The development of the above MLEs can be similarly extended to other continuous distributions, such as the Laplace distribution (see Exercise 4.1). In the following, the multinomial distribution, which is discrete, is considered, with the Bernoulli distribution as one of its special cases (see Exercise 4.2).

Example 2: Multinomial Distribution

Assume that $X_1, \ldots, X_n$ are i.i.d. with $X \sim \text{Multinomial}(x; p_1, \ldots, p_J)$, where $\mathbb{P}(X = x^j) = p_j$, $j = 1, \ldots, J$, and $\sum_{j=1}^{J} p_j = 1$.

The likelihood function is given by

$$L(p_1, \ldots, p_{J-1}; X_1, \ldots, X_n) = \prod_{i=1}^{n}\prod_{j=1}^{J-1} p_j^{I(X_i = x^j)} \left(1 - \sum_{j=1}^{J-1} p_j\right)^{I(X_i = x^J)}.$$

The corresponding log-likelihood function $\ell(p_1, \ldots, p_{J-1}; X_1, \ldots, X_n)$ is

$$\sum_{i=1}^{n} \left[\sum_{j=1}^{J-1} I(X_i = x^j) \log(p_j) + I(X_i = x^J) \log \left(1 - \sum_{j=1}^{J-1} p_j \right) \right].$$

The first derivative of the log-likelihood with respect to p_j is given by

$$\frac{\partial \ell(p_1, \ldots, p_{J-1})}{\partial p_j} = \sum_{i=1}^{n} \left[\frac{I(X_i = x^j)}{p_j} - \frac{I(X_i = x^J)}{1 - \sum_{j=1}^{J-1} p_j} \right], \quad j = 1, \ldots, J-1.$$

The second derivative of the log-likelihood with respect to p_j is

$$\frac{\partial^2 \ell}{\partial p_j^2} = -\frac{\sum_{i=1}^{n} I(X_i = x^j)}{p_j^2} - \frac{\sum_{i=1}^{n} I(X_i = x^J)}{p_J^2}, \quad j = 1, \ldots, J-1,$$

and for $j \neq k$, the mixed second derivative is

$$\frac{\partial^2 \ell}{\partial p_j \partial p_k} = -\frac{\sum_{i=1}^{n} I(X_i = x^J)}{p_J^2}.$$

Hence, the Hessian matrix of the log-likelihood function is given by

$$\begin{aligned} H = &-\sum_{i=1}^{n} \operatorname{diag} \left(\frac{I(X_i = x^1)}{p_1^2}, \ldots, \frac{I(X_i = x^{J-1})}{p_{J-1}^2} \right) \\ &- \frac{\sum_{i=1}^{n} I(X_i = x^J)}{p_J^2} \mathbf{1}_{J-1} \mathbf{1}_{J-1}^\top, \end{aligned}$$

where diag(a) denotes a diagonal matrix formed from a vector a, and $\mathbf{1}_{J-1}$ is the $(J-1)$-dimensional column vector of ones.

Since the diagonal matrix is positive definite and the matrix $\mathbf{1}_{J-1}\mathbf{1}_{J-1}^\top$ is non-negative definite, the Hessian matrix H is negative definite. Therefore, the MLEs of $p_1, \ldots, p_{J-1}$ are obtained by solving the following equations:

$$\sum_{i=1}^{n} \left[\frac{I(X_i = x^j)}{p_j} - \frac{I(X_i = x^J)}{1 - \sum_{j=1}^{J-1} p_j} \right] = 0, \quad j = 1, \ldots, J-1.$$

Solving the above yields the MLEs

$$\widehat{p}_{jn} = \frac{1}{n} \sum_{i=1}^{n} I(X_i = x^j), \quad j = 1, \ldots, J-1.$$

Consequently, the MLE of p_J is obtained as

$$\widehat{p}_{Jn} = 1 - \sum_{j=1}^{J-1} \widehat{p}_{jn} = \frac{1}{n} \sum_{i=1}^{n} I(X_i = x^J).$$

Thus, these MLEs correspond to the sample proportions. □

4.1.2 Nonparametric MLE

Assume that the estimand of interest is the density function $p(x)$ or the distribution function $P(x)$ itself. There are two scenarios in which the MLE method can be applied, depending on whether $p(x)$ is assumed to be parametric.

If it is assumed that $p(x) = p(x; \theta)$, where θ is a finite-dimensional parameter, then θ may be estimated by its MLE, denoted by $\widehat{\theta}_n$. Consequently, the distribution function $P(x) = \int_{-\infty}^{x} p(t)\, dt = \int_{-\infty}^{x} p(t; \theta)\, dt$ can be estimated by its plug-in MLE:

$$\widehat{P}_n(x) = \int_{-\infty}^{x} p(t; \widehat{\theta}_n)\, dt.$$

Alternatively, if no parametric form is imposed on $p(x)$, then the distribution function $P(x) = \int_{-\infty}^{x} p(t)\, dt$ can be estimated by the empirical distribution function:

$$\widehat{P}_n(x) = \frac{1}{n} \sum_{i=1}^{n} I(X_i \le x).$$

It can be shown that the empirical distribution function $\widehat{P}_n(x)$ constitutes the nonparametric MLE (NPMLE) of $P(x)$. To establish this, first assume that the observations $X_1, \ldots, X_n$ are all distinct. Under this setting, the likelihood function is given by

$$L(p(\cdot); X_1, \ldots, X_n) = \prod_{i=1}^{n} p(X_i),$$

subject to the constraints $0 \le p(X_i) \le 1$ for $i = 1, \ldots, n$, and $\sum_{i=1}^{n} p(X_i) \le 1$.

By the arithmetic mean–geometric mean inequality, it follows that

$$\prod_{i=1}^{n} p(X_i) \le \left[\frac{1}{n} \sum_{i=1}^{n} p(X_i) \right]^n \le 1,$$

where equality in the first inequality holds if and only if $p(X_1) = \cdots = p(X_n)$, and equality in the second holds if and only if $\sum_{i=1}^{n} p(X_i) = 1$. Thus, the likelihood is maximized when $p(X_i) = 1/n$ for all $i = 1, \ldots, n$.

If ties are present among $X_1, \ldots, X_n$, suppose there are J distinct values, denoted $x^1, \ldots, x^J$, with corresponding frequencies given by $m_j = \sum_{i=1}^{n} I(X_i = x^j)$ for $j = 1, \ldots, J$. Then, the likelihood function becomes

$$L(p(\cdot); X_1, \ldots, X_n) = \prod_{j=1}^{J} [p(x^j)]^{m_j},$$

subject to $0 \le p(x^j) \le 1$ for all j, and $\sum_{j=1}^{J} p(x^j) \le 1$.

Based on the findings in Example 2 (Multinomial distribution), the likelihood is maximized when

$$p(x^j) = \frac{m_j}{n} = \frac{1}{n}\sum_{i=1}^{n} I(X_i = x^j).$$

Combining these two cases, it follows that the empirical distribution $\widehat{P}_n(x)$ is the NPMLE of $P(x)$. □

Gold Coin # 13. *Assume that $X_1, \ldots, X_n$ are i.i.d. with $X \sim p(x)$. Two general approaches are available for estimating $p(x)$: the parametric and the nonparametric methods.*

Parametric MLE: *When it is assumed that $p(x)$ is parametrized by $p(x;\theta)$, the unknown parameter θ can be estimated using the maximum likelihood estimator (MLE), denoted $\widehat{\theta}_n$. The function $p(x)$ is then estimated by substituting $\widehat{\theta}_n$ into the parametric model, that is, $p(x;\widehat{\theta}_n)$.*

Nonparametric MLE: *When no parametric form is imposed on $p(x)$, its corresponding distribution function $P(x)$ can be estimated using the nonparametric maximum likelihood estimator (NPMLE), which in this case coincides with the empirical distribution function $\widehat{P}_n(x)$.*

Example 2 is now revisited. Since the sample space is given by $\{x^1, \ldots, x^J\}$ and the multinomial distribution involves $J-1$ free parameters, the parametric model under consideration is referred to as *saturated.*

It is known that a function may be represented in various forms: as a rule, a graph, or a table. For the multinomial distribution, Multinomial$(x; p_1, \ldots, p_J)$, it is convenient to represent it in tabular form, as shown in Table 4.1. If each p_j is replaced by its MLE $\widehat{p}_{nj}$, the collection $\{\widehat{p}_{1n}, \ldots, \widehat{p}_{Jn}\}$ serves as the MLE of the unknown distribution $\{p_1, \ldots, p_J\}$. By comparing this estimated distribution with the empirical distribution, we observe that, in this saturated setting, the parametric MLE and the nonparametric MLE coincide.

TABLE 4.1
Multinomial distribution

Sample space	x^1	x^2	$\cdots$	x^J
Distribution	p_1	p_2	$\cdots$	p_J
MLE	$\widehat{p}_{1n} = m_1/n$	$\widehat{p}_{2n} = m_2/n$	$\cdots$	$\widehat{p}_{Jn} = m_J/n$
NPMLE	$\widehat{p}_{1n} = m_1/n$	$\widehat{p}_{2n} = m_2/n$	$\cdots$	$\widehat{p}_{Jn} = m_J/n$

4.2 Non-asymptotic Theory of Likelihood

The theory of likelihood has been extensively developed in the context of parametric MLE. Suppose that $X_1, \ldots, X_n$ are i.i.d. with $X \sim p(x;\theta)$. Assume that the true value of the d-dimensional parameter θ is θ_0.

4.2.1 Score and Information

The log-likelihood function for a single observation X is defined as

$$\ell_\theta(X) = \log p(X;\theta).$$

The *score function*, denoted $s_\theta(X)$, is given by

$$\begin{aligned} s_\theta(X) &= \frac{\partial \ell_\theta(X)}{\partial \theta} \\ &= \frac{\partial \log p(X;\theta)}{\partial \theta} \\ &= \frac{\partial p(X;\theta)/\partial\theta}{p(X;\theta)}. \end{aligned}$$

The *Fisher information matrix* is defined as

$$\begin{aligned} \mathcal{I}_\theta &= \mathbb{E}_\theta\left[s_\theta(X) s_\theta(X)^\top\right] \\ &= \mathbb{E}_\theta\left[\frac{\partial \ell_\theta(X)}{\partial \theta}\frac{\partial \ell_\theta(X)}{\partial \theta^\top}\right], \end{aligned}$$

where $\mathbb{E}_\theta$ denotes expectation under the distribution $X \sim p(x;\theta)$.

Two key results concerning the score function and the information matrix can be established heuristically. The first result is:

$$\mathbb{E}_\theta\left[s_\theta(X)\right] = 0.$$

To show this, note that differentiating both sides of the identity

$$\int p(x;\theta)\, dx = 1$$

with respect to θ yields

$$\int \frac{\partial p(x;\theta)}{\partial \theta}\, dx = 0.$$

Hence, it follows that

$$\mathbb{E}_\theta\left[s_\theta(X)\right] = \int \frac{\partial p(x;\theta)/\partial\theta}{p(x;\theta)} p(x;\theta)\, dx = 0. \quad \square$$

The second result is:

$$\mathbb{E}_\theta\left[\frac{\partial \ell_\theta(X)}{\partial \theta}\frac{\partial \ell_\theta(X)}{\partial \theta^\top}\right] = -\mathbb{E}_\theta\left[\frac{\partial^2 \ell_\theta(X)}{\partial \theta \partial \theta^\top}\right].$$

To establish this, the derivative with respect to $\theta^\top$ is applied to both sides of the equation

$$\int \frac{\partial \ell_\theta(X)}{\partial \theta} p(x;\theta)\, dx = 0,$$

leading to

$$\int \left[\frac{\partial^2 \ell_\theta(X)}{\partial \theta \partial \theta^\top} p(x;\theta) + \frac{\partial \ell_\theta(X)}{\partial \theta}\frac{\partial p(x;\theta)}{\partial \theta^\top}\right] dx = 0.$$

This can be rewritten as

$$\int \left\{\frac{\partial^2 \ell_\theta(X)}{\partial \theta \partial \theta^\top} p(x;\theta) + \frac{\partial \ell_\theta(X)}{\partial \theta}\left[\frac{\partial \ell_\theta(X)}{\partial \theta}\right]^\top p(x;\theta)\right\} dx = 0.$$

Therefore, it follows that

$$\int \frac{\partial \ell_\theta(X)}{\partial \theta}\left[\frac{\partial \ell_\theta(X)}{\partial \theta}\right]^\top p(x;\theta)\, dx = -\int \frac{\partial^2 \ell_\theta(X)}{\partial \theta \partial \theta^\top} p(x;\theta)\, dx. \quad \square$$

4.2.2 Cramér–Rao Lower Bound

One of the most fundamental results in the theory of likelihood is: the inverse of the Fisher information matrix constitutes the Cramér–Rao lower bound (CRLB) on the covariance matrix of any unbiased estimator of θ.

An estimator $\widetilde{\theta}_n = \widetilde{\theta}(X_1, \ldots, X_n)$ is said to be unbiased if

$$\mathbb{E}_\theta[\widetilde{\theta}_n] = \theta,$$

for every θ within the parameter space. The condition can be written as

$$\int \widetilde{\theta}(x_1, \ldots, x_n) \prod_{i=1}^n p(x_i;\theta)\, dx_1 \ldots dx_n = \theta.$$

By differentiating both sides with respect to $\theta^\top$, the following is obtained:

$$\int \widetilde{\theta}(x_1, \ldots, x_n) \frac{\partial \prod_{i=1}^n p(x_i;\theta)}{\partial \theta^\top}\, dx_1 \ldots dx_n = I_{d\times d},$$

where d is the dimension of θ and $I_{d\times d}$ is the $d \times d$ identity matrix.

Since it holds that

$$\frac{\partial \prod_{i=1}^n p(x_i;\theta)}{\partial \theta} = \sum_{i=1}^n \prod_{i' \neq i} p(x_{i'};\theta) \frac{\partial p(x_i;\theta)}{\partial \theta} = \sum_{i=1}^n \prod_{i'=1}^n p(x_{i'};\theta) \frac{\partial \log p(x_i;\theta)}{\partial \theta},$$

it follows that

$$\int \widetilde{\theta}(x_1,\ldots,x_n) \left[\sum_{i=1}^n \frac{\partial \log p(x_i;\theta)}{\partial \theta^\top}\right] \prod_{i=1}^n p(x_i;\theta)\, dx_1 \ldots dx_n = I_{d\times d}.$$

Let the score function for the full sample be defined as

$$S_\theta = \sum_{i=1}^n \frac{\partial \log p(X_i;\theta)}{\partial \theta},$$

which satisfies the following properties:

$$\mathbb{E}_\theta[S_\theta] = 0 \quad \text{and} \quad \mathbb{E}_\theta[S_\theta S_\theta^\top] = n\mathcal{I}_\theta.$$

Hence, the previous identity can be rewritten as

$$\mathbb{E}_\theta[\widetilde{\theta}_n S_\theta^\top] = I_{d\times d}.$$

Given that $\mathbb{E}_\theta[S_\theta] = 0$, it can be further deduced that

$$\mathbb{E}_\theta[(\widetilde{\theta}_n - \theta) S_\theta^\top] = I_{d\times d}.$$

Now consider the non-negative definite matrix

$$\begin{pmatrix} \widetilde{\theta}_n - \theta \\ S_\theta \end{pmatrix} \begin{pmatrix} \widetilde{\theta}_n - \theta \\ S_\theta \end{pmatrix}^\top = \begin{pmatrix} (\widetilde{\theta}_n - \theta)(\widetilde{\theta}_n - \theta)^\top & (\widetilde{\theta}_n - \theta) S_\theta^\top \\ S_\theta(\widetilde{\theta}_n - \theta)^\top & S_\theta S_\theta^\top \end{pmatrix} \geq 0.$$

By taking expectations on both sides, the following inequality is obtained:

$$\mathbb{E}_\theta\left[\begin{pmatrix} (\widetilde{\theta}_n - \theta)(\widetilde{\theta}_n - \theta)^\top & (\widetilde{\theta}_n - \theta) S_\theta^\top \\ S_\theta(\widetilde{\theta}_n - \theta)^\top & S_\theta S_\theta^\top \end{pmatrix}\right] \geq 0.$$

By defining the covariance matrix of the estimator as

$$\mathrm{Cov}_\theta(\widetilde{\theta}_n) = \mathbb{E}_\theta\left[(\widetilde{\theta}_n - \theta)(\widetilde{\theta}_n - \theta)^\top\right],$$

it follows that

$$\begin{pmatrix} \mathrm{Cov}_\theta(\widetilde{\theta}_n) & I_{d\times d} \\ I_{d\times d} & n\mathcal{I}_\theta \end{pmatrix} \geq 0.$$

From the Schur complement condition for positive semi-definiteness, the following inequality is implied:

$$\mathrm{Cov}_\theta(\widetilde{\theta}_n) - I_{d\times d}\,(n\mathcal{I}_\theta)^{-1}I_{d\times d} \geq 0.$$

Therefore, it is concluded that

$$\mathrm{Cov}_\theta(\widetilde{\theta}_n) \geq \frac{1}{n}\mathcal{I}_\theta^{-1}. \quad \square$$

Gold Coin # 14. *Assume that $X_1, \ldots, X_n$ are i.i.d. with $X \sim p(x;\theta)$. Two key concepts in likelihood theory are the score function and the Fisher information matrix. The score function is defined as*

$$s_\theta(X) = \frac{\partial \ell_\theta(X)}{\partial \theta},$$

and the Fisher information matrix is defined as

$$\mathcal{I}_\theta = \mathbb{E}_\theta\left[\frac{\partial \ell_\theta(X)}{\partial \theta}\frac{\partial \ell_\theta(X)}{\partial \theta^\top}\right].$$

An important identity relates the first and second derivatives of the log-likelihood function:

$$\mathbb{E}_\theta\left[\frac{\partial \ell_\theta(X)}{\partial \theta}\frac{\partial \ell_\theta(X)}{\partial \theta^\top}\right] = -\mathbb{E}_\theta\left[\frac{\partial^2 \ell_\theta(X)}{\partial \theta \partial \theta^\top}\right].$$

Another fundamental result is the Cramér–Rao lower bound (CRLB), which connects unbiasedness to minimum variance. Specifically, for any unbiased estimator $\widetilde{\theta}_n$ of θ,

$$\mathrm{Cov}_\theta(\widetilde{\theta}_n) \geq \frac{1}{n}\mathcal{I}_\theta^{-1}.$$

Example 1 (continued): Normal Distribution

The Hessian matrix with respect to $\theta = (\mu, \sigma)^\top$ is

$$H = \begin{pmatrix} -\frac{1}{\sigma^2} & -\frac{1}{\sigma^4}(X-\mu) \\ -\frac{1}{\sigma^4}(X-\mu) & \frac{1}{2\sigma^4} - \frac{1}{\sigma^6}(X-\mu)^2 \end{pmatrix}.$$

Noting that

$$\mathbb{E}_\theta(X-\mu) = 0 \quad \text{and} \quad \mathbb{E}_\theta(X-\mu)^2 = \sigma^2,$$

the Fisher information matrix is obtained as

$$\mathcal{I}_\theta = -\mathbb{E}_\theta[H] = \begin{pmatrix} \frac{1}{\sigma^2} & 0 \\ 0 & \frac{1}{2\sigma^2} \end{pmatrix}.$$

Recall the MLE of μ is given by $\widehat{\mu}_n = \overline{X}_n$. It can be verified that

$$\mathbb{E}_\theta(\widehat{\mu}_n) = \mu \quad \text{and} \quad \mathbb{V}_\theta(\widehat{\mu}_n) = \frac{\sigma^2}{n}.$$

Hence, $\widehat{\mu}_n$ is shown to be an unbiased estimator of μ and the Cramér–Rao lower bound is achieved. Therefore, $\widehat{\mu}_n = \overline{X}_n$ is the uniformly minimum variance unbiased estimator (UMVUE) of μ. □

Example 2 (continued): Multinomial Distribution

The Hessian matrix of the log-likelihood function for a single observation is

$$H = -\text{diag}\left(\frac{I(X = x^1)}{p_1^2}, \ldots, \frac{I(X = x^{J-1})}{p_{J-1}^2}\right) - \frac{I(X = x^J)}{p_J^2}\mathbf{1}_{J-1}\mathbf{1}_{J-1}^\top.$$

Thus, the Fisher information matrix with respect to $\theta = (p_1, \ldots, p_{J-1})^\top$ is

$$\mathcal{I}_\theta = -\mathbb{E}_\theta[H] = \text{diag}\left(\frac{1}{p_1}, \ldots, \frac{1}{p_{J-1}}\right) + \frac{1}{1 - \sum_{j=1}^{J-1} p_j}\mathbf{1}_{J-1}\mathbf{1}_{J-1}^\top.$$

The inverse of the Fisher information matrix can be verified to be

$$\begin{aligned}
\mathcal{I}_\theta^{-1} =& \text{diag}(\theta) - \theta\theta^\top \\
=& \text{diag}(p_1, \ldots, p_{J-1}) - (p_1, \ldots, p_{J-1})^\top (p_1, \ldots, p_{J-1}) \\
=& \begin{pmatrix} p_1(1-p_1) & -p_1p_2 & \cdots & -p_1p_{J-1} \\ -p_2p_1 & p_2(1-p_2) & \cdots & -p_2p_{J-1} \\ \cdots & \cdots & \cdots & \cdots \\ -p_{J-1}p_1 & -p_{J-1}p_2 & \cdots & p_{J-1}(1-p_{J-1}) \end{pmatrix}.
\end{aligned}$$

In fact, it can be shown that

$$\begin{aligned}
\mathcal{I}_\theta\left[\text{diag}(\theta) - \theta\theta^\top\right] =& \left[\text{diag}(\theta^{-1}) + \frac{1}{p_J}\mathbf{1}_{J-1}\mathbf{1}_{J-1}^\top\right]\left[\text{diag}(\theta) - \theta\theta^\top\right] \\
=& \text{diag}(\theta^{-1})\text{diag}(\theta) - \text{diag}(\theta^{-1})\theta\theta^\top \\
&+ \frac{1}{p_J}\mathbf{1}_{J-1}\mathbf{1}_{J-1}^\top\text{diag}(\theta) - \frac{1}{p_J}\mathbf{1}_{J-1}\mathbf{1}_{J-1}^\top\theta\theta^\top \\
=& I_{(J-1)\times(J-1)} - \mathbf{1}_{J-1}\theta^\top + \frac{1}{p_J}\mathbf{1}_{J-1}\theta^\top - \frac{1-p_J}{p_J}\mathbf{1}_{J-1}\theta^\top \\
=& I_{(J-1)\times(J-1)}.
\end{aligned}$$

Recall that the MLE of θ is given by $\widehat{\theta}_n = (\widehat{p}_{1n}, \ldots, \widehat{p}_{J-1,n})^\top$, where $\widehat{p}_{jn} = \sum_{i=1}^n I(X_i = p^j)/n$. It can be verified that

$$\mathbb{E}(\widehat{p}_{jn}) = p_j, \quad \mathbb{V}(\widehat{p}_{jn}) = p_j(1-p_j)/n,$$

and, for $j_1 \neq j_2$,

$$\begin{aligned}
\mathrm{Cov}(\widehat{p}_{j_1 n}, \widehat{p}_{j_2 n}) &= \mathrm{Cov}\left(\frac{1}{n}\sum_{i=1}^n I(X_i = p^{j_1}), \frac{1}{n}\sum_{i=1}^n I(X_i = p^{j_2})\right) \\
&= \mathbb{E}\left(\frac{1}{n^2}\sum_{i=1}^n I(X_i = p^{j_1})\sum_{i=1}^n I(X_i = p^{j_2})\right) - p_{j_1}p_{j_2} \\
&= \frac{n(n-1)p_{j_1}p_{j_2}}{n^2} - p_{j_1}p_{j_2} \\
&= -\frac{p_{j_1}p_{j_2}}{n}.
\end{aligned}$$

Therefore, it is verified that

$$\mathrm{Cov}(\widehat{\theta}_n) = \mathrm{Cov}\begin{pmatrix} \widehat{p}_{1n} \\ \widehat{p}_{2n} \\ \vdots \\ \widehat{p}_{J-1,n} \end{pmatrix} = \frac{1}{n}\mathcal{I}_\theta^{-1}.$$

This implies that $\widehat{\theta}_n$ is an unbiased estimator of θ and that the CRLB is achieved. Consequently, $(\widehat{p}_{1n}, \widehat{p}_{2n}, \ldots, \widehat{p}_{J-1,n})^\top$ is the UMVUE for $(p_1, \ldots, p_{J-1})^\top$. □

As a by-product, the following result is obtained (see Exercise 4.3):

$$\mathrm{Cov}\begin{pmatrix} \widehat{p}_{1n} \\ \widehat{p}_{2n} \\ \vdots \\ \widehat{p}_{Jn} \end{pmatrix} = \frac{1}{n}\begin{pmatrix} p_1(1-p_1) & -p_1p_2 & \cdots & -p_1p_J \\ -p_2p_1 & p_2(1-p_2) & \cdots & -p_2p_J \\ \vdots & \vdots & \ddots & \vdots \\ -p_Jp_1 & -p_Jp_2 & \cdots & p_J(1-p_J) \end{pmatrix}.$$

4.3 Asymptotic Theory of Likelihood

4.3.1 Consistency and Asymptotic Normality

Assume that $X_1, \ldots, X_n$ are i.i.d. with $X \sim p(x; \theta_0)$. The estimand of interest is θ_0. To estimate θ_0, the MLE is considered:

$$\widehat{\theta}_n = \arg\max_\theta \frac{1}{n}\sum_{i=1}^n \ell_\theta(X_i),$$

which corresponds to the solution of the score equation:

$$\frac{1}{n}\sum_{i=1}^{n} s_{\widehat{\theta}_n}(X_i) = 0.$$

By the law of large numbers (LLN), the following convergence in probability holds:

$$\frac{1}{n}\sum_{i=1}^{n} \ell_\theta(X_i) \xrightarrow{p} \mathbb{E}_{\theta_0}[\ell_\theta(X)] = \int \log[p(x;\theta)]\, p(x;\theta_0)\, dx.$$

Hence, it is natural to expect that

$$\arg\max_\theta \frac{1}{n}\sum_{i=1}^{n} \ell_\theta(X_i) \xrightarrow{p} \arg\max_\theta \int \log[p(x;\theta)]\, p(x;\theta_0)\, dx.$$

Of course, regularity conditions on the model $p(x;\theta)$ are required to justify the above interchange of limit and maximization. Fortunately, these conditions are satisfied for the previously discussed distributions (normal, Laplace, Bernoulli, and multinomial).[12] For the purpose of this exposition, in this book, such technicalities are set aside, so that we can simply and safely assume that the above convergence holds, which implies that

$$\widehat{\theta}_n \xrightarrow{p} \theta_*,$$

where

$$\theta_* = \arg\max_\theta \int \log[p(x;\theta)]\, p(x;\theta_0)\, dx.$$

It can be shown that $\theta_* = \theta_0$. In fact, using the following inequality,[12]

$$\log x \leq 2(\sqrt{x} - 1),$$

where the equality holds if and only if $x = 1$, the following is obtained:

$$\begin{aligned}
\int \log\left[\frac{p(x;\theta)}{p(x;\theta_0)}\right] p(x;\theta_0)\, dx &\leq 2\int \left[\sqrt{\frac{p(x;\theta)}{p(x;\theta_0)}} - 1\right] p(x;\theta_0)\, dx \\
&= 2\int \sqrt{p(x;\theta)p(x;\theta_0)}\, dx - 2 \\
&= -\int \left[\sqrt{p(x;\theta)} - \sqrt{p(x;\theta_0)}\right]^2 dx \leq 0,
\end{aligned}$$

where the equality holds if and only if $p(x;\theta) = p(x;\theta_0)$ almost everywhere. Therefore, if $p(x;\theta_0) \neq p(x;\theta)$ on a set with positive probability, then

$$\int \log[p(x;\theta)]\, p(x;\theta_0)\, dx < \int \log[p(x;\theta_0)]\, p(x;\theta_0)\, dx.$$

Thus, $\theta_* = \theta_0$, and consequently,

$$\widehat{\theta}_n \xrightarrow{p} \theta_0. \quad \square$$

Once consistency of the MLE is established, a heuristic argument for asymptotic normality is outlined. Since $\widehat{\theta}_n \xrightarrow{p} \theta_0$, the Taylor expansion of the score function $s_{\widehat{\theta}_n}$ around θ_0 yields:

$$0 = \frac{1}{\sqrt{n}} \sum_{i=1}^{n} s_{\widehat{\theta}_n}(X_i) = \frac{1}{\sqrt{n}} \sum_{i=1}^{n} s_{\theta_0}(X_i) + \frac{1}{\sqrt{n}} \sum_{i=1}^{n} \frac{\partial^2 \ell_{\theta_0}(X_i)}{\partial \theta \, \partial \theta^\top} (\widehat{\theta}_n - \theta_0) + o_p(1),$$

where the following notation is used,

$$\frac{\partial^2 \ell_{\theta_0}(X_i)}{\partial \theta \, \partial \theta^\top} = \left. \frac{\partial^2 \ell_{\theta}(X_i)}{\partial \theta \, \partial \theta^\top} \right|_{\theta = \theta_0},$$

and the remainder term converges to zero in probability. (A rigorous justification for this order term is omitted here.)

By the LLN,

$$\frac{1}{n} \sum_{i=1}^{n} \frac{\partial^2 \ell_{\theta_0}(X_i)}{\partial \theta \, \partial \theta^\top} \xrightarrow{p} \mathbb{E}_{\theta_0} \left[\frac{\partial^2 \ell_{\theta_0}(X)}{\partial \theta \, \partial \theta^\top} \right].$$

Thus, applying the Slutsky's lemma,

$$\sqrt{n}(\widehat{\theta}_n - \theta_0) = - \left\{ \mathbb{E}_{\theta_0} \left[\frac{\partial^2 \ell_{\theta_0}(X)}{\partial \theta \, \partial \theta^\top} \right] \right\}^{-1} \cdot \frac{1}{\sqrt{n}} \sum_{i=1}^{n} s_{\theta_0}(X_i) + o_p(1).$$

Furthermore, by the central limit theorem (CLT),

$$\sqrt{n}(\widehat{\theta}_n - \theta_0) \xrightarrow{d} \mathcal{N}(0, \Sigma_{\theta_0}),$$

where

$$\Sigma_{\theta_0} = \left\{ \mathbb{E}_{\theta_0} \left[\frac{\partial^2 \ell_{\vartheta_0}(X)}{\partial \theta \, \partial \theta^\top} \right] \right\}^{-1} \mathcal{I}_{\theta_0} \left\{ \mathbb{E}_{\theta_0} \left[\frac{\partial^2 \ell_{\theta_0}(X)}{\partial \theta \, \partial \theta^\top} \right] \right\}^{-1}.$$

Recalling the identity that links the expected Hessian and the Fisher information:

$$\mathcal{I}_{\theta_0} = -\mathbb{E}_{\theta_0} \left[\frac{\partial^2 \ell_{\theta_0}(X)}{\partial \theta \, \partial \theta^\top} \right],$$

it follows that

$$\Sigma_{\theta_0} = \mathcal{I}_{\theta_0}^{-1}.$$

Hence, the asymptotic normality of the MLE is established:

$$\sqrt{n}(\widehat{\theta}_n - \theta_0) \xrightarrow{d} \mathcal{N}(0, \mathcal{I}_{\theta_0}^{-1}),$$

where the asymptotic covariance matrix is equal to the CRLB. $\square$

Gold Coin # 15. *Assume $X_1, \ldots, X_n$ are i.i.d. with $X \sim p(x; \theta_0)$. Let $\widehat{\theta}_n$ denote the maximum likelihood estimator of θ_0. Then, the following properties hold:*

$$\widehat{\theta}_n \xrightarrow{p} \theta_0, \quad \text{as } n \to \infty,$$

and

$$\sqrt{n}(\widehat{\theta}_n - \theta_0) \xrightarrow{d} \mathcal{N}(0, \mathcal{I}_{\theta_0}^{-1}), \quad \text{as } n \to \infty,$$

where $\mathcal{I}_{\theta_0}$ is the Fisher information matrix. The asymptotic covariance matrix achieves the Cramér–Rao lower bound.

Example 1 (continued): Normal Distribution

Let $\widehat{\theta}_n = (\widehat{\mu}_n, \widehat{\sigma}_n^2)^\top$ denote the MLE of $\theta = (\mu, \sigma^2)^\top$. By the CLT,

$$\sqrt{n}\begin{pmatrix} \widehat{\mu}_n - \mu_0 \\ \widehat{\sigma}_n^2 - \sigma_0^2 \end{pmatrix} \xrightarrow{d} \mathcal{N}\left(0, \begin{pmatrix} \sigma_0^2 & 0 \\ 0 & 2\sigma_0^2 \end{pmatrix}\right),$$

where the asymptotic covariance matrix corresponds to the CRLB. □

Example 2 (continued): Multinomial Distribution

Let $\widehat{\theta}_n = (\widehat{p}_{1n}, \ldots, \widehat{p}_{J-1,n})^\top$ denote the MLE of $\theta = (p_1, \ldots, p_{J-1})^\top$. By the CLT, it is shown that

$$\sqrt{n}\begin{pmatrix} \widehat{p}_{1n} - p_{10} \\ \vdots \\ \widehat{p}_{J-1,n} - p_{J-1,0} \end{pmatrix}$$

$$\xrightarrow{d} \mathcal{N}\left\{0, \begin{pmatrix} p_{10}(1-p_{10}) & -p_{10}p_{20} & \cdots & -p_{10}p_{J-1,0} \\ -p_{20}p_{10} & p_{20}(1-p_{20}) & \cdots & -p_{20}p_{J-1,0} \\ \vdots & \vdots & \ddots & \vdots \\ -p_{J-1,0}p_{10} & -p_{J-1,0}p_{20} & \cdots & p_{J-1,0}(1-p_{J-1,0}) \end{pmatrix}\right\},$$

where the asymptotic covariance matrix coincides with the non-asymptotic covariance matrix and corresponds to the CRLB. □

4.3.2 Asymptotic Efficiency

Two important results have been established. On the one hand, the CRLB for the covariance of any unbiased estimator $\widetilde{\theta}_n$ is given by

$$\text{Cov}(\widetilde{\theta}_n) \geq \mathcal{I}_{\theta_0}^{-1}.$$

On the other hand, the asymptotic normality of the MLE has been shown as

$$\sqrt{n}(\widehat{\theta}_n - \theta_0) \xrightarrow{d} \mathcal{N}\left(0, \mathcal{I}_{\theta_0}^{-1}\right).$$

The appearance of the same Fisher information matrix $\mathcal{I}_{\theta_0}$ in both results is notable. Therefore, several remarks are made based on this observation:

- If the MLE $\widehat{\theta}_n$ is unbiased and its finite-sample covariance matrix attains the CRLB, then it is the UMVUE.
- Since asymptotic normality does not necessarily imply the convergence of moments such as $\mathbb{E}_\theta(\widehat{\theta}_n)$ and $\mathbb{V}_\theta(\widehat{\theta}_n)$, it cannot be concluded that the MLE is asymptotically UMVUE. Nonetheless, this terminology is often used informally to describe the properties of the MLE.
- An asymptotic version of the CRLB (known as the generalized CRLB) will be developed in Part III of this book, where the asymptotic efficiency of the MLE will be formally established.

4.4 Right Model or Wrong Model

Assume that $X_1, \ldots, X_n$ are i.i.d. with $X \sim p_0(x)$, where $p_0(x)$ denotes the true but unknown probability mass function (for discrete variables) or probability density function (for continuous variables).

In the preceding two sections, it has been assumed that $p_0(x)$ is correctly specified by a parametric model $p(x; \theta_0)$, where θ_0 is the true but unknown parameter. In this setting, $p(x; \theta)$ is regarded as the correctly specified model, and the MLE of θ_0 has been shown to possess desirable properties.

However, by Box's famous saying, "all models are wrong, but some are useful," the question arises: What is the estimand targeted by the MLE $\widehat{\theta}_n$ if $p(x; \theta)$ is misspecified?

To illustrate, consider the normal model $\mathcal{N}(\mu, \sigma^2)$ as the assumed model, although the true distribution $p_0(x)$ is nct normal.

Example 1 (continued): Normal Distribution

Under the (incorrect) normal model $\mathcal{N}(\mu, \sigma^2)$, obtain the following estimator

$$(\widehat{\mu}_n, \widehat{\sigma}_n^2)^\top = \arg\max_{\mu, \sigma^2} \sum_{i=1}^{n} \log\left[\frac{1}{\sqrt{2\pi\sigma^2}} \exp\left(-\frac{(X_i - \mu)^2}{2\sigma^2}\right)\right].$$

By defining

$$m(X; \mu, \sigma^2) = \log\left[\frac{1}{\sqrt{2\pi\sigma^2}} \exp\left(-\frac{(X - \mu)^2}{2\sigma^2}\right)\right],$$

this estimator can be equivalently written as

$$(\widehat{\mu}_n, \widehat{\sigma}_n^2)^\top = \arg\max_{\mu,\sigma^2} \frac{1}{n}\sum_{i=1}^{n} m(X_i; \mu, \sigma^2).$$

Since the objective function is no longer the log-likelihood function under the misspecified model, the estimator may be better referred to as an M-estimator,[12] where M stands for maximum.

Denoting the objective function by

$$M_n(\mu, \sigma^2) = \frac{1}{n}\sum_{i=1}^{n} m(X_i; \mu, \sigma^2),$$

the LLN implies that

$$M_n(\mu, \sigma^2) \xrightarrow{p} \mathbb{E}_{X\sim p_0(x)}\left[m(X; \mu, \sigma^2)\right] \triangleq M_0(\mu, \sigma^2),$$

where

$$\begin{aligned} M_0(\mu, \sigma^2) &= \int_{-\infty}^{\infty}\left[-\frac{1}{2}\log(2\pi\sigma^2) - \frac{(x-\mu)^2}{2\sigma^2}\right] p_0(x)\, dx \\ &= -\frac{1}{2}\log(2\pi\sigma^2) - \frac{1}{2\sigma^2}\int_{-\infty}^{\infty}(x-\mu)^2 p_0(x)\, dx. \end{aligned}$$

Hence, it is natural to expect that

$$(\widehat{\mu}_n, \widehat{\sigma}_n^2)^\top = \arg\max_{\mu,\sigma^2} M_n(\mu, \sigma^2) \xrightarrow{p} \arg\max_{\mu,\sigma^2} M_0(\mu, \sigma^2) \triangleq (\mu_*, \sigma_*^2)^\top.$$

Thus, the quantity $(\mu_*, \sigma_*^2)^\top$, defined as the maximizer of $M_0(\mu, \sigma^2)$, is the estimand targeted by the M-estimator $(\widehat{\mu}_n, \widehat{\sigma}_n^2)^\top$.

Fortunately, in this example, the explicit form of (μ_*, σ_*^2) can be derived as the solutions to

$$\begin{aligned} \frac{\partial M_0(\mu, \sigma^2)}{\partial \mu} &= \frac{1}{\sigma^2}\int_{-\infty}^{\infty}(x-\mu)p_0(x)\, dx = 0, \\ \frac{\partial M_0(\mu, \sigma^2)}{\partial \sigma^2} &= -\frac{1}{2\sigma^2} + \frac{1}{2\sigma^4}\int_{-\infty}^{\infty}(x-\mu)^2 p_0(x)\, dx = 0. \end{aligned}$$

Solving these equations yields

$$\mu_* = \int_{-\infty}^{\infty} x\, p_0(x)\, dx \quad \text{and} \quad \sigma_*^2 = \int_{-\infty}^{\infty}(x-\mu_*)^2 p_0(x)\, dx,$$

which correspond to the population mean and variance, respectively. □

4.4.1 Kullback–Leibler Divergence

A generalization of the previous example involving the normal model can now be considered for arbitrary working models. Assume that $X_1, \ldots, X_n$ are i.i.d. with $X \sim p_0(x)$. A parametric model $p(x;\theta)$ is introduced not as an assumed truth but as a working model for constructing an estimator:

$$\widehat{\theta}_n = \arg\max_\theta \frac{1}{n}\sum_{i=1}^n \log p(X_i;\theta).$$

By defining the function

$$m(X_i;\theta) = \log p(X_i;\theta),$$

the estimator above can be expressed as

$$\widehat{\theta}_n = \arg\max_\theta \frac{1}{n}\sum_{i=1}^n m(X_i;\theta),$$

which is referred to as an *M-estimator*.[12] Let the objective function be

$$M_n(\theta) = \frac{1}{n}\sum_{i=1}^n \log p(X_i;\theta).$$

By the LLN, it follows that

$$M_n(\theta) \xrightarrow{p} \mathbb{E}_{X\sim p_0(x)} \log p(X;\theta) \triangleq M_0(\theta),$$

where

$$M_0(\theta) = \int_{-\infty}^{\infty} \log p(x;\theta)\, p_0(x)\, dx.$$

It is natural to expect that the maximizer of $M_n(\theta)$ converges in probability to the maximizer of $M_0(\theta)$:

$$\widehat{\theta}_n = \arg\max_\theta M_n(\theta) \xrightarrow{p} \arg\max_\theta M_0(\theta) \triangleq \theta_*.$$

Thus, the quantity θ_*—the maximizer of $M_0(\theta)$—serves as the estimand targeted by the M-estimator $\widehat{\theta}_n$.

To better understand the nature of θ_*, observe that

$$\begin{aligned}
\theta_* &= \arg\max_\theta \mathbb{E}_{X\sim p_0(x)} \log p(X;\theta) \\
&= \arg\min_\theta \left[\mathbb{E}_{X\sim p_0(x)} \log p_0(X) - \mathbb{E}_{X\sim p_0(x)} \log p(X;\theta)\right] \\
&= \arg\min_\theta \mathbb{E}_{X\sim p_0(x)} \left[\log p_0(X) - \log p(X;\theta)\right] \\
&= \arg\min_\theta \mathbb{E}_{X\sim p_0(x)} \log\left[\frac{p_0(X)}{p(X;\theta)}\right].
\end{aligned}$$

The final expression represents the Kullback–Leibler (KL) divergence[6] between the true model $p_0(x)$ and the working model $p(x;\theta)$:

$$D_{\text{KL}}\{p_0(x)\,||\,p(x;\theta)\} = \mathbb{E}_{X\sim p_0(x)} \log\left[\frac{p_0(X)}{p(X;\theta)}\right].$$

Example 1 (continued): Normal Distribution

An illustrative example may aid in understanding the KL divergence. Suppose that the true distribution is exponential with the rate parameter $\lambda = 1$:

$$p_0(x) = e^{-x} I(x > 0).$$

If the working model is $\mathcal{N}(\mu, 1)$, then the KL divergence between this normal model and the true exponential distribution is given by

$$D_{\text{KL}}(\mu) = \int_0^{\infty} \log\left[\frac{e^{-x}}{\frac{1}{\sqrt{2\pi}}e^{-(x-\mu)^2/2}}\right] e^{-x}\,dx.$$

This expression can be simplified through direct integration (see Exercise 4.4). As shown in the left panel of Figure 4.1, the function $D_{\text{KL}}(\mu)$ attains its minimum at $\mu_* = 1$.

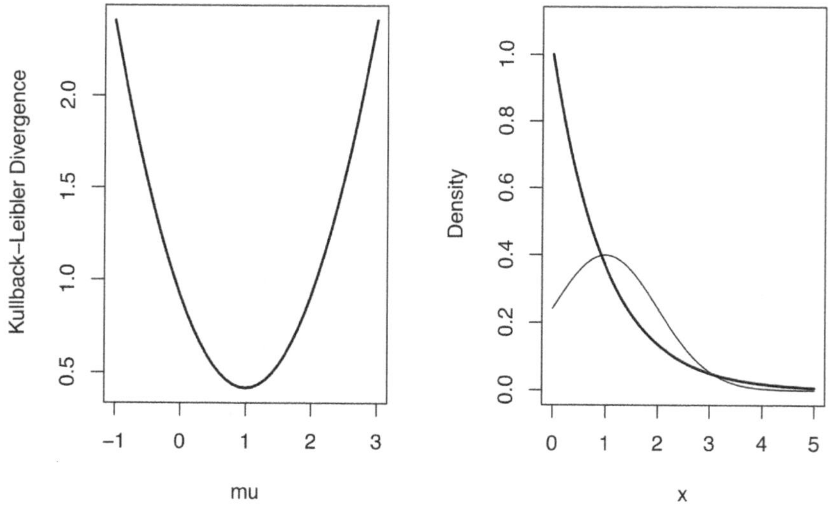

FIGURE 4.1
Left: the KL divergence between the working model $\mathcal{N}(\mu, 1)$ and the true model Exp(1). Right: the thicker line shows the density of Exp(1); the thinner line shows the density of $\mathcal{N}(1, 1)$

The right panel of Figure 4.1 compares the density of Exp(1) (i.e., true model) with that of $\mathcal{N}(\mu_*, 1)$ (i.e., the best approximation within the normal family), illustrating that $\mathcal{N}(\mu_*, 1)$ is the optimal normal approximation to Exp(1) in the KL sense. □

4.4.2 Consistency and Asymptotic Normality

Recall that, by the LLN, it holds that

$$M_n(\theta) \xrightarrow{p} M_0(\theta).$$

Hence, under regularity conditions imposed on the model $p(x;\theta)$—which can be ensured through the appropriate specification of the model—the maximizer of $M_n(\theta)$ is expected to converge in probability to the maximizer of $M_0(\theta)$:

$$\widehat{\theta}_n = \arg\max_\theta M_n(\theta) \xrightarrow{p} \arg\max_\theta M_0(\theta) = \theta_*.$$

This establishes the consistency of the estimator $\widehat{\theta}_n$ for θ_*.

To investigate the asymptotic normality, observe that the M-estimator $\widehat{\theta}_n$ solves the following estimating equation:

$$\frac{1}{\sqrt{n}} \sum_{i=1}^{n} \frac{\partial m(X_i; \widehat{\theta}_n)}{\partial \theta} = 0.$$

Given that $\widehat{\theta}_n \xrightarrow{p} \theta_*$, a Taylor expansion around θ_* yields:

$$0 = \frac{1}{\sqrt{n}} \sum_{i=1}^{n} \frac{\partial m(X_i; \theta_*)}{\partial \theta} + \frac{1}{\sqrt{n}} \sum_{i=1}^{n} \frac{\partial^2 m(X_i; \theta_*)}{\partial \theta \partial \theta^\top} (\widehat{\theta}_n - \theta_*) + o_p(1),$$

where the remainder term is of order $o_p(1)$ and is omitted for brevity.

Applying the LLN again, it holds that:

$$\frac{1}{n} \sum_{i=1}^{n} \frac{\partial^2 m(X_i; \theta_*)}{\partial \theta \partial \theta^\top} \xrightarrow{p} \mathbb{E}_0 \left[\frac{\partial^2 m(X; \theta_*)}{\partial \theta \partial \theta^\top} \right],$$

where $\mathbb{E}_0$ denotes the expectation with respect to $X \sim p_0(x)$. By the Slutsky's lemma, the following asymptotic linear representation holds:

$$\sqrt{n}(\widehat{\theta}_n - \theta_*) = -\left\{ \mathbb{E}_0 \left[\frac{\partial^2 m(X; \theta_*)}{\partial \theta \partial \theta^\top} \right] \right\}^{-1} \frac{1}{\sqrt{n}} \sum_{i=1}^{n} \frac{\partial m(X_i; \theta_*)}{\partial \theta} + o_p(1).$$

By the CLT, it follows that

$$\sqrt{n}(\widehat{\theta}_n - \theta_*) \xrightarrow{d} \mathcal{N}(0, \Sigma_{\theta_*}),$$

where the asymptotic covariance matrix is given by

$$\Sigma_{\theta_*} = \left\{ -\mathbb{E}_0 \left[\frac{\partial^2 m(X;\theta_*)}{\partial\theta\partial\theta^\top} \right] \right\}^{-1} \mathrm{Cov}_0 \left[\frac{\partial m(X;\theta_*)}{\partial\theta} \right] \left\{ -\mathbb{E}_0 \left[\frac{\partial^2 m(X;\theta_*)}{\partial\theta\partial\theta^\top} \right] \right\}^{-1},$$

with Cov_0 denoting the covariance under $X \sim p_0(x)$. The structure of the covariance matrix resembles a sandwich:

$$\text{Bread term} \times \text{Meat term} \times \text{Bread term}.$$

Gold Coin # 16. *Assume that $X_1, \ldots, X_n$ are i.i.d. with $X \sim p_0(x)$. The Kullback–Leibler divergence is used to quantify the discrepancy between the working model $p(x;\theta)$ and the true model $p_0(x)$. The estimand θ_* is defined as the minimizer of the KL divergence.*

Let $\widehat{\theta}_n$ denote the M-estimator constructed from the working model $p(x;\theta)$. Under suitable regularity conditions, it can be shown that

$$\widehat{\theta}_n \xrightarrow{p} \theta_*, \quad \text{as } n \to \infty,$$

and

$$\sqrt{n}(\widehat{\theta}_n - \theta_*) \xrightarrow{d} \mathcal{N}(0, \Sigma_{\theta_*}), \quad \text{as } n \to \infty,$$

where the asymptotic covariance matrix Σ_{θ_*} takes a "sandwich" form:

$$\Sigma_{\theta_*} = \textit{Bread–Meat–Bread}.$$

Example 1 (continued): Normal Distribution

An illustrative example is provided to aid in understanding the sandwich form of the asymptotic covariance matrix. Suppose the true distribution is exponential with rate parameter λ_0:

$$p_0(x) = \lambda_0 e^{-\lambda_0 x} I(x > 0).$$

If the working model is taken to be $\mathcal{N}(\mu, \sigma^2)$, then the KL divergence between this normal model and the true exponential distribution is given by:

$$D_{\mathrm{KL}}(\mu, \sigma^2) = \int_0^\infty \log \left[\frac{\lambda_0 e^{-\lambda_0 x}}{\frac{1}{\sqrt{2\pi\sigma^2}} e^{-(x-\mu)^2/(2\sigma^2)}} \right] \lambda_0 e^{-\lambda_0 x}\, dx.$$

It can be shown that the minimizer of the KL divergence is

$$\mu_* = \frac{1}{\lambda_0} \quad \text{and} \quad \sigma_*^2 = \frac{1}{\lambda_0^2},$$

since $\mathbb{E}_0(X) = 1/\lambda_0$ and $\mathbb{V}_0(X) = 1/\lambda_0^2$.

The M-estimators of μ and σ^2 are defined as:

$$(\widehat{\mu}_n, \widehat{\sigma}_n^2)^\top = \arg\max_{\mu,\sigma^2} \sum_{i=1}^{n} m(X_i; \mu, \sigma^2),$$

where the objective function is given by

$$m(X; \mu, \sigma^2) = -\frac{1}{2}\log(2\pi\sigma^2) - \frac{(X-\mu)^2}{2\sigma^2}.$$

In this case, $\widehat{\mu}_n$ and $\widehat{\sigma}_n^2$ correspond tc the sample mean and sample variance, respectively:

$$\widehat{\mu}_n = \frac{1}{n}\sum_{i=1}^{n} X_i \quad \text{and} \quad \widehat{\sigma}_n^2 = \frac{1}{n}\sum_{i=1}^{n}(X_i - \widehat{\mu}_n)^2.$$

To compute the asymptotic covariance matrix of $(\widehat{\mu}_n, \widehat{\sigma}_n^2)^\top$, the following derivatives are used:

$$\frac{\partial m(X; \mu, \sigma^2)}{\partial(\mu, \sigma^2)} = \left(\frac{X-\mu}{\sigma^2}, -\frac{1}{2\sigma^2} + \frac{(X-\mu)^2}{2\sigma^4}\right),$$

and

$$\frac{\partial^2 m(X; \mu, \sigma^2)}{\partial(\mu, \sigma^2)^\top \partial(\mu, \sigma^2)} = \begin{pmatrix} -\frac{1}{\sigma^2} & -\frac{X-\mu}{\sigma^4} \\ -\frac{X-\mu}{\sigma^4} & \frac{1}{2\sigma^4} - \frac{(X-\mu)^2}{\sigma^6} \end{pmatrix}.$$

Using these expressions (see Exercise 4.5), it can be verified that:

$$\mathrm{Cov}_0\left[\frac{\partial m(X; \mu_*, \sigma_*^2)}{\partial(\mu, \sigma^2)}\right] = \begin{pmatrix} \lambda_0^2 & \lambda_0^3 \\ \lambda_0^3 & 2\lambda_0^4 \end{pmatrix},$$

and

$$\mathbb{E}_0\left[\frac{\partial^2 m(X; \mu_*, \sigma_*^2)}{\partial(\mu, \sigma^2)^\top \partial(\mu, \sigma^2)}\right] = \begin{pmatrix} -\lambda_0^2 & 0 \\ 0 & -\lambda_0^4/2 \end{pmatrix}.$$

Thus, the asymptotic covariance matrix of $(\widehat{\mu}_n, \widehat{\sigma}_n^2)^\top$ is obtained as:

$$\begin{aligned}
&\mathbb{E}_0^{-1}\left[\frac{\partial^2 m(X; \mu_*, \sigma_*^2)}{\partial(\mu, \sigma^2)^\top \partial(\mu, \sigma^2)}\right] \cdot \mathrm{Cov}_0\left[\frac{\partial m(X; \mu_*, \sigma_*^2)}{\partial(\mu, \sigma^2)}\right] \cdot \mathbb{E}_0^{-1}\left[\frac{\partial^2 m(X; \mu_*, \sigma_*^2)}{\partial(\mu, \sigma^2)^\top \partial(\mu, \sigma^2)}\right] \\
&= \begin{pmatrix} -\lambda_0^{-2} & 0 \\ 0 & -2\lambda_0^{-4} \end{pmatrix} \begin{pmatrix} -\lambda_0^2 & 0 \\ 0 & -\lambda_0^4/2 \end{pmatrix} \begin{pmatrix} -\lambda_0^{-2} & 0 \\ 0 & -2\lambda_0^{-4} \end{pmatrix} \\
&= \begin{pmatrix} \lambda_0^{-2} & 2\lambda_0^{-3} \\ 2\lambda_0^{-3} & 8\lambda_0^{-4} \end{pmatrix}. \quad \square
\end{aligned}$$

4.5 Exercises

Exercise 4.1

Assume that $X_1, \ldots, X_n$ are i.i.d. with $X \sim p(x;\theta)$, where $\theta = (\mu, b)^\top$ and

$$p(x;\theta) = \frac{1}{2b} \exp\left(-\frac{|x-\mu|}{b}\right).$$

Show that the MLE of μ is the sample median,

$$\widehat{\mu}_n = \text{Median}\{X_1, \ldots, X_n\},$$

and the MLE of b is the mean absolute deviation,

$$\widehat{b}_n = \frac{1}{n}\sum_{i=1}^{n} |X_i - \widehat{\mu}_n|.$$

Exercise 4.2

Assume that $X_1, \ldots, X_n \stackrel{\text{i.i.d.}}{\sim} \text{Bernoulli}(p)$. Show that the MLE of p is

$$\widehat{p}_n = \frac{1}{n}\sum_{i=1}^{n} X_i = \frac{1}{n}\sum_{i=1}^{n} I(X_i = 1).$$

Exercise 4.3

Continue Example 2 of multinomial distribution. Show that

$$\text{Cov}\begin{pmatrix} \widehat{p}_{1n} \\ \widehat{p}_{2n} \\ \cdots \\ \widehat{p}_{Jn} \end{pmatrix} = \frac{1}{n}\begin{pmatrix} p_1(1-p_1) & -p_1p_2 & \cdots & -p_1p_J \\ -p_2p_1 & p_2(1-p_2) & \cdots & -p_2p_J \\ \cdots & \cdots & \cdots & \cdots \\ -p_Jp_1 & -p_Jp_2 & \cdots & p_J(1-p_J) \end{pmatrix}.$$

Exercise 4.4

Show that

$$D_{\text{KL}}(\mu) = \text{constant} + \frac{(1-\mu)^2}{2}.$$

Exercise 4.5

Suppose that the true distribution is the exponential distribution $\exp(\lambda_0)$ and the working model is $\mathcal{N}(\mu, \sigma^2)$. The M-estimators $\widehat{\mu}_n$ and $\widehat{\sigma}_n^2$ correspond to the sample mean and sample variance, respectively. Show that

$$n\mathrm{Cov}_0\left\{(\widehat{\mu}_n, \widehat{\sigma}_n^2)^\top\right\} \to \begin{pmatrix} \lambda_0^{-2} & 2\lambda_0^{-3} \\ 2\lambda_0^{-3} & 8\lambda_0^{-4} \end{pmatrix}.$$

5

Minimum Loss Estimation (MLE)

The concept of least squares has been around for more than two centuries, with its origins attributed to the work of Carl Friedrich Gauss in the late 18th century and Adrien-Marie Legendre in 1805. This method predates the maximum likelihood estimation technique, which was developed by R. A. Fisher in the early 20th century.

5.1 Dependent and Independent Variables

5.1.1 Relationship Between Two Variables

In pharmaceutical statistics, the relationship between two variables—such as the effect of a treatment on a patient's outcome—or, more generally, between one variable and multiple others, is often of primary interest. Accordingly, the distinction between dependent and independent variables is considered essential in the design of clinical studies.

A detailed discussion on the design of clinical studies and the classification of dependent and independent variables is deferred to Part II of this book. In this chapter, methods are introduced for analyzing data consisting of n independent and identically distributed (i.i.d.) pairs of observations $(X_1, Y_1), \ldots, (X_n, Y_n)$, with $(X, Y) \sim p_0(x, y)$, where $p_0(x, y)$ denotes the true but unknown joint distribution of (X, Y).

The Pearson correlation coefficient,* named after Karl Pearson (1857–1936), was developed in the late 19th century to quantify the strength and direction of the linear relationship between two variables. It is defined as

$$r = \frac{\sum_{i=1}^{n}(X_i - \overline{X})(Y_i - \overline{Y})}{\sqrt{\sum_{i=1}^{n}(X_i - \overline{X})^2}\sqrt{\sum_{i=1}^{n}(Y_i - \overline{Y})^2}},$$

where $\overline{X} = \sum_{i=1}^{n} X_i/n$ and $\overline{Y} = \sum_{i=1}^{n} Y_i/n$.

*It was developed by Karl Pearson from a related idea introduced by Francis Galton, who first used the term *correlation*, and for which the mathematical formula was derived by Auguste Bravais in 1844. This is another example of Stigler's Law.

DOI: 10.1201/9781003635468-6

The sign of r indicates the direction of the correlation, while the absolute value of r indicates its strength. By the Cauchy–Schwarz inequality,

$$\left[\sum_{i=1}^{n}(X_i-\overline{X})(Y_i-\overline{Y})\right]^2 \le \sum_{i=1}^{n}(X_i-\overline{X})^2\sum_{i=1}^{n}(Y_i-\overline{Y})^2,$$

with equality if and only if the vectors $(X_1-\overline{X},\dots,X_n-\overline{X})^\top$ and $(Y_1-\overline{Y},\dots,Y_n-\overline{Y})^\top$ are linearly dependent. Therefore, it follows that

$$r^2 \le 1,$$

with equality if and only if there exists a constant c such that

$$(Y_1-\overline{Y},\dots,Y_n-\overline{Y})^\top = c\cdot(X_1-\overline{X},\dots,X_n-\overline{X})^\top.$$

The Pearson correlation coefficient can be computed when X and Y are quantitative variables, ordinal variables, or binary variables coded as 0/1.

5.1.2 R^2

Suppose that multiple independent variables $X^{(1)},\dots,X^{(d)}$ are given. To abuse the notation, define the d-dimensional vector $X=(X^{(1)},\dots,X^{(d)})^\top$. To generalize the Pearson correlation coefficient for measuring the linear relationship between a response variable Y and the multivariate predictor X, a linear combination β is sought that maximizes the squared Pearson correlation between Y and $\beta^\top X$:

$$r^2(\beta)=\left\{\frac{\sum_{i=1}^{n}\beta^\top(X_i-\overline{X})(Y_i-\overline{Y})}{\sqrt{\sum_{i=1}^{n}[\beta^\top(X_i-\overline{X})]^2}\sqrt{\sum_{i=1}^{n}(Y_i-\overline{Y})^2}}\right\}^2.$$

Equivalently, the goal is to maximize the following ratio:

$$\frac{\left\{\beta^\top\left[\sum_{i=1}^{n}(X_i-\overline{X})(Y_i-\overline{Y})\right]\right\}^2}{\beta^\top\left[\sum_{i=1}^{n}(X_i-\overline{X})(X_i-\overline{X})^\top\right]\beta}.$$

Let the (scaled) sample covariance matrix of the predictors be denoted as

$$\Sigma_x=\sum_{i=1}^{n}(X_i-\overline{X})(X_i-\overline{X})^\top.$$

Under the transformation $\widetilde{\beta}=\Sigma_x^{1/2}\beta$, the optimization problem becomes one of maximizing

$$\frac{\left\{\widetilde{\beta}^\top\Sigma_x^{-1/2}\left[\sum_{i=1}^{n}(X_i-\overline{X})(Y_i-\overline{Y})\right]\right\}^2}{\widetilde{\beta}^\top\widetilde{\beta}}.$$

By the Cauchy–Schwarz inequality, the optimal value is attained when

$$\widetilde{\beta}_{\text{opt}} = c \cdot \Sigma_x^{-1/2} \sum_{i=1}^{n} (X_i - \overline{X})(Y_i - \overline{Y}), \quad \text{for some constant } c.$$

Therefore, the linear combination β that maximizes $r^2(\beta)$ is

$$\widehat{\beta}_{\text{opt}} = c \cdot \Sigma_x^{-1} \sum_{i=1}^{n} (X_i - \overline{X})(Y_i - \overline{Y}), \quad \text{for some constant } c.$$

Let R^2 be the maximum value of $r^2(\beta)$, that is, $R^2 = r^2(\widehat{\beta}_{\text{opt}})$, which is expressed as

$$R^2 = \frac{\left[\Sigma_x^{-1} \sum_{i=1}^{n} (X_i - \overline{X})(Y_i - \overline{Y})\right]^\top \sum_{i=1}^{n} (X_i - \overline{X})(Y_i - \overline{Y})}{\sum_{i=1}^{n} (Y_i - \overline{Y})^2}.$$

5.1.3 Least Squares Estimator

The least squares method is used to find the optimal linear combination β that minimizes the following sum of squared residuals:

$$\sum_{i=1}^{n} \left[Y_i - \overline{Y} - \beta^\top (X_i - \overline{X})\right]^2.$$

This minimization is motivated by the goal of optimally fitting the centered response $Y_i - \overline{Y}$ using the linear predictor $\beta^\top (X_i - \overline{X})$ for $i = 1, \ldots, n$. By taking the derivative of the objective function with respect to β and setting it to zero, the normal equation is obtained:

$$\sum_{i=1}^{n} 2 \left[Y_i - \overline{Y} - \beta^\top (X_i - \overline{X})\right] (X_i - \overline{X}) = 0,$$

which is equivalent to

$$\sum_{i=1}^{n} (Y_i - \overline{Y})(X_i - \overline{X}) = \sum_{i=1}^{n} (X_i - \overline{X})(X_i - \overline{X})^\top \beta.$$

The solution to this equation is given by

$$\widehat{\beta} = \left[\sum_{i=1}^{n} (X_i - \overline{X})(X_i - \overline{X})^\top\right]^{-1} \sum_{i=1}^{n} (X_i - \overline{X})(Y_i - \overline{Y}).$$

Thus, the estimator $\widehat{\beta}$ derived here coincides with the maximizer of $r^2(\beta)$ discussed in the previous subsection (with $c = 1$). Furthermore, using $\widehat{\beta}$, R^2

can be re-expressed as:

$$R^2 = \frac{\left[\Sigma_x^{-1} \sum_{i=1}^n (X_i - \overline{X})(Y_i - \overline{Y})\right]^\top \sum_{i=1}^n (X_i - \overline{X})(Y_i - \overline{Y})}{\sum_{i=1}^n (Y_i - \overline{Y})^2}$$

$$= \frac{\sum_{i=1}^n \widehat{\beta}^\top (X_i - \overline{X})(Y_i - \overline{Y})}{\sum_{i=1}^n (Y_i - \overline{Y})^2}$$

$$= 1 - \frac{\sum_{i=1}^n \left[Y_i - \overline{Y} - \widehat{\beta}^\top (X_i - \overline{X})\right]^2}{\sum_{i=1}^n (Y_i - \overline{Y})^2},$$

where the last equality follows from the normal equation (see Exercise 5.1).

This formulation of R^2 provides an alternative interpretation: it represents the proportion of variance in the dependent variable that is explained by the independent variables.

Gold Coin # 17. *The following two optimization problems are equivalent: finding β that maximizes the squared Pearson correlation between Y and $\beta^\top X$ (with the maximum value denoted by R^2),*

$$\widehat{\beta} = \arg\max_\beta \left\{ \frac{\sum_{i=1}^n \beta^\top (X_i - \overline{X})(Y_i - \overline{Y})}{\sqrt{\sum_{i=1}^n \left[\beta^\top (X_i - \overline{X})\right]^2} \sqrt{\sum_{i=1}^n (Y_i - \overline{Y})^2}} \right\}^2,$$

is equivalent to finding β that minimizes the sum of squared residuals (as used in the least squares method),

$$\widehat{\beta} = \arg\min_\beta \sum_{i=1}^n \left[Y_i - \overline{Y} - \beta^\top (X_i - \overline{X})\right]^2.$$

5.2 Minimum Loss Estimator

5.2.1 Two Main Types of Dependent Variables

Several common types of variables are encountered in statistical analyses, including continuous variables, discrete variables, quantitative variables, categorical variables, ordinal variables, nominal variables, binary variables, count variables, and time-to-event variables. These categories are not mutually exclusive: for example, continuous variables are a subset of quantitative variables, binary variables are a subset of categorical variables, and time-to-event variables are continuous variables subject to censoring. Furthermore, categorical variables can be classified as either ordinal or nominal.

In this chapter, two main types of dependent variables that are often encountered in clinical studies are considered: (1) continuous variables (in particular, bounded continuous variables) and (2) binary variables.

5.2.2 Law of Iterative Expectations

The Law of Iterated Expectations (LIE) states that, for any two random variables X and Y, the following identity holds:

$$\mathbb{E}[\mathbb{E}(Y \mid X)] = \mathbb{E}(Y).$$

A proof is provided for the case in which X and Y are discrete random variables (for continuous variables, the summations may be replaced by integrals). Let $p_X(x)$ and $p_Y(y)$ denote the probability mass functions of X and Y, respectively. Let $p_{X,Y}(x, y)$ denote the joint probability mass function, and let $p_{Y|x}(y) = p_{X,Y}(x, y)/p_X(x)$ denote the conditional probability mass function of Y given $X = x$. Then, the following derivation is obtained:

$$\begin{aligned}
\mathbb{E}[\mathbb{E}(Y \mid X)] &= \sum_x \mathbb{E}(Y \mid X = x)\, p_X(x) \\
&= \sum_x \sum_y y\, p_{Y|x}(y \mid x)\, p_X(x) \\
&= \sum_x \sum_y y\, p_{X,Y}(x, y) \\
&= \sum_y y \sum_x p_{X,Y}(x, y) \\
&= \sum_y y\, p_Y(y) = \mathbb{E}(Y). \quad \square
\end{aligned}$$

By applying the LIE, the following identity for variance can be established:

$$\mathbb{V}(Y) = \mathbb{E}[\mathbb{V}(Y \mid X)] + \mathbb{V}[\mathbb{E}(Y \mid X)].$$

In fact, this decomposition can be derived as follows:

$$\begin{aligned}
&\mathbb{E}[\mathbb{V}(Y \mid X)] + \mathbb{V}[\mathbb{E}(Y \mid X)] \\
=&\mathbb{E}[\mathbb{E}(Y^2 \mid X) - \mathbb{E}^2(Y \mid X)] + \mathbb{E}[\mathbb{E}^2(Y \mid X)] - \{\mathbb{E}[\mathbb{E}(Y \mid X)]\}^2 \\
=&\mathbb{E}[\mathbb{E}(Y^2 \mid X)] - \{\mathbb{E}[\mathbb{E}(Y \mid X)]\}^2 \\
=&\mathbb{E}(Y^2) - (\mathbb{E}Y)^2 = \mathbb{V}(Y). \quad \square
\end{aligned}$$

5.2.3 $\mathcal{L}_2$ Loss

Estimand

A continuous dependent variable Y is considered. Let $X = (X^{(1)}, \ldots, X^{(d)})^\top$ denote a vector of independent variables. Without loss of generality, assume that each independent variable is either quantitative or binary, given that a categorical variable with $J > 2$ levels can be represented by $J - 1$ binary (dummy) variables.

The least squares method is to identify a function of the independent variables which minimizes its "distance" from the dependent variable in terms of the sum of squared differences. In general, this distance is measured by a *loss function*. In particular, the sum of squares corresponds to the $\mathcal{L}_2$ loss.

Although linear functions are sometimes used, the function relating X to Y need not be linear. In general, let $Q(X)$ denote an arbitrary function of X used to model the relationship between X and Y. The quality of this function is evaluated by how close $Q(X)$ is to Y, measured by the $\mathcal{L}_2$ loss:

$$L(Y, Q(X)) = (Y - Q(X))^2.$$

The expected $\mathcal{L}_2$ loss is given by

$$\mathbb{E}[L(Y, Q(X))] = \mathbb{E}(Y - Q(X))^2.$$

By the LIE,

$$\mathbb{E}(Y - Q(X))^2 = \mathbb{E}\left\{\mathbb{E}\left[(Y - Q(X))^2 \mid X\right]\right\}.$$

Applying the bias–variance decomposition for mean squared error, the conditional loss can be expressed as:

$$\mathbb{E}\left[(Y - Q(X))^2 \mid X\right] = \mathbb{V}(Y \mid X) + |Q(X) - Q_0(X)|^2,$$

where

$$Q_0(X) = \mathbb{E}(Y \mid X)$$

is referred to as the *regression function* of Y given X. Thus, among all possible functions $Q(X)$, the expected loss is minimized when $Q(X) = Q_0(X)$.

A parametric regression function of the form $Q(X;\beta)$ is now considered, where β is a vector of finite parameters. A typical example is $Q(X;\beta) = \beta_0 + \beta_1^\top X$. The value of β that minimizes the expected loss is denoted by β_*:

$$\begin{aligned}\beta_* &= \arg\min_\beta \mathbb{E}[L(Y, Q(X;\beta))] \\ &= \arg\min_\beta \mathbb{E}\left[(Y - Q(X;\beta))^2\right].\end{aligned}$$

By differentiating the expected loss with respect to β and setting the derivative equal to zero, the following first-order condition is obtained:

$$\mathbb{E}\left[\frac{\partial Q(X;\beta_*)}{\partial \beta}(Y - Q(X;\beta_*))\right] = 0.$$

To simplify notation, define the *estimating function*:

$$m(X,Y;\beta) = \frac{\partial Q(X;\beta)}{\partial \beta}(Y - Q(X;\beta)).$$

The expectation of this function equals zero at β_*:

$$\mathbb{E}\left[m(X,Y;\beta_*)\right] = 0.$$

M-estimator

To estimate β_*, the empirical analog of the expected loss is minimized:

$$\widehat{\beta}_n = \arg\min_{\beta} \frac{1}{n}\sum_{i=1}^{n}(Y_i - Q(X_i;\beta))^2.$$

Defining the empirical estimating function as

$$M_n(\beta) \triangleq \frac{1}{n}\sum_{i=1}^{n} m(X_i, Y_i;\beta),$$

$\widehat{\beta}_n$ solves the following estimating equation,

$$M_n(\widehat{\beta}_n) = \frac{1}{n}\sum_{i=1}^{n} m(X_i, Y_i;\widehat{\beta}_n) = 0.$$

The estimator $\widehat{\beta}_n$ is referred to as the *minimum loss estimator* (MLE). It should be noted that the acronym MLE is also used to denote the *maximum likelihood estimator* (MLE), introduced in the preceding chapter.

In this book, this dual use is allowed for two reasons. First, both estimators are special cases of the M-estimator framework.[12] Second, when the loss function is taken to be the negative log-likelihood, the two estimators coincide. For example, the $\mathcal{L}_2$ loss corresponds to the log-likelihood under the normal distribution (see Exercise 5.2).

Linear Model

Consider the linear working model:

$$Q(X;\beta) = \beta_0 + \beta_1^\top X = \widetilde{X}^\top \beta,$$

where

$$\beta = \begin{pmatrix} \beta_0 \\ \beta_1 \end{pmatrix}, \quad \widetilde{X} = \begin{pmatrix} 1 \\ X \end{pmatrix}.$$

The parameter of interest, β_*, is given by:

$$\beta_* = \arg\min_{\beta} \mathbb{E}(Y - \widetilde{X}^\top \beta)^2,$$

and satisfies the moment condition:

$$\mathbb{E}\left[\widetilde{X}(Y - \widetilde{X}^\top \beta_*)\right] = 0.$$

An explicit form of β_* can be derived:

$$\beta_* = \left[\mathbb{E}(\widetilde{X}\widetilde{X}^\top)\right]^{-1} \mathbb{E}(\widetilde{X}Y).$$

The estimator $\widehat{\beta}_n$ is defined as the solution to the empirical version of the above estimating equation:

$$\frac{1}{n}\sum_{i=1}^{n} \widetilde{X}_i(Y_i - \widetilde{X}_i^\top \widehat{\beta}_n) = 0,$$

which yields the closed-form solution:

$$\widehat{\beta}_n = \left[\frac{1}{n}\sum_{i=1}^{n} \widetilde{X}_i \widetilde{X}_i^\top\right]^{-1} \left(\frac{1}{n}\sum_{i=1}^{n} \widetilde{X}_i Y_i\right).$$

Because in this example both $\widehat{\beta}_n$ and β_* admit explicit forms, the consistency and asymptotic normality of $\widehat{\beta}_n$ can be easily established. By the law of large numbers (LLN),

$$\frac{1}{n}\sum_{i=1}^{n} \widetilde{X}_i \widetilde{X}_i^\top \xrightarrow{p} \mathbb{E}(\widetilde{X}\widetilde{X}^\top) \quad \text{and} \quad \frac{1}{n}\sum_{i=1}^{n} \widetilde{X}_i Y_i \xrightarrow{p} \mathbb{E}(\widetilde{X}Y),$$

which implies

$$\widehat{\beta}_n \xrightarrow{p} \beta_*.$$

To establish asymptotic normality, note that by the Slutsky's lemma,

$$\sqrt{n}(\widehat{\beta}_n - \beta_*) \quad \text{and} \quad \sqrt{n}\left(\frac{1}{n}\sum_{i=1}^{n} [\mathbb{E}(\widetilde{X}\widetilde{X}^\top)]^{-1} \widetilde{X}_i Y_i - \beta_*\right)$$

have the same asymptotic distribution. By the central limit theorem (CLT),

$$\sqrt{n}(\widehat{\beta}_n - \beta_*) \xrightarrow{d} \mathcal{N}\left(0, [\mathbb{E}(\widetilde{X}\widetilde{X}^\top)]^{-1} \mathrm{Cov}(\widetilde{X}Y)\, [\mathbb{E}(\widetilde{X}\widetilde{X}^\top)]^{-1}\right),$$

where the asymptotic covariance has the "sandwich" form.

To conduct statistical inference (e.g., confidence intervals or hypothesis tests), an estimator of the asymptotic covariance matrix is required:

$$\left[\widehat{\mathbb{E}}(\widetilde{X}\widetilde{X}^\top)\right]^{-1} \widehat{\mathrm{Cov}}(\widetilde{X}Y) \left[\widehat{\mathbb{E}}(\widetilde{X}\widetilde{X}^\top)\right]^{-1},$$

where

$$\widehat{\mathbb{E}}(\widetilde{X}\widetilde{X}^\top) = \frac{1}{n}\sum_{i=1}^{n} \widetilde{X}_i\widetilde{X}_i^\top,$$

$$\widehat{\mathrm{Cov}}(\widetilde{X}Y) = \frac{1}{n}\sum_{i=1}^{n}\left(\widetilde{X}_iY_i - \frac{1}{n}\sum_{j=1}^{n}\widetilde{X}_jY_j\right)\left(\widetilde{X}_iY_i - \frac{1}{n}\sum_{j=1}^{n}\widetilde{X}_jY_j\right)^\top.$$

5.2.4 Binary Cross-Entropy Loss

Estimand

A binary dependent variable Y, encoded as $0/1$, is considered along with a vector of independent variables $X = (X^{(1)}, \ldots, X^{(d)})^\top$.

Although Y may be treated as a quantitative variable and analyzed via the $\mathcal{L}_2$ loss for its regression function (see Exercise 5.3),

$$Q_0(X) = \mathbb{E}(Y \mid X) = \mathbb{P}(Y = 1 \mid X),$$

it is often more natural for the negative log-likelihood of the Bernoulli distribution to be employed as the loss function. This leads to the so-called *binary cross-entropy (BCE) loss*. Under the BCE loss, the minimizer of the expected loss coincides with the maximum likelihood estimator (see Exercise 5.4).

The BCE loss between $Q(X)$ and Y is defined by

$$L\left(Y, Q(X)\right) = -\left[Y \log Q(X) + (1 - Y)\log(1 - Q(X))\right].$$

The expected loss is expressed as

$$\mathbb{E}[L(Y, Q(X))] = -\mathbb{E}\left[Y \log Q(X) + (1 - Y)\log(1 - Q(X))\right].$$

By the LIE,

$$\begin{aligned}\mathbb{E}[L(Y, Q(X))] &= -\mathbb{E}\left\{\mathbb{E}\left[Y \log Q(X) + (1 - Y)\log(1 - Q(X)) \mid X\right]\right\} \\ &= -\mathbb{E}\left\{Q_0(X)\log Q(X) + (1 - Q_0(X))\log(1 - Q(X))\right\}.\end{aligned}$$

Recall that the function

$$p \mapsto p_0 \log p + (1 - p_0)\log(1 - p)$$

achieves its maximum when $p = p_0$. Therefore, among all functions $Q(X)$, the regression function $Q_0(X)$ minimizes the expected loss.

A tentative parametric regression function $Q(X;\beta)$ is now considered, where β is a finite-dimensional parameter vector. Given that $0 < Q_0(X) < 1$, the model $Q(X;\beta)$ is chosen to ensure $0 < Q(X;\beta) < 1$. One such model is the logistic regression model to be introduced shortly.

The parameter β_* is defined as the value minimizing the expected loss:

$$\beta_* = \arg\min_{\beta} \left\{-\mathbb{E}\left[Y \log Q(X;\beta) + (1-Y)\log(1-Q(X;\beta))\right]\right\}.$$

By differentiating the expected loss with respect to β and equating to zero, the following estimating equation is obtained:

$$\mathbb{E}\left[\frac{\partial Q(X;\beta_*)}{\partial\beta} \cdot \frac{Y - Q(X;\beta_*)}{Q(X;\beta_*)(1-Q(X;\beta_*))}\right] = 0.$$

Define

$$m(X,Y;\beta) = \frac{\partial Q(X;\beta)/\partial\beta}{Q(X;\beta)(1-Q(X;\beta))}(Y - Q(X;\beta)).$$

Then,

$$\mathbb{E}[m(X,Y;\beta_*)] = 0.$$

M-Estimator

To estimate β_*, the minimizer of the empirical loss is employed as an estimator:

$$\widehat{\beta}_n = \arg\min_{\beta} \left\{-\frac{1}{n}\sum_{i=1}^{n}\left[Y_i \log Q(X_i;\beta) + (1-Y_i)\log(1-Q(X_i;\beta))\right]\right\}.$$

Defining the empirical estimating function as

$$M_n(\beta) \triangleq \frac{1}{n}\sum_{i=1}^{n} m(X_i, Y_i;\beta),$$

$\widehat{\beta}_n$ solves the estimating equation,

$$M_n(\widehat{\beta}_n) = \frac{1}{n}\sum_{i=1}^{n} m(X_i, Y_i;\widehat{\beta}_n) = 0.$$

Logistic Model

The logistic model is now introduced as the working model:

$$Q(X;\beta) = \frac{\exp(\beta_0 + \beta_1^\top X)}{1 + \exp(\beta_0 + \beta_1^\top X)} = \frac{\exp(\beta^\top \widetilde{X})}{1 + \exp(\beta^\top \widetilde{X})},$$

where

$$\beta = \begin{pmatrix}\beta_0\\ \beta_1\end{pmatrix}, \quad \widetilde{X} = \begin{pmatrix}1\\ X\end{pmatrix}.$$

The estimand β_* is characterized as:

$$\begin{aligned}\beta_* &= \arg\min_{\beta} \left\{-\mathbb{E}\left[Y \log Q(X;\beta) + (1-Y)\log(1-Q(X;\beta))\right]\right\} \\ &= \arg\min_{\beta} \left\{-\mathbb{E}\left[\beta^\top \widetilde{X} Y - \log(1+\exp(\beta^\top \widetilde{X}))\right]\right\},\end{aligned}$$

leading to the estimating equation:

$$\mathbb{E}\left[\widetilde{X}\left(Y - \frac{\exp(\widetilde{X}^\top \beta_*)}{1+\exp(\widetilde{X}^\top \beta_*)}\right)\right] = 0.$$

An estimator $\widehat{\beta}_n$ is defined as the solution to:

$$\frac{1}{n}\sum_{i=1}^{n} \widetilde{X}_i \left(Y_i - \frac{\exp(\widetilde{X}_i^\top \widehat{\beta}_n)}{1+\exp(\widetilde{X}_i^\top \widehat{\beta}_n)}\right) = 0.$$

Although a closed-form solution for $\widehat{\beta}_n$ is not available, it can be computed via the Newton-Raphson or iteratively reweighted least squares methods.[20] By the LLN, the consistency can be established:

$$\widehat{\beta}_n \xrightarrow{p} \beta_*.$$

To establish asymptotic normality, a Taylor expansion yields:

$$\begin{aligned}0 &= \frac{1}{n}\sum_{i=1}^{n} \widetilde{X}_i \left(Y_i - \frac{\exp(\widetilde{X}_i^\top \widehat{\beta}_n)}{1+\exp(\widetilde{X}_i^\top \widehat{\beta}_n)}\right) \\ &= \frac{1}{n}\sum_{i=1}^{n} \widetilde{X}_i \left(Y_i - \frac{\exp(\widetilde{X}_i^\top \beta_*)}{1+\exp(\widetilde{X}_i^\top \beta_*)}\right) \\ &\quad + \frac{1}{n}\sum_{i=1}^{n} \widetilde{X}_i \widetilde{X}_i^\top \frac{\exp(\widetilde{X}_i^\top \beta_*)}{\left(1+\exp(\widetilde{X}_i^\top \beta_*)\right)^2}(\widehat{\beta}_n - \beta_*) + o_p\left(\frac{1}{\sqrt{n}}\right).\end{aligned}$$

Hence, $\sqrt{n}(\widehat{\beta}_n - \beta_*)$ is expressed as:

$$-\left\{\mathbb{E}\left[\widetilde{X}\widetilde{X}^\top \frac{\exp(\widetilde{X}^\top \beta_*)}{\left(1+\exp(\widetilde{X}^\top \beta_*)\right)^2}\right]\right\}^{-1} \cdot \frac{1}{\sqrt{n}}\sum_{i=1}^{n} \widetilde{X}_i \left(Y_i - \frac{\exp(\widetilde{X}_i^\top \beta_*)}{1+\exp(\widetilde{X}_i^\top \beta_*)}\right).$$

Therefore, by the CLT,

$$\sqrt{n}(\widehat{\beta}_n - \beta_*) \xrightarrow{d} \mathcal{N}(0, B_*^{-1} M_* B_*^{-1}),$$

where

$$B_* = \mathbb{E}\left[\widetilde{X}\widetilde{X}^\top \frac{\exp(\widetilde{X}^\top \beta_*)}{\left(1+\exp(\widetilde{X}^\top \beta_*)\right)^2}\right] \quad \text{and}$$

$$M_* = \operatorname{Cov}\left[\widetilde{X}\left(Y - \frac{\exp(\widetilde{X}^\top \beta_*)}{1+\exp(\widetilde{X}^\top \beta_*)}\right)\right],$$

which can be estimated by

$$\widehat{B}_n = \frac{1}{n}\sum_{i=1}^{n}\left[\widetilde{X}_i\widetilde{X}_i^\top \frac{\exp(\widetilde{X}_i^\top \widehat{\beta}_n)}{\left(1+\exp(\widetilde{X}_i^\top \widehat{\beta}_n)\right)^2}\right] \quad \text{and}$$

$$\widehat{M}_n = \frac{1}{n}\sum_{i=1}^{n}\widetilde{X}_i\widetilde{X}_i^\top \left(Y_i - \frac{\exp(\widetilde{X}_i^\top \widehat{\beta}_n)}{1+\exp(\widetilde{X}_i^\top \widehat{\beta}_n)}\right)^2.$$

5.2.5 A Unified Approach

A quantitative dependent variable Y—in particular, bounded continuous variable or 0/1-coded binary variable—is considered along with a vector of independent variables $X = (X^{(1)}, \ldots, X^{(d)})^\top$.

The regression function of Y given X is defined as

$$Q_0(X) = \mathbb{E}(Y \mid X).$$

A loss function $L(Y, Q(X))$—such as the $\mathcal{L}_2$ loss, the BCE loss, or other valid alternatives—is assumed to satisfy the below condition:

$$Q_0(\cdot) = \arg\min_{Q(\cdot)} \mathbb{E}\left[L(Y, Q(X))\right].$$

See Exercise 5.5 for an example of using the BCE loss for bounded continuous dependent variable.

If a tentative parametric regression function $Q(X;\beta)$ is considered, where β is a finite-dimensional parameter vector, β_* is defined as the value that minimizes the expected loss:

$$\beta_* = \arg\min_{\beta} \mathbb{E}\left[L(Y, Q(X;\beta))\right],$$

which satisfies the following condition:

$$\mathbb{E}\left[\frac{\partial L(Y, Q(X;\beta_*))}{\partial Q}\frac{\partial Q(X;\beta_*)}{\partial \beta}\right] = 0.$$

Motivated by the above, the function

$$m(X,Y;\beta) = \frac{\partial L(Y, Q(X;\beta))}{\partial Q} \frac{\partial Q(X;\beta)}{\partial \beta}$$

is defined. The condition

$$\mathbb{E}\left[m(X,Y;\beta_*)\right] = 0$$

holds by construction.

To estimate β_*, consider the solution to the following estimating equation:

$$\frac{1}{n}\sum_{i=1}^{n} m(X_i, Y_i; \widehat{\beta}_n) = 0.$$

By applying a Taylor expansion, the following is obtained:

$$\begin{aligned}
0 &= \frac{1}{n}\sum_{i=1}^{n} m(X_i, Y_i; \widehat{\beta}_n) \\
&= \frac{1}{n}\sum_{i=1}^{n} m(X_i, Y_i; \beta_*) + \frac{1}{n}\sum_{i=1}^{n} \frac{\partial m(X_i, Y_i; \beta_*)}{\partial \beta^\top}(\widehat{\beta}_n - \beta_*) + o_p\left(\frac{1}{\sqrt{n}}\right).
\end{aligned}$$

It follows that

$$\sqrt{n}(\widehat{\beta}_n - \beta_*) = -\left[\mathbb{E}\frac{\partial m(X,Y;\beta_*)}{\partial \beta^\top}\right]^{-1} \cdot \frac{1}{\sqrt{n}}\sum_{i=1}^{n} m(X_i, Y_i; \beta_*) + o_p(1).$$

Therefore, by the CLT,

$$\sqrt{n}(\widehat{\beta}_n - \beta_*) \xrightarrow{d} \mathcal{N}(0, B_*^{-1} M_* B_*^{-1}),$$

where

$$B_* = \mathbb{E}\left[\frac{\partial m(X,Y;\beta_*)}{\partial \beta^\top}\right] \quad \text{and} \quad M_* = \mathrm{Cov}\left[m(X,Y;\beta_*)\right],$$

which can be estimated using the sample analogues:

$$\widehat{B}_n = \frac{1}{n}\sum_{i=1}^{n} \frac{\partial m(X_i, Y_i; \widehat{\beta}_n)}{\partial \beta^\top} \quad \text{and} \quad \widehat{M}_n = \frac{1}{n}\sum_{i=1}^{n} m(X_i, Y_i; \widehat{\beta}_n) m^\top(X_i, Y_i; \widehat{\beta}_n).$$

Gold Coin # 18. *The two methods sharing the acronym MLE—namely, maximum likelihood estimation and minimum loss estimation—can both be interpreted as special cases of M-estimation.*

In an M-estimation framework, a function $m(X,Y;\beta)$ is constructed such that its expectation is zero at β_*:

$$\mathbb{E}\left[m(X,Y;\beta_*)\right] = 0.$$

An M-estimator is defined as the solution to the estimating equation:

$$\frac{1}{n}\sum_{i=1}^{n} m(X_i,Y_i;\widehat{\beta}_n) = 0.$$

Under certain regularity conditions, the M-estimator is consistent:

$$\widehat{\beta}_n \xrightarrow{p} \beta_*,$$

and is asymptotically normal:

$$\sqrt{n}(\widehat{\beta}_n - \beta_*) \xrightarrow{d} \mathcal{N}(0, B_*^{-1}M_*B_*^{-1}),$$

where the asymptotic covariance has the sandwich form.

In particular, if the model $Q(X;\beta)$ correctly specifies $Q_0(X)$, then a true parameter value β_0 exists such that $Q_0(X) = Q(X;\beta_0)$ and

$$\mathbb{E}[m(X,Y;\beta_0)] = 0.$$

Under this condition, the M-estimator is consistent:

$$\widehat{\beta}_n \xrightarrow{p} \beta_0,$$

and is asymptotically normal:

$$\sqrt{n}(\widehat{\beta}_n - \beta_0) \xrightarrow{d} \mathcal{N}(0, B_0^{-1}M_0B_0^{-1}),$$

where

$$B_0 = \mathbb{E}\left[\frac{\partial m(X,Y;\beta_0)}{\partial \beta^\top}\right] \quad \text{and} \quad M_0 = \text{Cov}\left[m(X,Y;\beta_0)\right].$$

5.3 Two Purposes

Before an estimator is proposed, the target estimand of interest should be explicitly defined. In the preceding section, the estimand was taken to be β_*, which is associated with a tentative parametric regression function $Q(X;\beta)$, and, importantly, no assumption was made that the true regression function $Q_0(X)$ belongs to the parametric family represented by $Q(X; \beta)$. As a result, concerns about model misspecification are not relevant when the objective is solely to estimate β_*, which minimizes the expected loss.

However, if the estimand of interest is the true regression function $Q_0(X)$ itself, a different approach is warranted. In practice, the estimation of $Q_0(X)$ may serve at least two distinct purposes. One is *explanation*, where the estimated regression function $\widehat{Q}(X)$ is used to elucidate the relationship between the dependent variable Y and the independent variables X. The other is *prediction*, in which $\widehat{Q}(X)$ is employed as a predictive model to predict the value of Y given a new input X.

5.3.1 Explanation

To proceed with estimating $Q_0(X)$, assume that $Q_0(X)$ can be represented as $Q(X; \beta)$ with some true parameter value β_0 such that

$$Q(X; \beta_0) = Q_0(X).$$

The parameter β_0 can then be estimated using the M-estimation framework outlined previously, with the key distinction that the estimand now refers to β_0. Given an estimator $\widehat{\beta}_n$ of β_0, an estimator of $Q_0(X)$ can be formed as

$$\widehat{Q}_n(X) = Q(X; \widehat{\beta}_n).$$

Once the estimated regression function $\widehat{Q}_n(X) = Q(X; \widehat{\beta}_n)$ is obtained, it can be used to interpret the relationship between the dependent variable and each of the independent variables.

For instance, suppose a linear model of the form $Q(X; \beta) = \beta_0 + \beta_1^\top X$ is employed as the working model. Consider the coefficient $\beta^{(j)}$ corresponding to the jth independent variable. If $\widehat{\beta}^{(j)} > 0$, then, conditional on the remaining variables, a positive association between $X^{(j)}$ and Y is implied. Based on the estimated standard error—corresponding to the $(j, j)^{\text{th}}$ component of the estimated covariance matrix—a confidence interval for $\beta^{(j)}$ can be constructed, and a p-value for testing the null hypothesis $H_0 : \beta^{(j)} = 0$ can be computed.

5.3.2 Prediction

Following the estimation of the regression function $\widehat{Q}_n(X) = Q(X; \widehat{\beta}_n)$, it can be utilized as a predictive model for predicting future outcomes. Specifically, for a new input X^0, a prediction for the corresponding output Y^0 is made using $\widehat{Q}_n(X^0) = Q(X^0; \widehat{\beta}_n)$ prior to observing Y^0.

The accuracy of this prediction may be assessed via the expected loss conditional on $\widehat{\beta}_n$,

$$\mathbb{E}\left[L(Y^0, Q(X^0; \widehat{\beta}_n)) \,\middle|\, \widehat{\beta}_n \right],$$

or by the marginal (unconditional) expected loss,

$$\mathbb{E}\left[L(Y^0, Q(X^0; \widehat{\beta}_n))\right] = \mathbb{E}\left\{\mathbb{E}\left[L(Y^0, Q(X^0; \widehat{\beta}_n)) \,\Big|\, \widehat{\beta}_n\right]\right\}.$$

As a concrete example, if the $\mathcal{L}_2$ loss is adopted, the predictive accuracy may be quantified by the conditional expected loss

$$\mathbb{E}\left[(Y^0 - Q(X^0, \widehat{\beta}_n))^2 \,\Big|\, \widehat{\beta}_n\right],$$

or by the unconditional expected loss,

$$\mathbb{E}\left[(Y^0 - Q(X^0, \widehat{\beta}_n))^2\right] = \mathbb{E}\left\{\mathbb{E}\left[(Y^0 - Q(X^0, \widehat{\beta}_n))^2 \,\Big|\, \widehat{\beta}_n\right]\right\}.$$

If the BCE loss is used, the expected loss conditional on $\widehat{\beta}_n$ becomes

$$-\mathbb{E}\left[Y^0 \log Q(X^0; \widehat{\beta}_n) + (1 - Y^0)\log(1 - Q(X^0; \widehat{\beta}_n)) \,\Big|\, \widehat{\beta}_n\right],$$

while the unconditional expected loss is given by

$$\begin{aligned}
&- \mathbb{E}\left[Y^0 \log Q(X^0; \widehat{\beta}_n) + (1 - Y^0)\log(1 - Q(X^0; \widehat{\beta}_n))\right] \\
&= -\mathbb{E}\left\{\mathbb{E}\left[Y^0 \log Q(X^0; \widehat{\beta}_n) + (1 - Y^0)\log(1 - Q(X^0; \widehat{\beta}_n)) \,\Big|\, \widehat{\beta}_n\right]\right\}.
\end{aligned}$$

Gold Coin # 19. *Two main purposes of modeling are explanation and prediction. Explanation helps us understand the underlying structure or causal relationships, while prediction focuses on accurately forecasting future outcomes based on input data.*

5.3.3 Summary

The first five chapters are devoted to several key results in asymptotic statistics, which will be applied in Part II of this book. For a comprehensive study of asymptotic statistics, readers are referred to the excellent textbook *Asymptotic Statistics* by van der Vaart (2000).[12]

In Chapter 1, we argue that the probability density function of the normal distribution $\mathcal{N}(\mu, \sigma^2)$,

$$f(x) = \frac{1}{\sqrt{2\pi}\sigma} \exp\left[-\frac{1}{2}\left(\frac{x - \mu}{\sigma}\right)^2\right],$$

is the most useful equation in statistics. We also discuss why the distribution has the number 2 in the exponent as the order of its tail.

Chapter 2 introduces two fundamental types of convergence: convergence in probability and convergence in distribution. They are closely related to two cornerstone results in probability theory, the law of large numbers and the central limit theorem. In turn, these results underlie two desirable properties in asymptotic statistics: *consistency* and *asymptotic normality.*

In Chapter 3, we acknowledge the profound contributions of two pioneers of modern statistics—Fisher and Neyman—through their foundational work on hypothesis testing and confidence intervals. Of course, these represent only a small fraction of their extensive contributions on statistical theory. Chapter 4 is devoted to Fisher's likelihood theory, while Neyman's potential outcomes framework forms the foundation of Part II.

Chapters 4 and 5 cover two major estimation principles: maximum likelihood estimation (MLE) and minimum loss estimation (MLE), respectively. The dual use of the acronym "MLE" is intentional, emphasizing that both approaches can be unified under the broader framework of *M-estimation.*

With these fundamental tools in asymptotic statistics established, we are now prepared to explore causal inference in pharmaceutical statistics in Part II.

5.4 Exercises

Exercise 5.1

Show that

$$R^2 = 1 - \frac{\sum_{i=1}^n \left[Y_i - \overline{Y} - \widehat{\beta}^\top (X_i - \overline{X})\right]^2}{\sum_{i=1}^n (Y_i - \overline{Y})^2},$$

where

$$\widehat{\beta} = \left[\sum_{i=1}^n (X_i - \overline{X})(X_i - \overline{X})^\top\right]^{-1} \sum_{i=1}^n (X_i - \overline{X})(Y_i - \overline{Y}).$$

Exercise 5.2

Assume the following model:

$$Y_i = Q(X_i; \beta) + \varepsilon_i,$$

where ε_i, $i = 1, \ldots, n$, are i.i.d. with $\mathcal{N}(0, \sigma^2)$. Develop the maximum likelihood estimator of β and relate it to the minimum $\mathcal{L}_2$-loss estimator of β.

Exercise 5.3

For a 0/1 coded binary outcome variable Y, assume a model $Q(X;\beta)$ for $\mathbb{P}(Y = 1 \mid X)$. Develop the minimum $\mathcal{L}_2$-loss estimator of β.

Exercise 5.4

For a 0/1 coded binary outcome variable Y, assume a model $Q(X;\beta)$ for $\mathbb{P}(Y = 1 \mid X)$. Develop the maximum likelihood estimator of β and relate it to the minimum BCE-loss estimator of β.

Exercise 5.5

For a bounded outcome variable $Y \in [a, b]$, assume a model $Q(X;\beta)$ for $\mathbb{E}(Y \mid X)$. Develop the minimum BCE-loss estimator of β.

Part II

An Introduction to Causal Inference

6

Potential Outcomes

6.1 Two Sets of Guidance Documents

In the pharmaceutical industry, guidance documents are highly valued by statisticians. In this book, reference will be made to two sets of such documents: those issued by the International Council for Harmonisation of Technical Requirements for Pharmaceuticals for Human Use (ICH), available at `www.ich.org`, and those issued by the U.S. Food and Drug Administration (FDA), available at `www.fda.gov/regulatory-information`, and other agencies such as the European Medicines Agency (EMA).

ICH guidance documents are categorized according to efficiency (E), safety (S), quality (Q), and multidisciplinary (M). For instance, in this chapter, reference will be made to ICH E9, *Statistical Principles for Clinical Trials*, and ICH E9(R1), *Addendum on Estimands and Sensitivity Analysis in Clinical Trials to the Guideline on Statistical Principles for Clinical Trials.*

FDA guidance documents are generally organized by topic. In later chapters, reference will be made to the FDA guidance document *Adjusting for Covariates in Randomized Clinical Trials for Drugs and Biological Products*, released in May 2023, and the FDA guidance document *Real-World Evidence: Considerations Regarding Non-Interventional Studies for Drug and Biological Products*, released in March 2024.

6.2 Central Questions

6.2.1 Cause and Effect

Cause and effect refers to the relationship—hereafter referred to as the *causal relationship*—between two events, wherein the occurrence of one event directly leads to, or alters the probability of, the occurrence of the other event.

Generally, a cause is defined as an action that initiates an event, modifies the probability of its occurrence, or influences the distribution of a variable. The cause is considered the "why" behind an event's occurrence,[21] while the effect constitutes the resulting outcome.

DOI: 10.1201/9781003635468-8

Two primary perspectives have been taken in understanding causal relationships: the philosophical and the practical. In this book, we bypass the philosophical perspective to focus on the practical perspective. In the pharmaceutical industry, particular attention is given to assessing the effect of a drug or medical device—more broadly, a treatment, exposure, or intervention. This interest is generally framed as the investigation of *treatment effects*.

The ICH E9(R1) guidance document has articulated the following central questions for drug development and regulatory approval:

> "Central questions for drug development and licensing are to establish the existence, and to estimate the magnitude, of treatment effects: how the outcome of treatment compares to what would have happened to the same subjects under alternative treatment (i.e., had they not received the treatment, or had they received a different treatment)."

A more detailed examination of these central questions is warranted.[22] First, recall that two commonly used statistical inference tools are hypothesis testing and confidence intervals. Thus, hypothesis testing is employed to establish the existence of treatment effects, while confidence intervals are used to estimate their magnitude.

Second, a precise definition of treatment effects has been provided: "how the outcome of treatment compares to what would have happened to the same subjects under alternative treatment."

Third, the definition of treatment effects incorporates four key components, known by the acronym PICO:[23] P—Population/Patients/Participants, I—Intervention, C—Comparator, and O—Outcome variable.

Fourth, the definition of treatment effects relies on counterfactual reasoning—specifically, considering "what would have happened to the same subjects under alternative treatment (i.e., had they not received the treatment, or had they received a different treatment)."

Gold Coin # 20. *The central questions for the development of medical products are to uncover cause-and-effect relationships.*

Throughout Part II of this book, counterfactual thinking will be employed as a foundational concept. The phrase "*what would have happened to the same subjects under alternative treatment*" will be referred to as the *counterfactual outcome* or the *potential outcome* hereafter.

It should be noted that various terms are used in the literature to describe the outcome variable. In Chapter 5, it was referred to as the *dependent variable*, to distinguish it from independent variables. Other terms include *response variable*, which contrasts with explanatory variables, and *endpoint*, as commonly used by clinicians in the pharmaceutical industry.

Multiple outcome variables may be of interest in a given study. In practice, these are typically classified as primary, secondary, or exploratory outcome variables. However, for the purposes of this book, only a single outcome variable will be considered. In clinical studies, the presence of multiple outcome variables introduces the issue of multiple comparisons, for which appropriate statistical adjustments must be made.[24]

Before exploring potential outcomes in detail, the role of time in defining outcome variables will be discussed in the next subsection.

6.2.2 Baseline and Endpoint

By incorporating a fifth element—T (i.e., Time)—into the PICO framework, the extended PICOT framework is formed.[25] Time plays a critical role in defining the study duration, the baseline timepoint, the duration of treatment, and the measurement timepoint(s) of the outcome variable.

The focus here will be on two key timepoints: the baseline timepoint and the timepoint(s) at which the outcome variable is measured.

The baseline is defined as the initial set of measurements collected prior to the initiation of treatment. These typically include baseline covariates and may also include measurements of the outcome variable at the baseline.

An endpoint is typically defined as the measurement of the outcome variable measured at a prespecified post-treatment timepoint and is used for the assessment of treatment efficacy, effectiveness, and safety.

Study designs can be categorized into two types based on the number of post-baseline measurement timepoints. If only one such timepoint is investigated, the design is termed a *point-treatment study* or a *point-exposure study*. If multiple timepoints are included, the design is referred to as a *longitudinal study*. Part II of this book is devoted to point-exposure studies, while longitudinal studies will be addressed in Part III and Part IV.

6.3 Potential Outcomes

6.3.1 Two Spans of 50 Years

1923–1974

The concept of potential outcomes was first introduced by Jerzy Neyman in his 1923 Master's thesis. In 1974, the concept was extended by Donald Rubin into a more general framework for conducting causal inference and estimating treatment effects in both observational and experimental studies.[26]

1974–Present

"What are the most important statistical ideas of the past 50 years?"

In a 2021 paper bearing this title,[27] Andrew Gelman and Aki Vehtari presented a list of influential developments: counterfactual causal inference, bootstrapping and simulation-based inference, overparameterized models and regularization, Bayesian multilevel models, generic computation algorithms, adaptive decision analysis, robust inference, and exploratory data analysis.

Notably, *counterfactual causal inference* appeared at the top of the list. Since the 1970s, there has been "a cluster of different ideas that have appeared in statistics, econometrics, psychometrics, epidemiology, and computer science, all revolving around the challenges of causal inference."[27] Among these, the Rubin Causal Model (RCM),[26] which will be adopted in this book, has emerged as a prominent framework for the statistical analysis of cause-and-effect relationships based on the concept of potential outcomes.

The connections between the RCM and the Structural Causal Model (SCM)[28] will also be explored in this book.

6.3.2 Two Arms

In accordance with the PICO framework, studies are designed to compare two treatments: I (i.e., Intervention or Investigational treatment) and C (i.e., Comparator or Control treatment).

In the next chapter, two primary types of study designs will be discussed: interventional studies—particularly randomized controlled clinical trials (RCTs)—and non-interventional studies (NISs).

Study designs may involve two arms, three arms, four arms, or more. In a two-arm study, one investigational treatment is compared with one control treatment. In a three-arm study, either one investigational treatment is compared with two control treatments, or two investigational treatments are compared with one control treatment. Similar configurations are possible for studies involving additional arms.

Part II of this book is devoted to two-arm studies, whereas multi-arm studies will be addressed in Part IV.

6.3.3 Two Worlds

The Potential World

Let $i = 1, \ldots, N$ index the individuals in the population, where N denotes the population size. In certain studies, healthy individuals may be enrolled; hence, the terms "individuals," "subjects," "patients," and "participants" are used interchangeably.

Let the investigational treatment be denoted by 1 and the control treatment (e.g., placebo, standard of care, or alternative active treatment) by 0. The outcome variable, denoted by Y, may be continuous, binary, or time-to-event in nature.

A hypothetical world, referred to as the *potential world*, is imagined—one in which each individual is assumed to possess two potential outcomes prior to any study being conducted or any outcome being observed.

Let Y_i^a represent the potential outcome that would be observed had the individual i been treated with treatment a. In particular, when two treatments, 1 and 0, are compared, the potential outcomes are:

- Y_i^1 denotes the potential outcome had the individual i been treated with treatment 1,
- Y_i^0 denotes the potential outcome had the individual i been treated with treatment 0.

Thus, in the potential world, each individual in the population is assumed to have two potential outcomes, Y_i^1 and Y_i^0, as if "two hats" were worn.

The Real World

In the real world, the individual i is treated with treatment A_i, which takes the value of either 1 or 0. The corresponding outcome observed following this assignment is denoted by Y_i. Therefore, the observed outcomes in the real world are given by Y_i, for $i = 1, \ldots, N$.

While the potential world is envisioned to contain data of the form $\{(Y_i^{a=1}, Y_i^{a=0}), i = 1, \ldots, N\}$, the real world contains data of the form $\{(A_i, Y_i), i = 1, \ldots, N\}$.

An illustration of the two worlds is provided in Figure 6.1. The potential world exists only hypothetically, wherein each individual is assumed to "wear

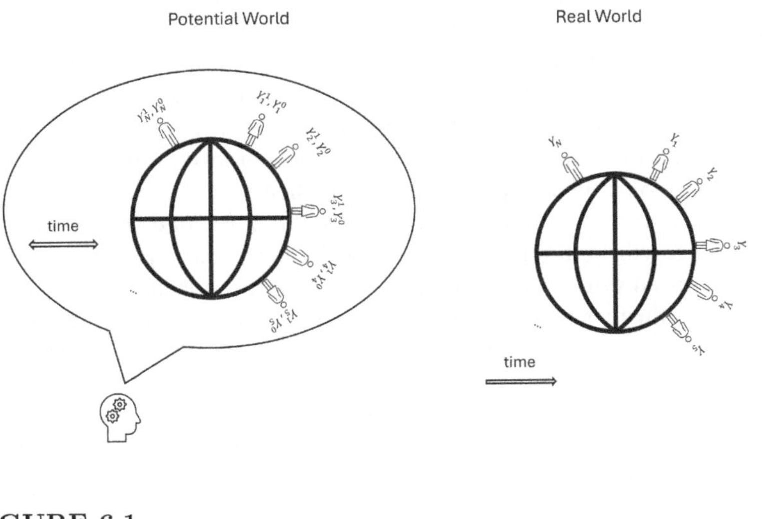

FIGURE 6.1
The two worlds—potential world and real world

two hats." (If $k > 2$ treatments are compared, then k potential outcomes would be assigned to each individual, metaphorically corresponding to "wearing k hats.") In this imagined world, time is considered bidirectional, allowing movement both forward and backward. The real world, in contrast, represents a single realization of the potential world, in which only one treatment is assigned (i.e., either treatment 1 or treatment 0) and only one outcome is observed—each individual metaphorically "wears only one hat." In the real world, time is moving only in the forward direction.

Gold Coin # 21. *Two worlds: the potential world and the real world.*

6.3.4 Tea Tasting Example Revisited

The hypothetical tea-tasting example introduced in Chapter 3 is revisited in this section, with slight modifications to align with the PICO framework (see Exercise 6.1) and to envision the potential outcomes, as detailed below.

The population is composed of 20 participants. The two interventions compared are: pouring milk into tea (M2T) and pouring tea into milk (T2M). The outcome variable is an ordinal score ranging from 1 to 10, representing the participant's evaluation of the milk tea's taste (with higher scores indicating greater tastefulness).

Prior to the intervention, participants are asked to taste a milk-free tea and provide a baseline score. In the potential world, each participant is assumed to possess two potential outcomes: Y^1, the score that the participant would give after tasting a cup of M2T milk tea; and Y^0, the score that the participant would give after tasting a cup of T2M milk tea.

Table 6.1 shows a simulated full dataset (see Exercise 6.2). The first column contains participant IDs. The second column indicates sex (1 for male, 0 for female). The third column shows the baseline score, which is observed in the real world. Columns four and five represent the potential outcomes Y^0 and Y^1, which exist only in the potential world—hence, the dataset is referred to as *the full dataset*. The final column shows *the individual treatment effect*, $Y^1 - Y^0$, which will be further discussed in the next chapter.

In the potential world, the arrow of time is conceptualized as being double-headed. As an illustrative example, consider the second participant. A score of 5 would be given after tasting a cup of T2M milk tea. Then, hypothetically returning to the moment just before that tasting and instead tasting a cup of M2T milk tea, a score of 6 would be given.

The hypothetical study can be generalized to one with a larger sample size (e.g., $n = 200$; see Exercise 6.3).

TABLE 6.1
The tea tasting example revisited

Participant	Sex (1-M; 0-F)	Baseline	Y^0	Y^1	$Y^1 - Y^0$
1	1	6	8	8	0
2	0	4	5	6	1
3	1	4	7	6	−1
4	0	4	5	7	2
5	0	4	6	6	0
6	1	7	10	10	0
7	1	4	7	5	−2
8	0	7	8	10	2
9	1	5	6	7	1
10	1	4	6	6	0
11	0	6	6	8	2
12	1	4	5	5	0
13	0	5	5	7	2
14	1	5	9	6	−3
15	1	5	8	5	−3
16	0	7	7	9	2
17	1	5	6	7	1
18	1	4	5	6	1
19	1	6	8	9	1
20	0	6	7	10	3
mean	0.60	5.10	6.70	7.15	0.45

6.4 Assumptions

6.4.1 SUTVA

The conceptualization of the potential world is not without constraints. One such constraint is the *stable unit treatment value assumption* (SUTVA) proposed by Donald Rubin:[29]

> "The potential outcomes for any unit do not vary with the treatments assigned to other units, and, for each unit, there are no different forms or versions of each treatment level, which lead to different potential outcomes."

The SUTVA comprises two components. First, it is assumed that "the potential outcomes for any unit do not vary with the treatments assigned to other units," which implies that the potential outcomes (Y_i^1, Y_i^0), $i = 1, \ldots, N$, are independent across units. Second, it is assumed that "for each unit, there are no different forms or versions of each treatment level, which lead to different potential outcomes," implying that (Y_i^1, Y_i^0), $i = 1, \ldots, N$, are identically distributed according to a common distribution. This common distribution is defined as that of (Y^1, Y^0), representing the potential outcomes for a randomly selected subject from a conceptually constructed superpopulation.

In summary, under the SUTVA and the superpopulation framework, it is assumed that

$$(Y_i^1, Y_i^0),\ i = 1, \ldots, N, \text{ are i.i.d. with the distribution of } (Y^1, Y^0).$$

In the real world, the subject i is treated with treatment $A_i \in \{0, 1\}$. The outcome observed for the subject i after being treated with treatment A_i is denoted by Y_i. Thus, in the real world, the observed data take the form

$$(A_i, Y_i),\ i = 1, \ldots, N.$$

The issues related to the study design and the treatment assignment mechanism will be addressed in the next chapter. Thus far, the two worlds have been discussed independently. The natural question that arises is: How can the two worlds be connected?

6.4.2 Connecting the Two Worlds

Two worlds are considered: the potential world (defined under the SUTVA) and the real world (characterized by observed data). To connect these two worlds, an assumption known as the *consistency assumption*[30] is introduced:

$$Y_i = Y_i^{A_i}, \quad i = 1, \ldots, N.$$

Gold Coin # 22. *The consistency assumption connects the two worlds—the potential world and the real world:*

$$Y_i = Y_i^1, \quad \text{if } A_i = 1,$$
$$Y_i = Y_i^0, \quad \text{if } A_i = 0.$$

The consistency assumption posits that the observed outcome is equal to the potential outcome corresponding to the treatment actually received. If this assumption is violated, the two worlds become disconnected, making it impossible to infer any information about the potential outcomes from the observed data. As a side note, it is important to emphasize that the consistency assumption is distinct from the asymptotic consistency (see Exercise 6.4).

The consistency assumption is inherently untestable. If the full dataset were available, a test of whether the consistency assumption is satisfied would be possible (see Exercise 6.5).

Additionally, let $B_i = 1 - A_i$ denote the alternative treatment not received by the subject i. If $A_i = 1$, then $B_i = 0$; if $A_i = 0$, then $B_i = 1$. Under this

TABLE 6.2
A (partial) full dataset for Exercise 6.5

Participant	Sex	Baseline	Y^0	Y^1	A	Y
1	1	6	8	8	0	8
2	0	4	5	6	1	6
3	1	4	7	6	1	6
4	0	4	5	7	0	8
5	0	4	6	6	0	6

notation, $Y_i = Y_i^{A_i}$ represents the *factual outcome*, corresponding to the treatment that was actually assigned, whereas, $Y_i^{B_i}$ represents the *counterfactual outcome*, corresponding to the treatment that was not assigned.

6.5 Exercises

Exercise 6.1

List the PICO components of the tea tasting example.

Exercise 6.2

Generate Table 6.1 in R.

Exercise 6.3

Generalize the above dataset to a larger one with $n = 200$.

Exercise 6.4

What is the difference between the consistency assumption and the asymptotic consistency?

Exercise 6.5

Examine whether the consistency assumption is satisfied by inspecting the (partial) full dataset presented in Table 6.2.

7

Study Designs

7.1 Interventional or Non-Interventional

For each study participant, two potential outcomes, Y^1 and Y^0, are defined in the potential world. However, in the real world, only one of these potential outcomes, either Y^1 or Y^0, is observed. The outcome that is observed depends on the treatment, 1 or 0, that the participant receives.

This raises the question: Who determines which treatment is received by the participant? Two scenarios are possible.

In the first scenario, treatments are assigned to participants by the study investigator. Studies conducted under this scheme are referred to as *interventional studies* or *experimental studies*. In the second scenario, treatments are not assigned but rather observed as they naturally occur; such studies are referred to as *non-interventional studies* or *observational studies*.

> **Gold Coin # 23.** *An essential difference between interventional and non-interventional studies is that, in interventional studies, the assignment of treatment is actively manipulated by the investigator, whereas in non-interventional studies, it is not.*

The two hypothetical studies presented in Chapter 3—in Tables 3.1 and 3.2, respectively—are examples of interventional studies. In the smaller study on tea tasting ($n = 8$), equal randomization was applied, while in the larger study ($n = 20$), simple randomization was used.

In the smaller study (Table 3.1), 4 of the 8 repetitions of the task were randomly selected to be served tea poured into milk (T2M), while the remaining 4 were served milk poured into tea (M2T). In general, when equal randomization is employed, one half of the participants are randomly assigned to one treatment arm, and the other half to the alternative arm.

In the larger study (Table 3.2), treatment assignment for each participant was determined by a coin flip: heads resulted in assignment to T2M, and tails to M2T. Upon completion of the randomization, 9 participants had been assigned to T2M and 11 to M2T. Generally, when simple randomization is

DOI: 10.1201/9781003635468-9

used, each participant is independently assigned to one of the treatment arms with equal probability.

To better understand the relationship between treatment assignment and pretreatment covariates, we next examine how participants from the population described in Table 6.1 would have been assigned to T2M or M2T.

7.1.1 Interventional Study

Consider the population of 20 participants described in Table 6.1, each associated with potential outcomes Y^1 and Y^0 in the potential world.

An interventional study can be designed using simple randomization (see Exercise 7.1). To implement this, 20 random binary numbers are generated, each with an equal probability (50%) of being 0 or 1. For the ith participant, a cup of tea is then prepared according to the ith random number: if the number is 1, milk is poured into tea (M2T); if 0, tea is poured into milk (T2M). As a result, the dataset shown in Table 7.1 is obtained.

For instance, for the second participant, the random number was 1, leading to an M2T assignment. Accordingly, the observed outcome was $Y_2 = 6$ in Table 7.1, since the second participant had potential outcomes $Y_2^0 = 5$ and $Y_2^1 = 6$ in Table 6.1.

TABLE 7.1
The tea tasting example: an interventional study

Participant	X_1 (Sex)	X_2 (Baseline)	A (1–M2T, 0–T2M)	Y
1	1	6	0	8
2	0	4	1	6
3	1	4	1	6
4	0	4	0	5
5	0	4	0	6
6	1	7	1	10
7	1	4	1	5
8	0	7	1	10
9	1	5	1	7
10	1	4	0	6
11	0	6	0	6
12	1	4	0	5
13	0	5	0	5
14	1	5	1	6
15	1	5	0	8
16	0	7	0	7
17	1	5	0	6
18	1	4	1	6
19	1	6	1	9
20	0	6	1	10

The concepts of *estimand* and *estimator* will be formally introduced in the following two chapters. However, a brief illustration can be provided here using the hypothetical interventional study above.

If the full dataset in Table 6.1 were available, the estimand (unknown in reality) would be identified as:

$$\theta_0 = 0.45,$$

which corresponds to the mean of the last column.

Based on the observed data in Table 7.1, an estimate of this estimand is given by (see Exercise 7.2):

$$\widehat{\theta} = \frac{\sum_{i=1}^{20} A_i Y_i}{\sum_{i=1}^{20} A_i} - \frac{\sum_{i=1}^{20}(1 - A_i)Y_i}{\sum_{i=1}^{20}(1 - A_i)} = 1.3.$$

7.1.2 Non-Interventional Study

Consider the population of 20 participants described in Table 6.1, each associated with potential outcomes Y^1 and Y^0 in the potential world.

A non-interventional study can be simulated (see Exercise 7.3) as follows. First, it is assumed that the ith participant chooses to prepare the tea by M2T or T2M according to the following probability (to be called the propensity score in the next chapter), which remains unknown to the investigator:

$$\mathbb{P}(A_i = 1 \mid X_{1i}, X_{2i}) = \frac{\exp\{1 - X_{1i} - 0.8I(X_{2i} > 5)\}}{1 + \exp\{1 - X_{1i} - 0.8I(X_{2i} > 5)\}}.$$

Next, 20 random binary numbers are generated, each based on the corresponding participant's propensity score. It is then assumed that tea is prepared by the participants themselves according to these generated values. As a result, a simulated dataset is obtained, as shown in Table 7.2.

For example, for the second participant, the generated random number was 1, indicating that the participant chose M2T. Accordingly, the observed outcome was $Y_2 = 6$ in Table 7.2, since this participant had potential outcomes $Y_2^0 = 5$ and $Y_2^1 = 6$ in Table 6.1.

If the full dataset in Table 6.1 were available, the estimand would be identified as $\theta_0 = 0.45$, which is the same as the one in the preceding scenario, because the estimand would not change due to different study designs.

Based on the observed data in Table 7.2, a naive estimate of this estimand is calculated as (see Exercise 7.4):

$$\widehat{\theta} = \frac{\sum_{i=1}^{20} A_i Y_i}{\sum_{i=1}^{20} A_i} - \frac{\sum_{i=1}^{20}(1 - A_i)Y_i}{\sum_{i=1}^{20}(1 - A_i)} = -0.28.$$

As will be discussed later, in the context of non-interventional studies, the presence of confounding makes the above unadjusted estimate undesirable.

TABLE 7.2
The tea tasting example: a non-interventional study

Participant	X_1 (Sex)	X_2 (Baseline)	A (1–M2T, 0–T2M)	Y
1	1	6	1	8
2	0	4	1	6
3	1	4	0	7
4	0	4	1	7
5	0	4	1	6
6	1	7	0	10
7	1	4	0	7
8	0	7	0	8
9	1	5	0	6
10	1	4	1	6
11	0	6	1	8
12	1	4	0	5
13	0	5	1	7
14	1	5	0	9
15	1	5	0	8
16	0	7	1	9
17	1	5	1	7
18	1	4	1	6
19	1	6	0	8
20	0	6	1	10

7.2 Randomized Controlled Trials

7.2.1 Randomization and Blinding

Randomized Controlled Trials (RCTs) are regarded as the gold standard in clinical research and regulatory submissions. According to ICH E9, *Statistical Principles for Clinical Trials*,

> "The most important design techniques for avoiding bias in clinical trials are blinding and randomization, and these should be normal features of most controlled clinical trials intended to be included in a marketing application."

Blinding is employed to minimize both conscious and unconscious bias during the conduct and interpretation of a clinical trial. Without blinding, subjects may respond differently, or investigators may treat subjects differently, based on knowledge of treatment assignments. In a double-blind trial, neither the subject nor the investigator is aware of the treatment being administered. In cases where blinding is not feasible, open-label RCTs may be conducted, in which both researchers and participants are aware of the assigned treatments.

Randomization is used to ensure that treatment arms are comparable, such that the distributions of both known and unknown prognostic factors and effect modifiers are balanced across treatment arms. When combined with blinding, randomization helps prevent bias by eliminating predictability in treatment assignments.

Gold Coin # 24. *"The most important design techniques for avoiding bias in clinical trials are blinding and randomization."—ICH E9*

7.2.2 Efficacy and Safety

ICH E9 opens with the following statement: "The efficacy and safety of medicinal products should be demonstrated by clinical trials which follow the guidance in 'Good Clinical Practice: Consolidated Guideline' (ICH E6)."

In any RCT of a medicinal product, the primary objective is typically the demonstration of efficacy and safety. To support this objective, ICH E9 provides guidance on how the primary outcome variable should be specified:

> "The primary variable ('target' variable, primary endpoint) should be the variable capable of providing the most clinically relevant and convincing evidence directly related to the primary objective of the trial. There should generally be only one primary variable. This will usually be an efficacy variable, because the primary objective of most confirmatory trials is to provide strong scientific evidence regarding efficacy. Safety/tolerability may sometimes be the primary variable, and will always be an important consideration. Measurements relating to quality of life and health economics are further potential primary variables."

According to this guideline, a single primary variable is typically selected to ensure that the conclusions of the trial are based on a well-defined and clinically meaningful outcome. (In some cases, co-primary variables or multiple primary variables are selected.) In most cases, an efficacy variable is chosen, although in some cases safety or tolerability may be designated as the primary focus. Regardless of the chosen primary endpoint, safety is always expected to be evaluated as a critical component of the trial.

7.2.3 Internal Validity and External Validity

In the context of an RCT, it is essential to understand the relationship among three populations: the target population, the intention-to-treat (ITT) population, and the superpopulation.

The *target population* is defined as the group of individuals for whom the treatment is intended and from whom the trial's conclusions are meant to be generalized. This population is typically specified using a set of inclusion and exclusion criteria determined prior to trial enrollment.

The *ITT population* consists of the participants enrolled in the trial who are assigned to either the investigational treatment or the control treatment. In the hypothetical study presented in Table 7.1, the ITT population consists of 20 participants.

The *superpopulation* is a conceptual construct from which the ITT population is assumed to have been drawn as a simple random sample. The observations from participants in the ITT population are assumed to be independently and identically distributed (i.i.d.) according to the common distribution of the superpopulation.

The definitions of target population, ITT population, and superpopulation are the key to understand internal validity and external validity. *Internal validity* refers to the extent to which the causal relationship between the treatment variable (A) and the outcome variable (Y) is accurately estimated within the trial, free from the influence of confounding or bias. In contrast, *external validity* refers to the extent to which the trial findings can be generalized beyond the ITT population to the broader target population.

As illustrated in Figure 7.1, internal validity concerns the inference from the ITT population to the superpopulation, while external validity concerns the generalizability from the ITT population to the target population.

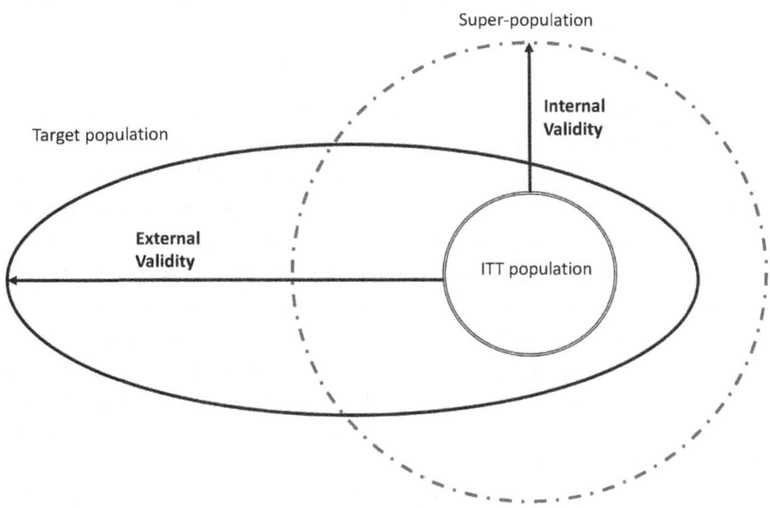

FIGURE 7.1
Internal validity and external validity

7.3 Observational Studies

7.3.1 Simpson's Paradox and Berkson's Paradox

In *The Book of Why*,[21] Judea Pearl devotes a chapter to some of the most well-known and perplexing paradoxes in probability and statistics, as these are not only intellectually stimulating but also illustrative of key concepts. Here, two such paradoxes—Simpson's paradox and Berkson's paradox—are briefly described, as they serve to clarify the notions of confounding bias and selection bias, respectively, which will be further discussed in the next subsection.

Simpson's paradox[31] refers to a statistical phenomenon in which an association between two variables is observed within several separate groups but disappears or reverses when the groups are aggregated. In some cases, no association may be observed within any subgroup, while a significant association may emerge when the groups are combined.

For example, let A denote a treatment variable (with $A = 1$ indicating investigational treatment and $A = 0$ indicating control treatment), and let Y be a binary outcome variable (1 for response, 0 for nonresponse). Among 100 male patients, 20 were treated with $A = 1$ and 80 with $A = 0$, and a response rate of 0.5 was observed in both treatment arms. Among 200 female patients, 160 received $A = 1$ and 40 received $A = 0$, with a response rate of 0.8 in both arms. When these groups are combined, the response rate among the 180 patients who received $A = 1$ is calculated as $[20(0.5) + 160(0.8)]/180 = 76.7\%$, while the response rate among the 120 patients who received $A = 0$ is $[80(0.5) + 40(0.8)]/120 = 60\%$. Thus, although A and Y are conditionally independent given C (sex), they are not marginally independent.

Berkson's paradox[32] describes a phenomenon in which a conditional association appears counterintuitive or misleading due to selection bias.

For instance, let A represent the treatment variable (1 for investigational treatment, 0 for control), and let Y denote the binary outcome (1 for response, 0 for nonresponse). Among 100 patients treated with $A = 1$, 80 responded; likewise, among 100 patients treated with $A = 0$, 80 also responded, indicating a marginal response rate of 0.8 for both arms. Suppose that among patients with $A = 1$ and $Y = 1$, a 50% missing rate is observed, while a 25% missing rate is observed among patients with $A = 0$ and $Y = 1$. As a result, the observed (nonmissing) data consist of 60 patients in the $A = 1$ arm (40 responders and 20 nonresponders) and 80 patients in the $A = 0$ arm (60 responders and 20 nonresponders). Let $E = 1$ denote the nonmissingness. Consequently, the response rate among the patients with $E = 1$ is $40/60 = 66.7\%$ for those treated with $A = 1$ and $60/80 = 75\%$ for those treated with $A = 0$. Therefore, while A and Y are marginally independent, they are not conditionally independent given $E = 1$.

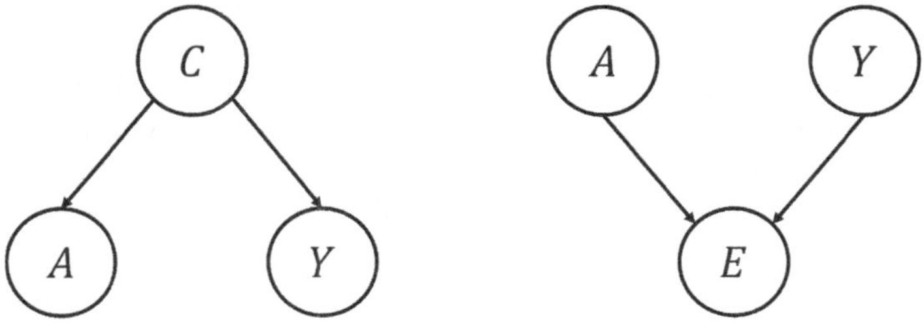

FIGURE 7.2
Common cause and common effect

In Figure 7.2, the impact of a common cause and a common effect is illustrated. In the left panel, a common cause C is shown, where A and Y are not marginally independent but become conditionally independent given any level c of C. This explains the reason behind Simpson's paradox. In the right panel, a common effect E is shown, where A and Y are marginally independent but may become conditionally dependent given any level e of E. This explains the reason behind Berkson's paradox.

7.3.2 Confounding Bias and Selection Bias

Demystifying Simpson's paradox and Berkson's paradox facilitates an understanding of confounding bias and selection bias.

By adding a directed edge from A to Y, Figure 7.2 is transformed into Figure 7.3. The causal relationship between A and Y is depicted by the direct path $A \to Y$ in Figure 7.3. In the left panel, the marginal association between A and Y is influenced both by the true causal effect $A \to Y$ and by a spurious association induced by the common cause C through the path $A \leftarrow C \to Y$. Therefore, when the marginal association between A and Y is used to estimate the causal effect, a confounding bias is introduced.

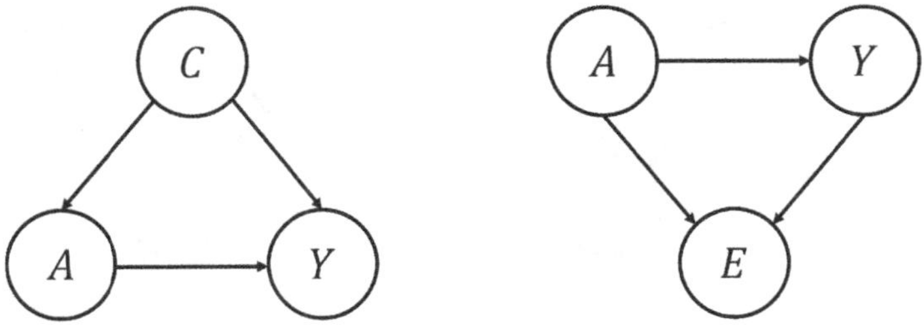

FIGURE 7.3
Common cause and common effect (continued)

In the right panel of Figure 7.3, the marginal association between A and Y reflects the true causal effect. However, when conditioning on $E = e$, where e denotes any level of E, the conditional association between A and Y becomes a mixture of the true causal effect and a spurious association arising from conditioning on the collider E. Consequently, if the conditional association given $E = e$ is used to estimate the causal effect, a selection bias is introduced.

7.3.3 DAG and SCM

The FDA guidance document, *Real-World Evidence: Considerations Regarding Non-Interventional Studies for Drug and Biological Products*, emphasizes that "(e)ach protocol should concisely describe each of the critical elements," in which the first element suggested by the guidance is:

> "Schema to describe overall study design as well as a causal diagram to specify the theorized causal relationship."

According to the guidance, both a schema of the study design and a causal diagram are essential. The most widely adopted causal diagrams are *directed acyclic graphs* (DAGs).[33,34] Four such DAGs have already been presented in Figures 7.2 and 7.3.

In Figure 7.4, three basic DAG structures—chain, fork, and collider—are illustrated in panels (a), (b), and (c), respectively, to represent three distinct patterns of relationships among three variables.

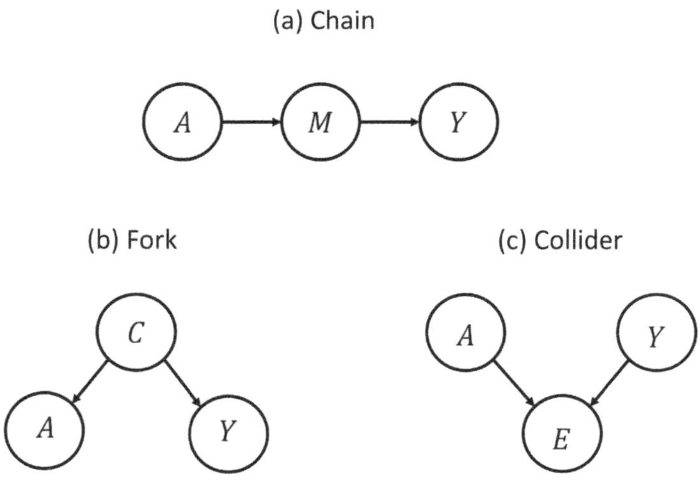

FIGURE 7.4
Chain, fork, and collider

Chain, Fork, and Collider

In the chain structure shown in Figure 7.4(a), the following relationships are implied (see Exercise 7.5):

- A and M are likely to be dependent;
- M and Y are likely to be dependent;
- A and Y are likely to be dependent;
- Conditional on M, A and Y are independent.

In the fork structure shown in Figure 7.4(b), the following relationships are implied:

- A and C are likely to be dependent;
- Y and C are likely to be dependent;
- A and Y are likely to be dependent;
- Conditional on C, A and Y are independent.

In the collider structure illustrated in Figure 7.4(c), the following relationships are implied:

- A and E are likely to be dependent;
- Y and E are likely to be dependent;
- A and Y are independent;
- Conditional on E, A and Y are likely to be dependent.

These three basic DAG structures may appear as subgraphs within more complex DAGs. For instance, the DAG depicted in Figure 7.5 includes a fork

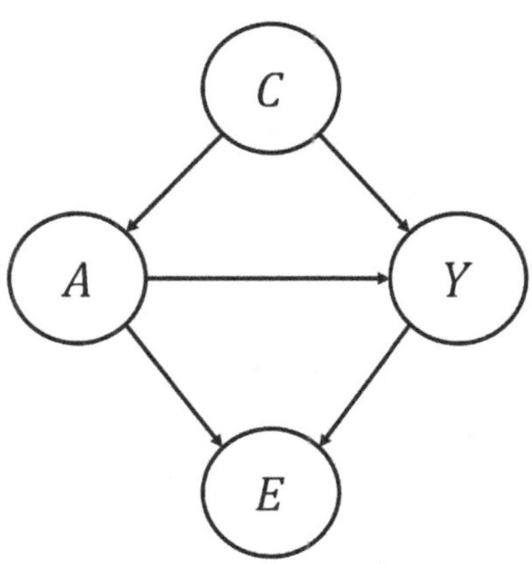

FIGURE 7.5
An example of a DAG

($A \leftarrow C \rightarrow Y$), a chain ($X \rightarrow Y \rightarrow E$), and a collider ($A \rightarrow E \leftarrow Y$). This foundational understanding of DAGs is sufficient for the material covered in the remaining chapters. For more comprehensive discussions, the readers are referred to Pearl's famous book *Causality*.[28]

Structural Causal Model

While DAGs provide a visual representation of causal structure, *structural causal models* (SCMs) extend this framework by assigning functional relationships that describe how variables are generated.[28] The DAG in Figure 7.5 is used as an example to construct an SCM.

In this DAG, the following parental relationships are assumed: C has no parent; A has C as a parent; Y has C and A as parents; and E has A and Y as parents. Accordingly, the corresponding SCM is defined as:

$$
\begin{aligned}
C &= f_C(U_C), \\
A &= f_A(C, U_A), \\
Y &= f_Y(C, A, U_Y), \\
E &= f_E(A, Y, U_E),
\end{aligned}
$$

where f_C, f_A, f_Y, and f_E denote unknown deterministic functions, and U_C, U_A, U_Y, and U_E represent mutually independent exogenous variables (random disturbances). Mutual independence is assumed for the exogenous variables since no edges among them are represented in the DAG.

One application of the SCM is to formally connect the real world with the potential world. Under the assumption of consistency, potential outcomes $Y^{a=1}$ and $Y^{a=0}$ can be expressed as:

$$
\begin{aligned}
Y^{a=1} &= f_Y(C, 1, U_Y), \\
Y^{a=0} &= f_Y(C, 0, U_Y).
\end{aligned}
$$

Gold Coin # 25. *DAGs and SCMs are two essential tools for specifying theorized causal relationships.*

7.3.4 Time-Independent and Time-Varying Confounding

Numerous types of observational study designs exist, including, but not limited to, case-control studies, cross-sectional studies, and cohort studies. In this book, the focus is placed on cohort studies.

Within the framework of cohort studies, two primary scenarios are considered. In one scenario, only a single follow-up time point is of interest (even though multiple time points may exist, only the primary endpoint is analyzed). In the other scenario, multiple follow-up time points are of interest.

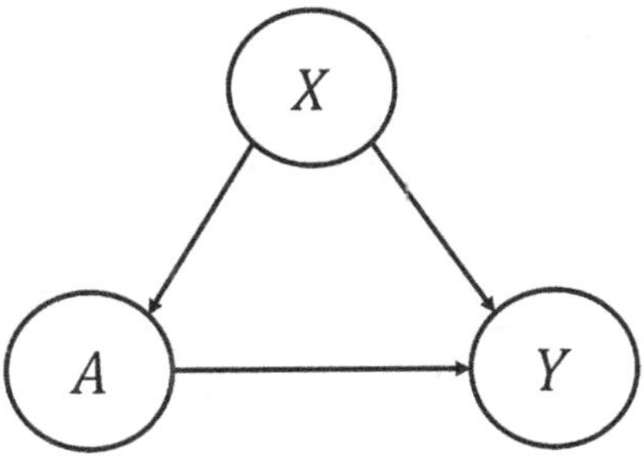

FIGURE 7.6
DAG for a typical cohort study

To distinguish between these two scenarios, studies in the former case are referred to as cohort studies with time-independent confounding (or simply, cohort studies), while those in the latter case are referred to as cohort studies with time-varying confounding (or simply, longitudinal cohort studies).

Distinguishing between time-independent and time-varying confounding in cohort studies can be challenging. However, with the aid of two powerful tools—DAGs and SCMs—this distinction can be clarified.

Figure 7.6 illustrates a DAG for a cohort study involving time-independent confounding. In this DAG, X denotes a vector of baseline confounders measured prior to treatment. The corresponding SCM is specified as follows:

$$\begin{aligned} X &= f_X(U_X), \\ A &= f_A(X, U_A), \\ Y &= f_Y(X, A, U_Y), \end{aligned}$$

where U_X, U_A, and U_Y are mutually independent exogenous variables. Under this SCM, the potential outcomes $Y^{a=1}$ and $Y^{a=0}$ can be expressed as:

$$\begin{aligned} Y^{a=1} &= f_Y(X, 1, U_Y), \\ Y^{a=0} &= f_Y(X, 0, U_Y). \end{aligned}$$

Figure 7.7 presents a DAG for a longitudinal cohort study involving $T = 2$ follow-up visits. In this diagram, L_0 denotes a vector of baseline confounders measured at time $t = 0$, A_0 is the treatment administered after $t = 0$, L_1 represents time-varying confounders measured at $t = 1$, A_1 is the treatment administered after $t = 1$, and Y is the primary outcome measured at $t = 2$.

In a general longitudinal cohort study with T follow-up visits, the temporal ordering of variables is represented as:

$$L_0 \to A_0 \to L_1 \to A_1 \to \cdots \to L_{T-1} \to A_{T-1} \to Y,$$

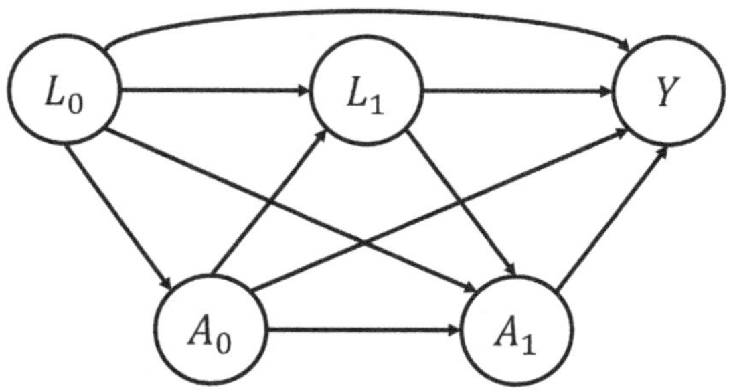

FIGURE 7.7
DAG for a longitudinal cohort study with $T = 2$

where each treatment variable A_t may take values such as 1 (investigational treatment), 0 (control treatment), `NULL` (no treatment), 2 (alternative treatment), etc. As the corresponding DAG becomes increasingly complex with higher T, it is often more convenient to represent the study using the corresponding SCM:

$$\begin{aligned}
L_0 &= f_{L_0}(U_{L_0}), \\
A_0 &= f_{A_0}(L_0, U_{A_0}), \\
L_1 &= f_{L_1}(L_0, A_0, U_{L_1}), \\
A_1 &= f_{A_1}(L_0, A_0, L_1, U_{A_1}), \\
&\vdots \\
L_{T-1} &= f_{L_{T-1}}(L_0, A_0, \ldots, L_{T-2}, A_{T-2}, U_{L_{T-1}}), \\
A_{T-1} &= f_{A_{T-1}}(L_0, A_0, \ldots, A_{T-2}, L_{T-1}, U_{A_{T-1}}), \\
Y &= f_Y(L_0, A_0, \ldots, L_{T-1}, A_{T-1}, U_Y),
\end{aligned}$$

where all exogenous variables are assumed to be mutually independent, unless otherwise indicated.

The potential outcome $Y^{\overline{a}}$ corresponding to any specific treatment sequence $\overline{a} = (a_0, \ldots, a_{T-1})$ can then be expressed as:

$$f_Y(L_0, a_0, f_{L_1}(L_0, a_0, U_{L_1}), a_1, \ldots, f_{T_{T-1}}(L_0, a_0, \ldots, a_{T-2}, U_{L_{T-1}}), a_{T-1}, U_Y).$$

In particular,

$$\begin{aligned}
Y^{(1,\ldots,1)} &= f_Y(L_0, 1, f_{L_1}(L_0, 1, U_{L_1}), 1, \ldots, f_{T_{T-1}}(L_0, 1, \ldots, 1, U_{L_{T-1}}), 1, U_Y), \\
Y^{(0,\ldots,0)} &= f_Y(L_0, 0, f_{L_1}(L_0, 0, U_{L_1}), 0, \ldots, f_{T_{T-1}}(L_0, 0, \ldots, 0, U_{L_{T-1}}), 0, U_Y).
\end{aligned}$$

7.4 Concurrent Control and External Control

In the ICH E10 guideline, *Choice of Control Group and Related Issues in Clinical Trials*, five types of control groups are described: (1) placebo concurrent control, (2) no-treatment concurrent control, (3) dose-response concurrent control, (4) active treatment concurrent control, and (5) external control.

From an analytical standpoint, these control types are commonly classified into two broader categories: (1) concurrent controls and (2) external controls.

The control arm in a previously discussed RCT is considered an example of concurrent control. According to the definition provided in ICH E10,

> "A concurrent control group is one chosen from the same population as the test group and treated in a defined way as part of the same trial that studies the test treatment, and over the same period of time. The test and control groups should be similar with regard to all baseline and on-treatment variables that could influence outcome, except for the study treatment."

Furthermore, ICH E10 states that an external control group

> "[...] can be a group of patients treated at an earlier time (historical control) or a group treated during the same time period but in another setting."

When external controls are used in place of concurrent controls, various sources of bias—such as confounding bias—may be introduced. To address these concerns, the FDA released in February 2023 the guidance document *Considerations for the Design and Conduct of Externally Controlled Trials for Drug and Biological Products*.

As a result, two major types of clinical trials are distinguished: (1) randomized controlled clinical trials (RCTs) and (2) externally controlled clinical trials (ECTs). As will be shown in later chapters, the analytical methods applicable to ECTs are equivalent to those used in observational cohort studies.

Gold Coin # 26. *Two main types of control groups—concurrent control and external control—correspond to two major clinical trial designs: randomized controlled clinical trials (RCTs) and externally controlled clinical trials (ECTs).*

In addition to these two primary designs, other hybrid designs have also been explored. For instance, an RCT may be augmented with an external control arm in settings where the concurrent placebo arm is underpowered due to ethical considerations (e.g., limiting placebo enrollment) or logistical constraints (e.g., in rare disease studies where patient enrollment is limited).

7.5 Exercises

Exercise 7.1

Generate the data in Table 7.1 in R.

Exercise 7.2

Verify that the naive estimate of the treatment effect based on the data in Table 7.1 is $\widehat{\theta} = 1.3$.

Exercise 7.3

Generate the data in Table 7.2 in R.

Exercise 7.4

Verify that the naive estimate of the treatment effect based on the data in Table 7.2 is $\widehat{\theta} = -0.28$.

Exercise 7.5

Prove the relationships in the three basic DAGs in Figure 7.4.

8

Estimand

8.1 Causal Estimand and Statistical Estimand

The ICH E9(R1) *Addendum on Estimands and Sensitivity Analysis in Clinical Trials* provides the following definition for an estimand:

> "A precise description of the treatment effect reflecting the clinical question posed by the trial objective. It summarizes at a population-level what the outcomes would be in the same patients under different treatment conditions being compared."

First, in each trial, the estimand of interest (also referred to as the target estimand) is considered to be the treatment effect reflecting the clinical question. Second, the treatment effect is characterized in terms of potential outcomes—that is, "what the outcomes would be in the same patients under different treatment conditions being compared." Third, an estimand in a trial is described by four attributes:

- Patient population
- Treatment conditions
- Outcome variable
- Population-level summary

Because an estimand defined according to the above definition relies on potential outcomes, it is regarded as existing only in the potential world. To distinguish between the estimand existing in the potential world and its counterpart projected onto the real world, the term *causal estimand* is used for the former and *statistical estimand* for the latter. The statistical estimand is understood to be "estimable" based on the data collected in the study. The process by which the causal estimand is translated into a statistical estimand is referred to as *identification*.

DOI: 10.1201/9781003635468-10

Gold Coin # 27. *An estimand is defined as a precise description of the treatment effect reflecting the clinical question posed by the trial objective. It is referred to as the causal estimand when defined in terms of potential outcomes, and as the statistical estimand when defined in terms of observed outcomes.*

8.2 Identification for an RCT

8.2.1 Causal Estimand

In the potential world, the population $\mathcal{P}$ is assumed to consist of

$$(X_i, Y_i^{z=1}, Y_i^{z=0}), \quad i = 1, \ldots, N,$$

which are considered to be independently and identically distributed (i.i.d.) as $(X, Y^{z=1}, Y^{z=0})$, under the Stable Unit Treatment Value Assumption (SUTVA) and the conceptual construction of a superpopulation from which these N sets of potential outcomes are sampled.

A 1:1 randomized controlled trial (RCT) with simple randomization is considered, in which data of the form (X_i, Z_i, Y_i) are generated for $i = 1, \ldots, N$, where N denotes the population size. Due to simple randomization, the treatment assignments Z_i, for $i = 1, \ldots, N$, are i.i.d. copies of a random variable Z, for which

$$\mathbb{P}(Z = 1) = \mathbb{P}(Z = 0) = 1/2.$$

Under the consistency assumption that the observed outcome Y is equal to Y^Z, i.e., a projection of the pair $(Y^{z=1}, Y^{z=0})$ from the potential world to the real world. Accordingly, the observed data $(X_i, Z_i, Y_i), i = 1, \ldots, N$, are regarded as the projections of the full data $(X_i, Y_i^{z=1}, Y_i^{z=0}), i = 1, \ldots, N$, from the potential world to the real world.

Since $(X_i, Y_i^{z=1}, Y_i^{z=0}), i = 1, \ldots, N$, are i.i.d. copies of $(X, Y^{z=1}, Y^{z=0})$, and the treatment indicators Z_i are i.i.d. copies of Z, it follows that the observed data $(X_i, Z_i, Y_i), i = 1, \ldots, N$, are i.i.d. copies of (X, Z, Y).

The outcome variable may be continuous, binary, or time-to-event. Time-to-event outcomes will be discussed in Part IV of this book. Parts II and III will focus on continuous and binary outcomes.

For a continuous outcome, the mean is used as the population-level summary, and the following causal estimand is defined:

$$\theta^* = \mathbb{E}\left(Y^{z=1} - Y^{z=0}\right),$$

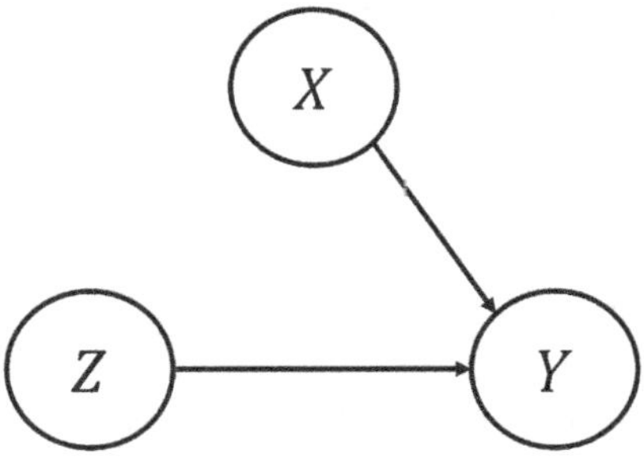

FIGURE 8.1
DAG for a typical RCT

where the expectation is taken over the superpopulation. This estimand is commonly referred to as the *average treatment effect (ATE)*.

For a binary outcome coded as 0/1, the proportion is used as the population-level summary, and the ATE estimand is defined as

$$\theta^* = \mathbb{E}\left[I(Y^{z=1} = 1) - I(Y^{z=0} = 1)\right] = \mathbb{E}\left(Y^{z=1} - Y^{z=0}\right),$$

where $I(\cdot)$ denotes the indicator function.

8.2.2 Statistical Estimand

The directed acyclic graph (DAG) representing the RCT described above is shown in Figure 8.1. Due to randomization, no arrow is included from X to Z. The corresponding structural causal model (SCM) is defined as

$$\begin{aligned} X &= f_X(U_X), \\ Z &= f_Z(U_Z), \\ Y &= f_Y(X, Z, U_Y), \end{aligned}$$

where U_X, U_Y, and U_Z are assumed to be mutually independent. Consequently, the potential outcomes can be represented as

$$\begin{aligned} Y^{z=1} &= f_Y(f_X(U_X), 1, U_Y), \\ Y^{z=0} &= f_Y(f_X(U_X), 0, U_Y). \end{aligned}$$

Because independence is assumed between (U_X, U_Y) and U_Z, the potential outcomes $(Y^{z=1}, Y^{z=0})$ are independent of the treatment assignment Z.

The process of identification can now be carried out. The causal estimand is expressed as

$$\begin{aligned}\theta^* &= \mathbb{E}\left(Y^{z=1} - Y^{z=0}\right) \\ &\overset{(a)}{=} \mathbb{E}\left(Y^{z=1} \mid Z=1\right) - \mathbb{E}\left(Y^{z=0} \mid Z=0\right) \\ &\overset{(b)}{=} \mathbb{E}\left(Y \mid Z=1\right) - \mathbb{E}\left(Y \mid Z=0\right),\end{aligned}$$

where (a) follows from the independence between $(Y^{z=1}, Y^{z=0})$ and Z, and (b) follows by the consistency assumption. Therefore, the causal estimand θ^* is translated into the following statistical estimand:

$$\theta = \mathbb{E}(Y \mid Z=1) - \mathbb{E}(Y \mid Z=0).$$

Although such identification is straightforward in RCTs, it is instructive to consider an alternative approach using weighting. This yields:

$$\begin{aligned}\theta^* &= \mathbb{E}(Y^{z=1}) - \mathbb{E}(Y^{z=0}) \\ &\overset{(a)}{=} \mathbb{E}\left[\frac{I(Z=1)}{\mathbb{P}(Z=1)}Y^{z=1}\right] - \mathbb{E}\left[\frac{I(Z=0)}{\mathbb{P}(Z=0)}Y^{z=0}\right] \\ &\overset{(b)}{=} \mathbb{E}\left[\frac{I(Z=1)Y}{\mathbb{P}(Z=1)}\right] - \mathbb{E}\left[\frac{I(Z=0)Y}{\mathbb{P}(Z=0)}\right] \\ &\overset{(c)}{=} \mathbb{E}\left[\frac{ZY}{\mathbb{P}(Z=1)}\right] - \mathbb{E}\left[\frac{(1-Z)Y}{\mathbb{P}(Z=0)}\right],\end{aligned}$$

where (a) follows from the independence between $(Y^{z=1}, Y^{z=0})$ and Z, allowing the expectation to be factorized as

$$\mathbb{E}\left[\frac{I(Z=j)}{\mathbb{P}(Z=j)}Y^{z=j}\right] = \mathbb{E}\left[\frac{I(Z=j)}{\mathbb{P}(Z=j)}\right]\mathbb{E}\left[Y^{z=j}\right] = \mathbb{E}\left[Y^{z=j}\right], \quad j=1,0.$$

In addition, (b) uses the consistency assumption, and (c) holds because $I(Z=1)=Z$ and $I(Z=0)=1-Z$.

Thus, a second expression for the statistical estimand is obtained:

$$\theta = \mathbb{E}\left[\frac{ZY}{\mathbb{P}(Z=1)}\right] - \mathbb{E}\left[\frac{(1-Z)Y}{\mathbb{P}(Z=0)}\right].$$

These two formulations of the statistical estimand are equivalent. In fact,

$$\begin{aligned}\mathbb{E}\left[\frac{ZY}{\mathbb{P}(Z=1)}\right] &= \mathbb{E}\left[\frac{ZY}{\mathbb{P}(Z=1)} \,\middle|\, Z=1\right]\mathbb{P}(Z=1) = \mathbb{E}(Y \mid Z=1), \\ \mathbb{E}\left[\frac{(1-Z)Y}{\mathbb{P}(Z=0)}\right] &= \mathbb{E}\left[\frac{(1-Z)Y}{\mathbb{P}(Z=0)} \,\middle|\, Z=0\right]\mathbb{P}(Z=0) = \mathbb{E}(Y \mid Z=0).\end{aligned}$$

Gold Coin # 28. *In an RCT, if the causal estimand is the average treatment effect (ATE),*

$$\theta^* = \mathbb{E}\left(Y^{z=1} - Y^{z=0}\right),$$

then the corresponding statistical estimand can be expressed in the following two equivalent forms:

$$\theta = \mathbb{E}(Y \mid Z = 1) - \mathbb{E}(Y \mid Z = 0),$$
$$\theta = \mathbb{E}\left[\frac{ZY}{\mathbb{P}(Z = 1)}\right] - \mathbb{E}\left[\frac{(1 - Z)Y}{\mathbb{P}(Z = 0)}\right].$$

8.2.3 Estimator

A more general framework for estimator development will be introduced in the next chapter. Here, a simple estimator for the RCT setting is considered as a special case of that framework.

Given the statistical estimand

$$\theta = \mathbb{E}(Y \mid Z = 1) - \mathbb{E}(Y \mid Z = 0),$$

a straightforward estimator can be constructed by replacing population means with sample means:

$$\widehat{\theta} = \frac{\sum_{i=1}^{N} Z_i Y_i}{\sum_{i=1}^{N} Z_i} - \frac{\sum_{i=1}^{N}(1 - Z_i)Y_i}{\sum_{i=1}^{N}(1 - Z_i)}.$$

This estimator may be undefined if either $\sum_{i=1}^{N} Z_i = 0$ or $\sum_{i=1}^{N}(1-Z_i) = 0$. As a result, inference cannot be based on the unconditional mean $\mathbb{E}[\widehat{\theta}]$ or variance $\mathbb{V}[\widehat{\theta}]$. Instead, the conditional mean $\mathbb{E}[\widehat{\theta} \mid \mathcal{Z}]$ and conditional variance $\mathbb{V}[\widehat{\theta} \mid \mathcal{Z}]$ are used, where

$$\mathcal{Z} = \{Z_1, \ldots, Z_N\}.$$

The conditional unbiasedness of $\widehat{\theta}$ can be verified:

$$\begin{aligned}
\mathbb{E}[\widehat{\theta} \mid \mathcal{Z}] &= \mathbb{E}\left[\frac{\sum_{i=1}^N Z_i Y_i}{\sum_{i=1}^N Z_i} - \frac{\sum_{i=1}^N (1-Z_i) Y_i}{\sum_{i=1}^N (1-Z_i)} \,\middle|\, \mathcal{Z}\right] \\
&\overset{(a)}{=} \frac{\sum_{i=1}^N Z_i \mathbb{E}(Y_i \mid Z_i = 1)}{\sum_{i=1}^N Z_i} - \frac{\sum_{i=1}^N (1-Z_i) \mathbb{E}(Y_i \mid Z_i = 0)}{\sum_{i=1}^N (1-Z_i)} \\
&\overset{(b)}{=} \frac{\sum_{i=1}^N Z_i \mathbb{E}(Y \mid Z = 1)}{\sum_{i=1}^N Z_i} - \frac{\sum_{i=1}^N (1-Z_i) \mathbb{E}(Y \mid Z = 0)}{\sum_{i=1}^N (1-Z_i)} \\
&\overset{(c)}{=} \mathbb{E}(Y \mid Z = 1) - \mathbb{E}(Y \mid Z = 0) = \theta,
\end{aligned}$$

where (a) holds because $Z_i \in \{0, 1\}$, (b) follows from the i.i.d. assumption, and (c) results from simplification.

The conditional variance of $\widehat{\theta}$ is given by:

$$\begin{aligned}
\mathbb{V}[\widehat{\theta} \mid \mathcal{Z}] &= \mathbb{V}\left[\frac{\sum_{i=1}^N Z_i Y_i}{\sum_{i=1}^N Z_i} - \frac{\sum_{i=1}^N (1-Z_i) Y_i}{\sum_{i=1}^N (1-Z_i)} \,\middle|\, \mathcal{Z}\right] \\
&\overset{(a)}{=} \mathbb{V}\left[\frac{\sum_{i=1}^N Z_i Y_i}{\sum_{i=1}^N Z_i} \,\middle|\, \mathcal{Z}\right] + \mathbb{V}\left[\frac{\sum_{i=1}^N (1-Z_i) Y_i}{\sum_{i=1}^N (1-Z_i)} \,\middle|\, \mathcal{Z}\right] \\
&\overset{(b)}{=} \sum_{i=1}^N \frac{\mathbb{V}(Z_i Y_i \mid Z_i)}{(\sum_{i=1}^N Z_i)^2} + \sum_{i=1}^N \frac{\mathbb{V}[(1-Z_i) Y_i \mid Z_i]}{(\sum_{i=1}^N (1-Z_i))^2} \\
&\overset{(c)}{=} \sum_{i=1}^N \frac{Z_i \mathbb{V}(Y_i \mid Z_i = 1)}{(\sum_{i=1}^N Z_i)^2} + \sum_{i=1}^N \frac{(1-Z_i) \mathbb{V}(Y_i \mid Z_i = 0)}{(\sum_{i=1}^N (1-Z_i))^2} \\
&\overset{(d)}{=} \sum_{i=1}^N \frac{Z_i \mathbb{V}(Y \mid Z = 1)}{(\sum_{i=1}^N Z_i)^2} + \sum_{i=1}^N \frac{(1-Z_i) \mathbb{V}(Y \mid Z = 0)}{(\sum_{i=1}^N (1-Z_i))^2} \\
&\overset{(e)}{=} \frac{\mathbb{V}(Y \mid Z = 1)}{\sum_{i=1}^N Z_i} + \frac{\mathbb{V}(Y \mid Z = 0)}{\sum_{i=1}^N (1-Z_i)},
\end{aligned}$$

where (a) holds because $Z_i(1-Z_i) = 0$, (b) follows from the i.i.d. assumption, (c) holds because $Z_i^2 = Z_i$ and $(1-Z_i)^2 = (1-Z_i)$, (d) also follows from the i.i.d. assumption, and (e) results from simplification.

Let the observed data be divided into treatment and control arms:

$$\text{Treatment Arm} : \{Y_{1i} : i = 1, \ldots, N_1\}; \quad \text{Control Arm} : \{Y_{0i} : i = 1, \ldots, N_0\}.$$

The conditional variances $\mathbb{V}(Y \mid Z = j)$, $j = 1, 0$, can be estimated using the corresponding sample variances:

$$\widehat{\sigma}_j^2 = \frac{1}{N_j - 1} \sum_{i=1}^{N_j} \left(Y_{ji} - \frac{1}{N_j} \sum_{i=1}^{N_j} Y_{ji} \right)^2, \quad j = 1, 0.$$

An estimator for the conditional standard error of $\widehat{\theta}$ is then given by:

$$\widehat{\mathrm{SE}} = \sqrt{\frac{\widehat{\sigma}_1^2}{N_1} + \frac{\widehat{\sigma}_0^2}{N_0}}.$$

This allows for the construction of a 95% confidence interval for θ and calculation of the corresponding p-value for testing $H_0 : \theta = 0$.

As an example, applying the estimator $\widehat{\theta}$ to the data in Table 7.1 (see Exercise 8.1), the following values are obtained:

$$\widehat{\theta} = 1.3, \quad \widehat{\sigma}_1^2 = 4.06, \quad \widehat{\sigma}_0^2 = 1.29, \quad \widehat{\mathrm{SE}} = 0.73, \quad 95\%\ \mathrm{CI} = 1.3 \pm 2 \times 0.73.$$

Although not essential for the RCT setting, it is instructive to consider an alternative estimator motivated by the weighting-based statistical estimand:

$$\theta = \mathbb{E}\left[\frac{ZY}{\mathbb{P}(Z=1)}\right] - \mathbb{E}\left[\frac{(1-Z)Y}{\mathbb{P}(Z=0)}\right].$$

This leads to the estimator:

$$\widetilde{\theta} = \frac{1}{N}\sum_{i=1}^{N}\frac{Z_i Y_i}{g_1} - \frac{1}{N}\sum_{i=1}^{N}\frac{(1-Z_i)Y_i}{1-g_1},$$

where $g_1 = \mathbb{P}(Z=1) = 0.5$.

The estimator $\widetilde{\theta}$ is unbiased:

$$\begin{aligned}
\mathbb{E}(\widetilde{\theta}) &= \mathbb{E}\left[\frac{\sum_{i=1}^{N} Z_i Y_i}{N g_1} - \frac{\sum_{i=1}^{N}(1-Z_i)Y_i}{N(1-g_1)}\right] \\
&= \frac{\sum_{i=1}^{N}\mathbb{E}[ZY]}{N g_1} - \frac{\sum_{i=1}^{N}\mathbb{E}[(1-Z)Y]}{N(1-g_1)} \\
&= \mathbb{E}\left[\frac{ZY}{g_1}\right] - \mathbb{E}\left[\frac{(1-Z)Y}{1-g_1}\right] = \theta.
\end{aligned}$$

However, the conditional mean of $\widetilde{\theta}$ given $\mathcal{Z}$ does not equal θ, and therefore the conditional variance cannot be used for inference. Instead, the unconditional variance must be employed:

$$\mathbb{V}(\widetilde{\theta}) = \mathbb{E}[\mathbb{V}(\widetilde{\theta} \mid \mathcal{Z})] + \mathbb{V}[\mathbb{E}(\widetilde{\theta} \mid \mathcal{Z})].$$

Each term can be expanded and simplified accordingly. Both are found to be of the same order, and together they imply that the unconditional variance of $\widetilde{\theta}$ is larger than the conditional variance of $\widehat{\theta}$. Therefore, despite knowledge of $g_1 = 1/2$ in an RCT, the empirical proportion $\widehat{g}_1 = N_1/N$ should be used in constructing the estimator to enable valid inference.

That is, if the empirical proportion $\widehat{g}_1 = N_1/N$ is used instead of the known theoretical proportion g_1, the following estimator is obtained:

$$\begin{aligned}
\widetilde{\theta} &= \frac{1}{N}\sum_{i=1}^{N}\frac{Z_iY_i}{\widehat{g}_1} - \frac{1}{N}\sum_{i=1}^{N}\frac{(1-Z_i)Y_i}{1-\widehat{g}_1} \\
&= \frac{1}{N}\sum_{i=1}^{N}\frac{Z_iY_i}{N_1/N} - \frac{1}{N}\sum_{i=1}^{N}\frac{(1-Z_i)Y_i}{N_0/N} \\
&= \sum_{i=1}^{N}\frac{Z_iY_i}{N_1} - \sum_{i=1}^{N}\frac{(1-Z_i)Y_i}{N_0} \\
&= \frac{\sum_{i=1}^{N} Z_iY_i}{\sum_{i=1}^{N} Z_i} - \frac{\sum_{i=1}^{N}(1-Z_i)Y_i}{\sum_{i=1}^{N}(1-Z_i)},
\end{aligned}$$

which coincides with the previously constructed estimator $\widehat{\theta}$.

8.3 Two Identifiability Assumptions for an NIS

A simple non-interventional study (NIS), specifically a cohort study, is considered in this section. A cohort study is assumed to generate data (X_i, A_i, Y_i), $i = 1, \ldots, n$, where n denotes the sample size. Here, X_i represents a vector of pretreatment covariates, A_i denotes the treatment variable taking the value 1 for the investigational treatment and 0 for the control treatment, and Y_i denotes the outcome variable.

Let $Y_i^{a=j}$ denote the potential outcome that would be observed had the subject i been treated with treatment j, where $j \in \{0, 1\}$. Accordingly, in the potential world, the full dataset is represented by

$$(X_i, A_i, Y_i^{a=1}, Y_i^{a=0}),\ i = 1, \ldots, n,$$

and is assumed to be i.i.d. with $(X, A, Y^{a=1}, Y^{a=0})$, under the SUTVA and the existence of a superpopulation from which these n observations are drawn.

Under the consistency assumption, the observed outcome satisfies $Y_i = Y_i^{A_i}$; that is, $Y_i = Y_i^{a=1}$ if $A_i = 1$, and $Y_i = Y_i^{a=0}$ if $A_i = 0$. Therefore, under both the SUTVA and the consistency assumption, the observed data

$$(X_i, A_i, Y_i),\ i = 1, \ldots, n,$$

are i.i.d. copies of (X, A, Y).

In addition to the SUTVA, which is assumed for the potential outcomes, and the consistency assumption, which connects the potential and observed outcomes, two further assumptions are required for identification. These assumptions pertain to the real world: one is the exchangeability assumption,

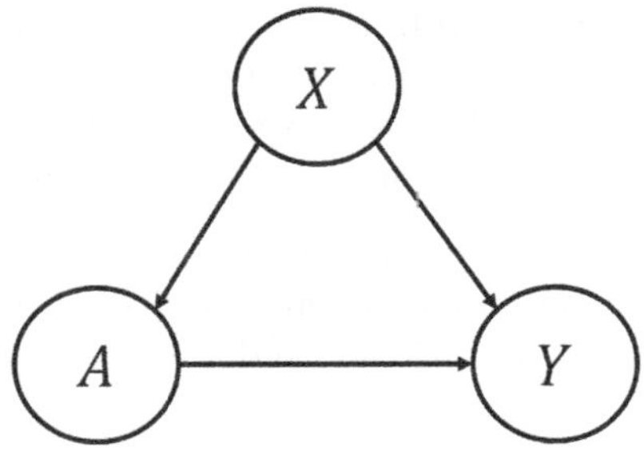

FIGURE 8.2
DAG for a simple cohort study

which is implied by the assumed DAG, and the other is the positivity assumption, which is not implied by the DAG.

8.3.1 An Assumption Implied by the DAG

For the cohort study under consideration, where (X_i, A_i, Y_i), $i = 1, \ldots, n$ are i.i.d. with (X, A, Y), the DAG depicted in Figure 8.2 is adopted as the structural representation of the assumed relationships. These relationships are equivalently described by the following SCM:

$$\begin{aligned} X &= f_X(U_X), \\ A &= f_A(X, U_A), \\ Y &= f_Y(X, A, U_Y), \end{aligned}$$

where U_X, U_A, and U_Y are mutually independent.

From this SCM, the exchangeability assumption is implied:

$$(Y^{a=1}, Y^{a=0}) \perp\!\!\!\perp A \mid X,$$

which states that, conditional on X, the potential outcomes $(Y^{a=1}, Y^{a=0})$ are independent of the treatment assignment A.

To see this, note that, conditional on $X = x$, the potential outcomes and treatment can be expressed as

$$\begin{aligned} (Y^{a=1}, Y^{a=0}) &= (f_Y(x, 1, U_Y),\ f_Y(x, 0, U_Y)), \\ A &= f_A(x, U_A). \end{aligned}$$

Given the assumption that $U_Y \perp\!\!\!\perp U_A$, it follows that

$$(f_Y(x, 1, U_Y),\ f_Y(x, 0, U_Y)) \perp\!\!\!\perp U_A,$$

which implies that $(Y^{a=1}, Y^{a=0}) \perp\!\!\!\perp A \mid X = x$. □

Moreover, it can be observed from the derivation that a weaker condition, namely $U_A \perp\!\!\!\perp U_Y \mid X$, is sufficient to establish the conditional independence $(Y^{a=1}, Y^{a=0}) \perp\!\!\!\perp A \mid X$. Here, the assumption $U_A \perp\!\!\!\perp U_Y \mid X$ is weaker than the original convenient assumption that $\{U_X, U_A, U_Y\}$ are mutually independent.

In the causal inference literature, the exchangeability assumption is also referred to as the no unmeasured confounding (NUC) assumption, the ignorability assumption, or the conditional randomization assumption.[30, 35]

8.3.2 An Assumption Not Implied by the DAG

In addition to the exchangeability assumption that is implied by the DAG, another assumption pertaining to the real world—which is not implied by the DAG—must be made. This is the *positivity assumption*, which states that

$$\mathbb{P}(A = j \mid X = x) > 0, \quad \text{for } j = 0, 1, \text{ and any } x \in \text{supp}(X).$$

That is, for every x in the support of X, the following holds:

$$\begin{aligned} 0 < \mathbb{P}(A = 1 \mid X = x) < 1, \\ 0 < \mathbb{P}(A = 0 \mid X = x) < 1. \end{aligned}$$

This assumption ensures that, conditional on any covariate level $X = x$, data are available for both treatment groups. In other words, for each possible covariate level, it is required that some subjects receive each treatment. A *violation of positivity* occurs when this condition is not satisfied—i.e., when, for certain values of X, no subjects are observed in one of the treatment arms. In such cases, the causal effect of treatment cannot be estimated due to the absence of comparable data.

The two assumptions concerning the real world, the exchangeability and positivity assumptions, are often in tension; See Chapter 10 for more details. Satisfying the exchangeability assumption typically requires adjustment for a comprehensive set of measured covariates. However, including an extensive number of covariates may increase the possibility of violating the positivity assumption. To mitigate such violations, variables or levels of categorical variables are sometimes collapsed. Yet, collapsing may reduce covariate detail and thereby jeopardize the exchangeability assumption.

Gold Coin # 29. *In addition to the SUTVA and the consistency assumption, two further assumptions are required for identification in a non-interventional study: the exchangeability assumption and the positivity assumption. Unfortunately, these two assumptions are often in tension.*

8.4 Two Identification Strategies

The goal is to estimate the following causal estimand,

$$\theta^*_{\text{ATE}} = \mathbb{E}(Y^{a=1}) - \mathbb{E}(Y^{a=0}).$$

To facilitate the identification, several assumptions are imposed: the SUTVA, the consistency assumption, the exchangeability assumption, and the positivity assumption. These assumptions are collectively referred to as the *identifiability assumptions* for the cohort study illustrated in Figure 8.2.

With these assumptions in place, identification of the causal estimand can proceed. Two common strategies are available for this purpose: the *standardization strategy* and the *weighting strategy*.[30,36,22]

8.4.1 Standardization Strategy

Under these identifiability assumptions, the following can be derived:

$$\begin{aligned}
\theta^*_{\text{ATE}} &= \mathbb{E}(Y^{a=1}) - \mathbb{E}(Y^{a=0}) \\
&\overset{(a)}{=} \mathbb{E}[\mathbb{E}(Y^{a=1} \mid X)] - \mathbb{E}[\mathbb{E}(Y^{a=0} \mid X)] \\
&\overset{(b)}{=} \mathbb{E}[\mathbb{E}(Y^{a=1} \mid A = 1, X)] - \mathbb{E}[\mathbb{E}(Y^{a=0} \mid A = 0, X)] \\
&\overset{(c)}{=} \mathbb{E}[\mathbb{E}(Y \mid A = 1, X)] - \mathbb{E}[\mathbb{E}(Y \mid A = 0, X)],
\end{aligned}$$

where (a) follows from the law of iterated expectations (LIE), (b) holds under the exchangeability and positivity assumptions, and (c) holds under the consistency assumption.

Thus, the causal estimand is translated into a statistical estimand:

$$\theta_{\text{ATE}} = \mathbb{E}[\mathbb{E}(Y \mid A = 1, X)] - \mathbb{E}[\mathbb{E}(Y \mid A = 0, X)].$$

8.4.2 Weighting Strategy

Under the same identifiability assumptions, the following can be derived:

$$\begin{aligned}
\mathbb{E}\left\{\frac{I(A=a)}{\mathbb{P}(A=a \mid X)}Y\right\} &\overset{(a)}{=} \mathbb{E}\left\{\frac{I(A=a)}{\mathbb{P}(A=a \mid X)}Y^a\right\} \\
&\overset{(b)}{=} \mathbb{E}\left[\mathbb{E}\left\{\frac{I(A=a)}{\mathbb{P}(A=a \mid X)}Y^a \,\middle|\, X\right\}\right] \\
&\overset{(c)}{=} \mathbb{E}\left[\mathbb{E}\left\{\frac{I(A=a)}{\mathbb{P}(A=a \mid X)} \,\middle|\, X\right\}\mathbb{E}(Y^a \mid X)\right] \\
&\overset{(d)}{=} \mathbb{E}[\mathbb{E}(Y^a \mid X)] \\
&\overset{(e)}{=} \mathbb{E}(Y^a),
\end{aligned}$$

where (a) holds because the left-hand side is well-defined under the positivity assumption and the right-hand side uses the consistency assumption, (b) follows from the LIE, (c) holds under the exchangeability assumption, (d) holds because $\mathbb{E}[I(A = a) \mid X] = \mathbb{P}(A = a \mid X)$, and (e) uses the LIE again.

Thus, the causal estimand is translated into an alternative but equivalent form of the statistical estimand:

$$\theta_{\text{ATE}} = \mathbb{E}\left[\frac{I(A=1)}{\mathbb{P}(A=1 \mid X)}Y\right] - \mathbb{E}\left[\frac{I(A=0)}{\mathbb{P}(A=0 \mid X)}Y\right].$$

This statistical estimand can be further rewritten as:

$$\theta_{\text{ATE}} = \mathbb{E}\left[\frac{2(A-1)Y}{\mathbb{P}(A \mid X)}\right].$$

Gold Coin # 30. *Two fundamental strategies are available for identification: the standardization strategy and the weighting strategy. Consequently, the causal estimand for the average treatment effect (ATE) corresponds to the following two equivalent forms:*

$$\theta_{ATE} = \mathbb{E}[\mathbb{E}(Y \mid A = 1, X)] - \mathbb{E}[\mathbb{E}(Y \mid A = 0, X)],$$

$$\theta_{ATE} = \mathbb{E}\left[\frac{I(A=1)}{\mathbb{P}(A=1 \mid X)}Y\right] - \mathbb{E}\left[\frac{I(A=0)}{\mathbb{P}(A=0 \mid X)}Y\right] = \mathbb{E}\left[\frac{2(A-1)Y}{\mathbb{P}(A \mid X)}\right].$$

8.5 Two Subpopulations

Recall that the first attribute of an estimand is the population. In the preceding discussion, the population has been defined as all subjects who received either treatment 1 or treatment 0. Based on this population, the causal estimand θ^*_{ATE} is defined and used to summarize, at the population level, the difference between the following two potential outcomes (expressed in the subjunctive mood):

- The outcomes that would have been observed if all subjects had been treated with treatment 1,
- The outcomes that would have been observed if all subjects had been treated with treatment 0.

As an exercise (see Exercise 8.2), suppose the subjects in Table 6.1 are taken to represent the target population—for instance, imagine each subject corresponds to one million individuals. Under this interpretation, the value of the ATE estimand, which is unknown in reality, is $\theta^*_{\text{ATE}} = 0.45$.

8.5.1 ATT and ATC

Target estimands may also be defined with respect to specific subpopulations. Two such subpopulations are commonly considered:

1. One that consists of all subjects who actually received treatment 1;
2. One that consistes of all subjects who actually received treatment 0.

The *average treatment effect on the treated (ATT)* is defined as

$$\theta^*_{\text{ATT}} = \mathbb{E}(Y^{a=1} - Y^{a=0} \mid A = 1).$$

The ATT estimand summarizes, at the population level, the difference between the following two potential outcomes (reflecting counterfactual thinking):

- The outcomes observed for those who actually received treatment 1;
- The outcomes that would have been observed if these same subjects had instead received treatment 0.

The *average treatment effect on the control (ATC)* is defined as

$$\theta^*_{\text{ATC}} = \mathbb{E}(Y^{a=1} - Y^{a=0} \mid A = 0).$$

The ATC estimand summarizes, at the population level, the difference between the following two potential outcomes:

- The outcomes that would have been observed if all subjects who received treatment 0 had instead been treated with treatment 1;
- The outcomes observed for those who actually received treatment 0.

Figure 8.3 provides a visual summary of the definitions of ATE, ATT, and ATC. The oval represents the treated population, and the rectangle represents the control population.

As an example, if the subjects in Table 6.1 are assumed to represent the population (if helpful, imagine that each subject represents one million individuals), and the distribution of A in the population is taken to match that in Table 7.2, then two subpopulations can be defined: the subpopulation of individuals treated with $A = 1$ and the subpopulation treated with $A = 0$. Given the full dataset in Table 6.1, the values of the ATT estimand and the ATC estimand are obtained as $\theta^*_{\text{ATT}} = 1.27$ (see Exercise 8.3) and $\theta^*_{\text{ATC}} = -0.56$ (see Exercise 8.4), respectively.

Gold Coin # 31. *In addition to the averaget treatment effect (ATE), two other estimands defined with respect to subpopulations may be of interest: the average treatment effect on the treated (ATT) and the average treatment effect on the control (ATC).*

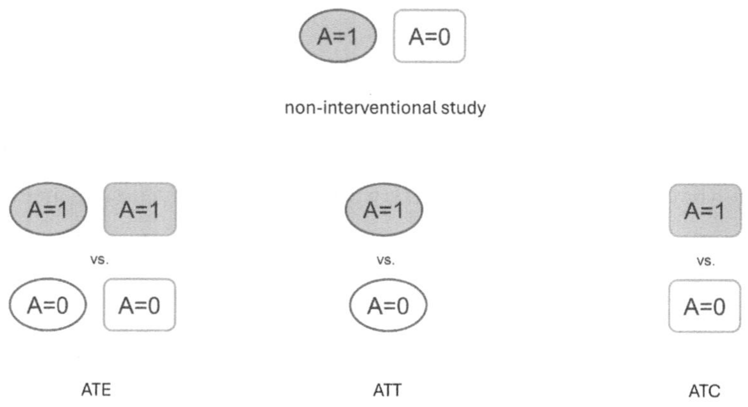

FIGURE 8.3
ATE, ATT, and ATC

Remark:

Consider an externally controlled trial (ECT), which consists of a single-arm clinical trial supplemented with an external control arm. Let $A = 1$ indicate inclusion in the single-arm trial (i.e., the treated arm), and $A = 0$ indicate inclusion in the external control group. If the population of the single-arm trial is considered the population of interest, then the appropriate target estimand is the ATT. Conversely, for purposes such as health technology assessment (HTA),[37] where the population of interest may correspond to the external control group, the ATC becomes the relevant estimand.

8.5.2 Identification

For the purpose of identifying θ^*_{ATT}, the same set of identifiability assumptions is adopted—namely, SUTVA, consistency, exchangeability, and positivity. Under these assumptions, two identification strategies can be employed.

The Standardization Strategy

Under the identifiability assumptions, the following derivation is established:

$$\begin{aligned}
\theta^*_{\text{ATT}} &= \mathbb{E}(Y^{a=1} \mid A = 1) - \mathbb{E}(Y^{a=0} \mid A = 1) \\
&\overset{(a)}{=} \mathbb{E}(Y \mid A = 1) - \mathbb{E}(Y^{a=0} \mid A = 1) \\
&\overset{(b)}{=} \mathbb{E}(Y \mid A = 1) - \mathbb{E}_{X|A=1}\left[\mathbb{E}(Y^{a=0} \mid X, A = 1)\right] \\
&\overset{(c)}{=} \mathbb{E}(Y \mid A = 1) - \mathbb{E}_{X|A=1}\left[\mathbb{E}(Y^{a=0} \mid X, A = 0)\right] \\
&\overset{(d)}{=} \mathbb{E}(Y \mid A = 1) - \mathbb{E}_{X|A=1}\left[\mathbb{E}(Y \mid X, A = 0)\right],
\end{aligned}$$

where (a) follows from the consistency assumption, (b) follows from the LIE, where $\mathbb{E}_{X|A=1}$ denotes expectation with respect to the conditional distribution of X given $A = 1$, (c) follows from the exchangeability and positivity assumptions, and (d) again follows from the consistency assumption.

Therefore, the ATT estimand θ^*_{ATT} is translated into the following statistical estimand:

$$\theta_{\text{ATT}} = \mathbb{E}(Y \mid A = 1) - \mathbb{E}_{X|A=1}\left[\mathbb{E}(Y \mid X, A = 0)\right].$$

The Weighting Strategy

Under the same identifiability assumptions, an alternative expression for θ^*_{ATT} can be derived using the weighting strategy:

$$\begin{aligned}
&\mathbb{E}\left[\frac{I(A=0)\,\mathbb{P}(A=1 \mid X)\,Y}{\mathbb{P}(A=1)\,\mathbb{P}(A=0 \mid X)}\right] \\
&\overset{(a)}{=}\mathbb{E}\left[\frac{I(A=0)\,\mathbb{P}(A=1 \mid X)\,Y^{a=0}}{\mathbb{P}(A=1)\,\mathbb{P}(A=0 \mid X)}\right] \\
&\overset{(b)}{=}\mathbb{E}\left\{\mathbb{E}\left[\frac{I(A=0)\,\mathbb{P}(A=1 \mid X)\,Y^{a=0}}{\mathbb{P}(A=1)\,\mathbb{P}(A=0 \mid X)} \,\middle|\, X\right]\right\} \\
&\overset{(c)}{=}\mathbb{E}\left\{\frac{\mathbb{P}(A=1 \mid X)\,\mathbb{E}(Y^{a=0} \mid X)}{\mathbb{P}(A=1)}\right\} \\
&\overset{(d)}{=}\mathbb{E}_{X|A=1}\left[\mathbb{E}(Y^{a=0} \mid X, A=1)\right] \\
&\overset{(e)}{=}\mathbb{E}(Y^{a=0} \mid A=1),
\end{aligned}$$

where (a) is valid under the positivity assumption and follows from the consistency assumption, (b) uses the LIE, (c) follows from the exchangeability assumption, (d) follows from Bayes' theorem (see Exercise 8.5), and (e) again follows from the LIE.

Accordingly, the following statistical estimand is obtained:

$$\theta_{\text{ATT}} = \mathbb{E}\left[\frac{I(A=1)\,Y}{\mathbb{P}(A=1)}\right] - \mathbb{E}\left[\frac{I(A=0)\,\mathbb{P}(A=1 \mid X)\,Y}{\mathbb{P}(A=1)\,\mathbb{P}(A=0 \mid X)}\right].$$

In summary, the two statistical estimands obtained through the standardization and weighting strategies are equivalent under the identifiability assumptions, as both are equal to the causal estimand:

$$\begin{aligned}
\theta_{\text{ATT}} &= \mathbb{E}(Y \mid A = 1) - \mathbb{E}_{X|A=1}\left[\mathbb{E}(Y \mid X, A = 0)\right] \\
&= \mathbb{E}\left[\frac{I(A=1)\,Y}{\mathbb{P}(A=1)}\right] - \mathbb{E}\left[\frac{I(A=0)\,\mathbb{P}(A=1 \mid X)\,Y}{\mathbb{P}(A=1)\,\mathbb{P}(A=0 \mid X)}\right].
\end{aligned}$$

Extension to ATC

Similarly, under the same identifiability assumptions, the causal estimand θ^*_{ATC} can be expressed as:

$$\begin{aligned}\theta_{\text{ATC}} &= \mathbb{E}_{X|A=0}\left[\mathbb{E}(Y \mid X, A=1)\right] - \mathbb{E}(Y \mid A=0)\\ &= \mathbb{E}\left[\frac{I(A=1)\,\mathbb{P}(A=0 \mid X)\,Y}{\mathbb{P}(A=0)\,\mathbb{P}(A=1 \mid X)}\right] - \mathbb{E}\left[\frac{I(A=0)\,Y}{\mathbb{P}(A=0)}\right].\end{aligned}$$

8.6 Exercises

Exercise 8.1

Analyze the data in Table 7.1 to obtain estimates for $\widehat{\theta}$, $\widehat{\sigma}_1^2$, $\widehat{\sigma}_0^2$, $\widehat{\text{SE}}$, and 95% CI.

Exercise 8.2

Consider the subjects in Table 6.1 as a superpopulation (imagine that one subject in the table represents one million subjects) and find the true value of the causal estimand of ATE.

Exercise 8.3

Consider the subjects in Table 6.1 as a superpopulation and assume that the distribution of A in the superpopulation is the same as that of A in Table 7.2. Find the value of ATT.

Exercise 8.4

Consider the subjects in Table 6.1 as a superpopulation and assume that the distribution of A in the superpopulation is the same as that of A in Table 7.2. Find the value of ATC.

Exercise 8.5

Using Bayes' theorem, show that

$$\mathbb{E}\left\{\frac{\mathbb{P}(A=1 \mid X)\mathbb{E}[Y^{a=0} \mid X]}{\mathbb{P}(A=1)}\right\} = \mathbb{E}_{X|A=1}\left\{\mathbb{E}[Y^{a=0} \mid X, A=1]\right\}.$$

9

Estimator

In the development of causal inference in statistics, two major schools of estimation methods have been established: the propensity score-based methods, pioneered by Donald Rubin and colleagues,[38, 35] and the g-methods, pioneered by James Robins and colleagues.[39, 30] Both schools of estimation methods are solid and well-established. However, this book places greater emphasis on g-methods, as they form the foundation for the doubly robust methods to be introduced in this chapter and are further extended to address time-varying confounding in later chapters.

9.1 Two Models

In the preceding chapter, the identification of the following causal estimand—the average treatment effect (ATE)—was considered:

$$\theta^*_{\text{ATE}} = \mathbb{E}(Y^{a=1}) - \mathbb{E}(Y^{a=0}).$$

To identify this estimand, two strategies were employed: standardization and weighting. As a result, two equivalent versions of the corresponding statistical estimand were obtained. The same identification approach was applied to the other two estimands—the average treatment effect on the treated (ATT) and the average treatment effect on the controls (ATC). The following two subsections will demonstrate that each version of the statistical estimand corresponds to a distinct underlying model.

9.1.1 Outcome Model

Under the identifiability assumptions, the standardization strategy yields the following version of the statistical estimand:

$$\theta_{\text{ATE}} = \mathbb{E}\left[\mathbb{E}(Y \mid A = 1, X)\right] - \mathbb{E}\left[\mathbb{E}(Y \mid A = 0, X)\right].$$

Define the conditional expectation function as

$$Q(X, A) = \mathbb{E}(Y \mid X, A).$$

DOI: 10.1201/9781003635468-11

This function $Q(X, A)$ is referred to as *the outcome regression function* and is derived from the following outcome model:

$$Y \sim X + A,$$

where Y is treated as the dependent variable, and X and A are considered the independent variables.

Accordingly, the statistical estimand can be rewritten as

$$\theta_{\text{ATE}} = \mathbb{E}[Q(X, 1)] - \mathbb{E}[Q(X, 0)].$$

9.1.2 Treatment Model

Under the identifiability assumptions, the weighting strategy yields the following version of the statistical estimand:

$$\theta_{\text{ATE}} = \mathbb{E}\left[\frac{I(A=1)}{\mathbb{P}(A=1 \mid X)}Y\right] - \mathbb{E}\left[\frac{I(A=0)}{\mathbb{P}(A=0 \mid X)}Y\right] = \mathbb{E}\left[\frac{2(A-1)Y}{\mathbb{P}(A \mid X)}\right].$$

Define

$$g(1 \mid X) = \mathbb{P}(A = 1 \mid X) \quad \text{and} \quad g(0 \mid X) = \mathbb{P}(A = 0 \mid X).$$

The function $g(A \mid X)$ is referred to as *the propensity score function*[38] and is derived from the following treatment model:

$$A \sim X,$$

where A is the dependent variable and X is the independent variable.

Accordingly, the statistical estimand can be rewritten as

$$\theta_{\text{ATE}} = \mathbb{E}\left[\frac{I(A=1)Y}{g(1 \mid X)}\right] - \mathbb{E}\left[\frac{I(A=0)Y}{g(0 \mid X)}\right] = \mathbb{E}\left[\frac{2(A-1)Y}{g(A \mid X)}\right].$$

Gold Coin # 32. *Typically, constructing the estimators requires specifying one or both of two models: the outcome model $Y \sim X + A$ and the treatment model $A \sim X$.*

9.2 Two Estimators

For each statistical estimand, two equivalent representations can be derived: one involving the outcome regression function and the other involving the propensity score function. Accordingly, if an estimator for the outcome regression function is available, an estimator for the statistical estimand can be obtained from the former representation, leading to the construction of *the g-computation estimator*.[39] Similarly, if an estimator for the propensity score function is available, a corresponding estimator can be derived from the latter representation, resulting in *the inverse probability of treatment weighted (IPW) estimator*.[40]

Gold Coin # 33. *Since each statistical estimand admits two equivalent representations, two singly robust estimators can be directly constructed: the g-computation estimator and the inverse probability of treatment weighted (IPW) estimator.*

9.2.1 G-Computation Estimator

Point Estimator

Consider the following statistical estimand expressed in terms of the outcome regression function $Q(X, A) = \mathbb{E}(Y \mid X, A)$:

$$\theta = \mathbb{E}[Q(X,1)] - \mathbb{E}[Q(X,0)],$$

where the subscript "ATE" is omitted hereafter for notational simplicity.

If an estimator $\widehat{Q}(x, a)$ for $Q(x, a)$ is available, and the population means are approximated by their empirical counterparts, the g-computation estimator can be defined as

$$\widehat{\theta}_{\text{g-comp}} = \frac{1}{n}\sum_{i=1}^{n} \widehat{Q}(X_i, 1) - \frac{1}{n}\sum_{i=1}^{n} \widehat{Q}(X_i, 0),$$

where the subscript "g-comp" refers to "g-computation."

To conduct inference based on $\widehat{\theta}_{\text{g-comp}}$, an estimator for its variance is needed. While the bootstrap method can be used without deriving an explicit formula, it is often desirable for analytical purposes to obtain a closed-form expression for the asymptotic variance, which will be derived below.

Parametric Outcome Model

For a continuous outcome Y, the following linear model is considered:

$$Q(X, A; \beta) = \beta(0) + \beta(1)A + \beta(2)^\top X + \beta(3)^\top AX,$$

where $\beta = (\beta(0), \beta(1), \beta(2)^\top, \beta(3)^\top)^\top$.

The minimum loss estimator (MLE) $\widehat{\beta}$ is defined as the minimizer of the $\mathcal{L}_2$ loss function:

$$\widehat{\beta} = \arg\min_{\beta} \sum_{i=1}^{n} \left[Y_i - Q(X_i, A_i; \beta)\right]^2,$$

which is equivalent to solving the estimating equation:

$$\sum_{i=1}^{n} \frac{\partial Q(X_i, A_i; \beta)}{\partial \beta} \left[Y_i - Q(X_i, A_i; \beta)\right] = 0,$$

where

$$\frac{\partial Q(X, A; \beta)}{\partial \beta} = \begin{pmatrix} 1 \\ A \\ X \\ AX \end{pmatrix}.$$

For a binary outcome Y, the following logistic model is used:

$$\text{logit}[Q(X, A; \beta)] = \beta(0) + \beta(1)A + \beta(2)^\top X + \beta(3)^\top AX,$$

where $\text{logit}(Q) = \log(Q/(1 - Q))$ and β is defined as above.

The MLE $\widehat{\beta}$ is obtained by minimizing the binary cross-entropy (BCE) loss function:

$$\widehat{\beta} = \arg\min_{\beta} \sum_{i=1}^{n} \left\{-Y_i \log[Q(X_i, A_i; \beta)] - (1 - Y_i) \log[1 - Q(X_i, A_i; \beta)]\right\},$$

which corresponds to solving the estimating equation:

$$\sum_{i=1}^{n} \frac{\partial Q(X_i, A_i; \beta)/\partial \beta}{Q(X_i, A_i; \beta)(1 - Q(X_i, A_i; \beta))} \left[Y_i - Q(X_i, A_i; \beta)\right] = 0,$$

where

$$\frac{\partial Q(X, A; \beta)/\partial \beta}{Q(X, A; \beta)(1 - Q(X, A; \beta))} = \begin{pmatrix} 1 \\ A \\ X \\ AX \end{pmatrix}.$$

After an estimator $\widehat{\beta}$ is obtained, either via the linear or logistic model, the fitted regression function $\widehat{Q}(X, A) = Q(X, A; \widehat{\beta})$ is used to construct the g-computation estimator:

$$\widehat{\theta}_{\text{g-comp}} = \frac{1}{n} \sum_{i=1}^{n} \left[Q(X_i, 1; \widehat{\beta}) - Q(X_i, 0; \widehat{\beta})\right].$$

Asymptotic Variance

To derive the asymptotic variance of $\widehat{\theta}_{\text{g-comp}}$, the framework of M-estimation discussed in Chapter 5 is employed. The derivation is presented for continuous outcomes, and the extension to binary outcomes follows similarly.

Let $\mu = (\theta, \beta^\top)^\top$ and define the stacked estimating function as

$$M(O; \mu) = (M_1(O, \mu), M_2^\top(O, \mu))^\top,$$

where

$$\begin{aligned}
M_1(O, \mu) &= Q(X, 1; \beta) - Q(X, 0; \beta) - \theta, \\
M_2(O, \mu) &= \frac{\partial Q(X, A; \beta)}{\partial \beta}[Y - Q(X, A; \beta)].
\end{aligned}$$

Let $\mu_0 = (\theta_0, \beta_0^\top)^\top$ denote the true value of μ, assuming the outcome model $Q(X, A; \beta)$ is correctly specified.

The meat (i.e., variance) term in the sandwich estimator is given by:

$$\mathbb{E}\left\{M^{\otimes 2}(O; \mu_0)\right\} = \begin{pmatrix} \mathbb{E}[Q(X, 1; \beta_0) - Q(X, 0; \beta_0) - \theta_0]^2 & 0^\top \\ 0 & \mathbf{A}_0 \end{pmatrix},$$

where $v^{\otimes 2} = vv^\top$ for any vector v and

$$\mathbf{A}_0 = \mathbb{E}\left\{[Y - Q(X, A; \beta_0)]^2 \left[\frac{\partial Q(X, A; \beta_0)}{\partial \beta}\right]^{\otimes 2}\right\}.$$

The bread (i.e., derivative) term is:

$$\mathbb{E}\left[\frac{\partial M(O, \mu_0)}{\partial \mu^\top}\right] = \begin{pmatrix} -1 & \mathbf{B}_0 \\ 0 & -\mathbf{C}_0 \end{pmatrix},$$

whose inverse is

$$\left\{\mathbb{E}\left[\frac{\partial M(O, \mu_0)}{\partial \mu^\top}\right]\right\}^{-1} = \begin{pmatrix} -1 & -\mathbf{B}_0\mathbf{C}_0^{-1} \\ 0 & -\mathbf{C}_0^{-1} \end{pmatrix},$$

where

$$\mathbf{B}_0 = \mathbb{E}\left\{\frac{\partial Q(X, 1; \beta_0)}{\partial \beta^\top} - \frac{\partial Q(X, 0; \beta_0)}{\partial \beta^\top}\right\} = (0, 1, 0, \mathbb{E}[X^\top])$$

and

$$\mathbf{C}_0 = \mathbb{E}\left\{\left[\frac{\partial Q(X, A; \beta_0)}{\partial \beta}\right]^{\otimes 2}\right\}.$$

The estimator $\widehat{\theta}_{\text{g-comp}}$ is thus consistent and asymptotically normal:

$$\widehat{\theta}_{\text{g-comp}} \xrightarrow{p} \theta_0,$$
$$\sqrt{n}(\widehat{\theta}_{\text{g-comp}} - \theta_0) \xrightarrow{d} \mathcal{N}(0, \sigma^2_{0,\text{g-comp}}),$$

where the asymptotic variance is

$$\sigma^2_{0,\text{g-comp}} = \mathbb{E}\left[Q(X,1;\beta_0) - Q(X,0;\beta_0) - \theta_0\right]^2 + \mathbf{B}_0\mathbf{C}_0^{-1}\mathbf{A}_0\mathbf{C}_0^{-1}\mathbf{B}_0^\top.$$

A consistent estimator of this variance is

$$\widehat{\sigma}^2_{\text{g-comp}} = \frac{1}{n}\sum_{i=1}^{n}\left[Q(X_i,1;\widehat{\beta}) - Q(X_i,0;\widehat{\beta}) - \widehat{\theta}_{\text{g-comp}}\right]^2 + \widehat{\mathbf{B}}\widehat{\mathbf{C}}^{-1}\widehat{\mathbf{A}}\widehat{\mathbf{C}}^{-1}\widehat{\mathbf{B}}^\top,$$

where

$$\widehat{\mathbf{A}} = \frac{1}{n}\sum_{i=1}^{n}\left\{[Y_i - Q(X_i,A_i;\widehat{\beta})]^2\left[\frac{\partial Q(X_i,A_i;\widehat{\beta})}{\partial\beta}\right]^{\otimes 2}\right\},$$
$$\widehat{\mathbf{B}} = \frac{1}{n}\sum_{i=1}^{n}\left\{\frac{\partial Q(X_i,1;\widehat{\beta})}{\partial\beta^\top} - \frac{\partial Q(X_i,0;\widehat{\beta})}{\partial\beta^\top}\right\},$$
$$\widehat{\mathbf{C}} = \frac{1}{n}\sum_{i=1}^{n}\left[\frac{\partial Q(X_i,A_i;\widehat{\beta})}{\partial\beta}\right]^{\otimes 2}.$$

9.2.2 IPW Estimator

Point Estimator

The following statistical estimand is considered in terms of the propensity score function $g(A \mid X) = \mathbb{P}(A \mid X)$:

$$\theta = \mathbb{E}\left[\frac{I(A=1)Y}{g(1 \mid X)}\right] - \mathbb{E}\left[\frac{I(A=0)Y}{g(0 \mid X)}\right].$$

An estimator of $g(a \mid x)$, denoted by $\widehat{g}(a \mid x)$, can be obtained, and the population mean can subsequently be estimated using the sample mean. Based on this approach, the inverse probability of treatment weighted (IPW) estimator is defined as

$$\widehat{\theta}_{\text{IPW}} = \frac{1}{n}\sum_{i=1}^{n}\frac{I(A_i=1)}{\widehat{g}(1 \mid X_i)}Y_i - \frac{1}{n}\sum_{i=1}^{n}\frac{I(A_i=0)}{\widehat{g}(0 \mid X_i)}Y_i.$$

To conduct statistical inference based on $\widehat{\theta}_{\text{IPW}}$, an estimator of its variance is needed. Similarly, it is often desirable to obtain a closed-form expression for the asymptotic variance, which will be derived below.

Parametric Treatment Model

For the treatment variable A, assume the following logistic regression model:

$$\text{logit}[g(1 \mid X; \gamma)] = \gamma(0) + \gamma(1)^\top X,$$

where $\text{logit}[g(1 \mid X; \gamma)] = \log[g(1 \mid X; \gamma)/g(0 \mid X; \gamma)]$, and $\gamma = (\gamma(0), \gamma(1)^\top)^\top$.

The MLE of γ is obtained as

$$\widehat{\gamma} = \arg\min \sum_{i=1}^{n} - \{A_i \log[g(1 \mid X_i; \gamma)] + (1 - A_i) \log[g(0 \mid X_i; \gamma)]\},$$

which corresponds to the solution of the following estimating equation:

$$\sum_{i=1}^{n} \widetilde{X}_i \left[A_i - g(1 \mid X_i; \gamma)\right] = 0,$$

where

$$\widetilde{X} = \begin{pmatrix} 1 \\ X \end{pmatrix}.$$

Once the estimator $\widehat{\gamma}$ is obtained from the logistic model, an estimator of g can be constructed as $\widehat{g}(1 \mid X) = g(1 \mid X; \widehat{\gamma})$. Consequently, the IPW estimator is expressed as

$$\widehat{\theta}_{\text{IPW}} = \frac{1}{n} \sum_{i=1}^{n} \frac{I(A_i = 1)}{g(1 \mid X_i; \widehat{\gamma})} Y_i - \frac{1}{n} \sum_{i=1}^{n} \frac{I(A_i = 0)}{g(0 \mid X_i; \widehat{\gamma})} Y_i.$$

Asymptotic Variance

To derive an explicit expression for the asymptotic variance of $\widehat{\theta}_{\text{IPW}}$, the M-estimation framework discussed in Chapter 5 is applied again.

Let $\mu = (\theta, \gamma^\top)^\top$ and define

$$M(O; \mu) = (M_1(O, \mu), M_2^\top(O, \mu))^\top,$$

where

$$M_1(O, \mu) = \frac{AY}{g(1 \mid X; \gamma)} - \frac{(1 - A)Y}{g(0 \mid X; \gamma)} - \theta,$$
$$M_2(O, \mu) = [A - g(1 \mid X; \gamma)]\widetilde{X}.$$

Let θ_0 and γ_0 denote the true values of θ and γ, respectively, under the assumption that the model $g(a \mid X; \gamma)$ is correctly specified. Then, $\mu_0 = (\theta_0, \gamma_0^\top)^\top$ represents the true value of μ.

The meat term of the sandwich variance formula is given by

$$\mathbb{E}\left\{M^{\otimes 2}(O;\mu_0)\right\} = \begin{pmatrix} \mathbb{E}\left[\frac{AY}{g(1|X;\gamma_0)} - \frac{(1-A)Y}{g(0|X;\gamma_0)} - \theta\right]^2 & 0^\top \\ 0 & \mathbf{D}_0 \end{pmatrix},$$

where

$$\mathbf{D}_0 = \mathbb{E}\left\{[A - g(1 \mid X;\gamma_0)]^2 \widetilde{X}^{\otimes 2}\right\}.$$

The bread term is given by

$$\mathbb{E}\left[\frac{\partial M(O,\mu_0)}{\partial \mu^\top}\right] = \begin{pmatrix} -1 & \mathbf{E}_0 \\ 0 & -\mathbf{F}_0 \end{pmatrix},$$

with its inverse expressed as

$$\left\{\mathbb{E}\left[\frac{\partial M(O,\mu_0)}{\partial \mu^\top}\right]\right\}^{-1} = \begin{pmatrix} -1 & -\mathbf{E}_0\mathbf{F}_0^{-1} \\ 0 & -\mathbf{F}_0^{-1} \end{pmatrix},$$

where

$$\mathbf{E}_0 = -\mathbb{E}\left\{\frac{[A - g(1 \mid X;\gamma_0)]^2\, Y}{g(1 \mid X;\gamma_0) g(0 \mid X;\gamma_0)} \widetilde{X}^\top\right\},$$

$$\mathbf{F}_0 = \mathbb{E}\left\{g(1 \mid X;\gamma_0) g(0 \mid X;\gamma_0) \widetilde{X}^{\otimes 2}\right\}.$$

Under the correct specification of $g(a \mid x;\gamma)$, the consistency and asymptotic normality of $\widehat{\theta}_{\text{IPW}}$ are established:

$$\widehat{\theta}_{\text{IPW}} \xrightarrow{p} \theta_0,$$

$$\sqrt{n}(\widehat{\theta}_{\text{IPW}} - \theta_0) \xrightarrow{d} \mathcal{N}(0, \sigma^2_{0,\text{IPW}}),$$

where

$$\sigma^2_{0,\text{IPW}} = \mathbb{E}\left[\frac{AY}{g(1 \mid X;\gamma_0)} - \frac{(1-A)Y}{g(0 \mid X;\gamma_0)} - \theta_0\right]^2 + \mathbf{E}_0\mathbf{F}_0^{-1}\mathbf{D}_0\mathbf{F}_0^{-1}\mathbf{E}_0^\top,$$

which can be estimated by

$$\widehat{\sigma}^2_{\text{IPW}} = \frac{1}{n}\sum_{i=1}^n \left[\frac{A_iY_i}{g(1 \mid X_i;\widehat{\gamma})} - \frac{(1-A_i)Y_i}{g(0 \mid X_i;\widehat{\gamma})} - \widehat{\theta}_{\text{IPW}}\right]^2 + \widehat{\mathbf{E}}\widehat{\mathbf{F}}^{-1}\widehat{\mathbf{D}}\widehat{\mathbf{F}}^{-1}\widehat{\mathbf{E}}^\top,$$

where

$$\widehat{\mathbf{D}} = \frac{1}{n}\sum_{i=1}^n [A_i - g(1 \mid X_i;\widehat{\gamma})]^2 \widetilde{X}_i^{\otimes 2},$$

$$\widehat{\mathbf{E}} = -\frac{1}{n}\sum_{i=1}^n \frac{[A_i - g(1 \mid X_i;\widehat{\gamma})]^2\, Y_i}{g(1 \mid X_i;\widehat{\gamma}) g(0 \mid X_i;\widehat{\gamma})} \widetilde{X}_i^\top,$$

$$\widehat{\mathbf{F}} = \frac{1}{n}\sum_{i=1}^n g(1 \mid X_i;\widehat{\gamma}) g(0 \mid X_i;\widehat{\gamma}) \widetilde{X}_i^{\otimes 2}.$$

9.3 Doubly-Robust Estimator

9.3.1 A Class of Point Estimators

It has been shown that $\widehat{\theta}_{\text{IPW}}$ is consistent and asymptotically normal when the propensity score function is correctly specified. However, it is often desirable to construct an estimator that is not only consistent and asymptotically normal, but also achieves the smallest possible asymptotic variance. Such an estimator is referred to as an *efficient estimator.*[41, 42]

A general class of estimators is considered, defined as

$$\widehat{\theta}(h) = \frac{1}{n}\sum_{i=1}^{n}\left\{\frac{A_i Y_i}{g(1 \mid X_i; \widehat{\gamma})} - \frac{(1-A_i)Y_i}{g(0 \mid X_i; \widehat{\gamma})} - [A_i - g(1 \mid X_i; \widehat{\gamma})]h(X_i)\right\},$$

where $h(X)$ denotes any measurable function of X. Notably, $\widehat{\theta}_{\text{IPW}}$ corresponds to the special case in which $h(X) \equiv 0$.

It can be shown that for any choice of $h(X)$, the estimator $\widehat{\theta}(h)$ remains consistent provided that the propensity score model is correctly specified.

In fact, under the assumption that γ_0 represents the true value of γ, the following holds:

$$\begin{aligned}\mathbb{E}\{[A - g(1 \mid X; \gamma_0)]h(X)\} &= \mathbb{E}\{\mathbb{E}([A - g(1 \mid X; \gamma_0)]h(X) \mid X)\} \\ &= \mathbb{E}\{h(X) \cdot \mathbb{E}[A - g(1 \mid X; \gamma_0) \mid X]\} = 0.\end{aligned}$$

9.3.2 Minimum-Variance "Estimator"

The asymptotic variance of $\widehat{\theta}(h)$ can be derived by applying the M-estimation framework. A logistic regression model is assumed for the propensity score function. Let $\mu = (\theta, \gamma^\top)^\top$ and define

$$M_h(O; \mu) = (M_{h1}(O; \mu), M_{h2}^\top(O; \mu))^\top,$$

where

$$\begin{aligned}M_{h1}(O; \mu) &= \frac{AY}{g(1 \mid X; \gamma)} - \frac{(1-A)Y}{g(0 \mid X; \gamma)} - [A - g(1 \mid X; \gamma)]h(X) - \theta, \\ M_{h2}(O; \mu) &= [A - g(1 \mid X; \gamma)]\widetilde{X}.\end{aligned}$$

Assuming that $g(a \mid x; \gamma)$ is correctly specified, asymptotic normality of $\widehat{\theta}(h)$ for any given h follows:

$$\sqrt{n}(\widehat{\theta}(h) - \theta_0) \xrightarrow{d} \mathcal{N}(0, \sigma_0^2(h)),$$

where a closed-form of the asymptotic variance can be derived following the sandwich formula.

By some tedious calculus,[22] we can find a closed-form for the optimal function $h^*(X)$ that minimizes $\sigma_0^2(h)$:

$$h^*(X) = \frac{\mathbb{E}(Y|X, A=1)}{g(1 \mid X; \gamma_0)} + \frac{\mathbb{E}(Y|X, A=0)}{g(0 \mid X; \gamma_0)}.$$

The resulting optimal "estimator", having the smallest asymptotic variance, is given by

$$\widehat{\theta}(h^*) = \frac{1}{n}\sum_{i=1}^{n}\left\{\frac{A_iY_i}{g(1 \mid X_i; \widehat{\gamma})} - \frac{(1-A_i)Y_i}{g(0 \mid X_i; \widehat{\gamma})} - [A_i - g(1 \mid X_i; \widehat{\gamma})]\right.$$
$$\left.\times\left(\frac{Q(X_i, 1)}{g(1 \mid X_i; \gamma_0)} + \frac{Q(X_i, 0)}{g(0 \mid X_i; \gamma_0)}\right)\right\}.$$

9.3.3 AIPW Estimator

The "estimator" introduced at the end of the previous subsection cannot be considered an estimator in the strict sense, as it depends on the unknown outcome regression function $Q(x, a)$. To address this, a parametric model, denoted by $Q(X, A; \beta)$, can be specified and the parameter β estimated via $\widehat{\beta}$. Based on this approach, *the augmented inverse probability of treatment weighted (AIPW) estimator* is constructed as follows:

$$\widehat{\theta}_{\text{AIPW}} = \frac{1}{n}\sum_{i=1}^{n}\left\{\frac{A_iY_i}{g(1 \mid X_i; \widehat{\gamma})} - \frac{(1-A_i)Y_i}{g(0 \mid X_i; \widehat{\gamma})} - [A_i - g(1 \mid X_i; \widehat{\gamma})]\right.$$
$$\left.\times\left(\frac{Q(X_i, 1; \widehat{\beta})}{g(1 \mid X_i; \widehat{\gamma})} + \frac{Q(X_i, 0; \widehat{\beta})}{g(0 \mid X_i; \widehat{\gamma})}\right)\right\}.$$

The construction of $\widehat{\theta}_{\text{AIPW}}$ is now placed within the framework of M-estimation. Let $\mu = (\theta, \beta^\top, \gamma^\top)^\top$ and define

$$M(O; \mu) = (M_1(O; \mu), M_2^\top(O; \mu), M_3^\top(O; \mu))^\top,$$

where

$$\begin{aligned}
M_1(O; \mu) &= \frac{AY}{g(1 \mid X; \gamma)} - \frac{(1-A)Y}{g(0 \mid X; \gamma)} \\
&\quad - [A - g(1 \mid X; \gamma)]\left[\frac{Q(X, 1; \beta)}{g(1 \mid X; \gamma)} + \frac{Q(X, 0; \beta)}{g(0 \mid X; \gamma)}\right] - \theta, \\
M_2(O; \mu) &= \frac{\partial Q(X, A; \beta)}{\partial \beta^\top}[Y - Q(X, A; \beta)], \\
M_3(O; \mu) &= \widetilde{X}[A - g(1 \mid X; \gamma)].
\end{aligned}$$

Under the assumption that the propensity score model g is correctly specified, the M-estimation theory implies that $\widehat{\gamma} \to \gamma_0$. Consequently, it can be verified that, by the law of iterative expectations (see Exercise 9.1),

$$\mathbb{E}[M_1(O;\theta_0,\beta,\gamma_0)] = 0,$$

for any β. In particular, $\mathbb{E}[M_1(O;\theta_0,\beta_*,\gamma_0)] = 0$, where β_* is the minimizer of the KL divergence between the working and true models of the Q function, that is, the one satisfying $\mathbb{E}[M_2(O;\theta_0,\beta_*,\gamma_0)] = 0$. Thus, the consistency of $\widehat{\theta}_{\text{AIPW}}$ is ensured when the model for g is correctly specified, regardless of whether the model for Q is correctly specified or not.

Similarly, if the outcome regression model Q is correctly specified, then the M-estimation theory implies $\widehat{\beta} \to \beta_0$. Consequently, it can be verified that, by the law of iterative expectations (see Exercise 9.2),

$$\mathbb{E}[M_1(O;\theta_0,\beta_0,\gamma)] = 0,$$

for any γ. In particular, $\mathbb{E}[M_1(O;\theta_0,\beta_0,\gamma_*)] = 0$, where γ_* is the minimizer of the KL divergence between the working and true models of the g function, that is, the one satisfying $\mathbb{E}[M_3(O;\theta_0,\beta_0,\gamma_*)] = 0$. Thus, the consistency of $\widehat{\theta}_{\text{AIPW}}$ is ensured when the model for Q is correctly specified, regardless of whether the model for g is correctly specified or not.

This property is referred to as *double robustness*,[43] as the consistency of $\widehat{\theta}_{\text{AIPW}}$ is ensured when either g or Q is correctly specified. This concept is illustrated by the parallel structure shown in Figure 9.1.

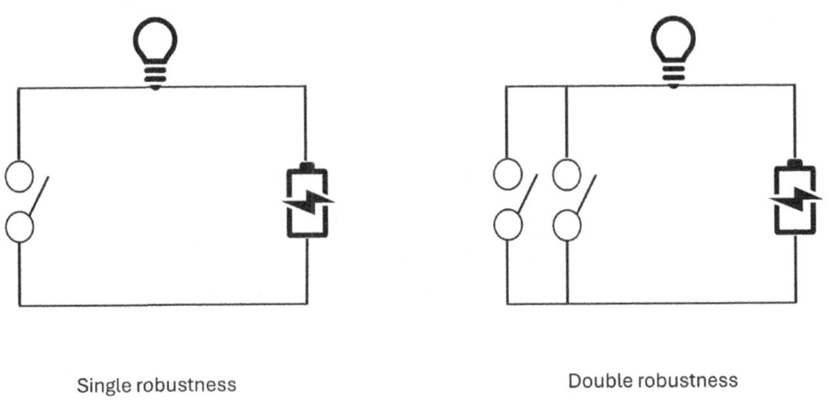

FIGURE 9.1
Single robustness and double robustness

Gold Coin # 34. *The augmented inverse probability of treatment weighted (AIPW) estimator is doubly robust, meaning that the asymptotic consistency is ensured if either of the two functions—the propensity score function or the outcome regression function—is correctly specified.*

When both Q and g are correctly specified, the asymptotic normality of $\widehat{\theta}_{\text{AIPW}}$ follows from the M-estimation theory (details omitted):

$$\sqrt{n}(\widehat{\theta}_{\text{AIPW}} - \theta_0) \xrightarrow{d} \mathcal{N}(0, \sigma^2_{0,\text{AIPW}}),$$

where the asymptotic variance is given by:

$$\sigma^2_{0,\text{AIPW}} = \mathbb{E}\left\{ \frac{AY}{g(1 \mid X; \gamma_0)} - \frac{(1-A)Y}{g(0 \mid X; \gamma_0)} - [A - g(1 \mid X; \gamma_0)] \times \left[\frac{Q(X,1;\beta_0)}{g(1 \mid X; \gamma_0)} + \frac{Q(X,0;\beta_0)}{g(0 \mid X; \gamma_0)} \right] - \theta_0 \right\}^2,$$

and can be estimated via:

$$\widehat{\sigma}^2_{\text{AIPW}} = \frac{1}{n}\sum_{i=1}^{n} \left\{ \frac{A_iY_i}{g(1 \mid X_i; \widehat{\gamma})} - \frac{(1-A_i)Y_i}{g(0 \mid X_i; \widehat{\gamma})} - [A_i - g(1 \mid X_i; \widehat{\gamma})] \times \left[\frac{Q(X_i,1;\widehat{\beta})}{g(1 \mid X_i; \widehat{\gamma})} + \frac{Q(X_i,0;\widehat{\beta})}{g(0 \mid X_i; \widehat{\gamma})} \right] - \widehat{\theta}_{\text{AIPW}} \right\}^2.$$

In summary, it has been established that $\widehat{\theta}_{\text{AIPW}}$ is an asymptotically linear estimator:

$$\sqrt{n}(\widehat{\theta}_{\text{AIPW}} - \theta_0) = \frac{1}{\sqrt{n}}\sum_{i=1}^{n} \phi_{\text{AIPW}}(O_i) + o_p(1),$$

where $\phi_{\text{AIPW}}(O)$ is the corresponding influence function, given by:

$$\phi_{\text{AIPW}}(O) = \frac{AY}{g(1 \mid X; \gamma_0)} - \frac{(1-A)Y}{g(0 \mid X; \gamma_0)} - [A - g(1 \mid X; \gamma_0)] \times \left[\frac{Q(X,1;\beta_0)}{g(1 \mid X; \gamma_0)} + \frac{Q(X,0;\beta_0)}{g(0 \mid X; \gamma_0)} \right] - \theta_0.$$

Remarks:

Since $\widehat{\theta}_{\text{AIPW}}$ achieves the semiparametric efficiency bound (in this chapter, roughly speaking, the minimum asymptotic variance) if both g and Q are correctly specified, the influence function $\phi_{\text{AIPW}}(O)$ is referred to as the *efficient influence function (EIF).*

Note that so far the EIF has been derived using only elementary calculus. A more rigorous and elegant treatment is available within the framework of semiparametric theory.[41,42] The essential role of the EIF in statistical inference will be further examined in Parts III and IV of this book.

In addition, this chapter focuses on non-interventional studies. For estimation purposes, interventional studies can be regarded as special cases of non-interventional studies in which the propensity score function is known or can be correctly specified with certainty. Consequently, in interventional studies, the consistency of the AIPW estimators is ensured because the propensity score function is known or can be correctly specified, regardless of whether the outcome regression function is correctly specified or not.

9.3.4 "One Plus One Equals Three"

The three estimators introduced above, along with their corresponding standard errors, 95% confidence intervals, and p-values, can be implemented using `PROC CAUSALTRT` in SAS.[44] Within this procedure, two model statements are provided: `model` for specifying the outcome model and `psmodel` for specifying the treatment (propensity score) model.

When only the `model` statement is specified, the g-computation estimator is produced. When only the `psmodel` statement is specified, the IPW estimator is obtained. When both statements are included, the AIPW estimator is computed. This property illustrates that "one plus one equals three." The structure of the corresponding SAS code is shown below.

```
/* g-computation */
proc causaltrt;
  model y = x1 x2;
  psmodel trt;
run;

/* IPW */
proc causaltrt;
  model y;
  psmodel trt = x1 x2;
run;

/* AIPW */
proc causaltrt;
  model y = x1 x2;
  psmodel trt = x1 x2;
run;
```

9.3.5 Tea Tasting Example Continued

The hypothetical non-interventional study described in Chapter 7 is revisited, with the data presented in Table 7.2. For simplicity, the variable `Baseline` is dichotomized as $X_2 = I(\texttt{Baseline} > 5)$. As a result, Table 7.2 is simplified and presented as Table 9.1.

TABLE 9.1
The tea tasting example: a non-interventional study (continued)

ID	X_1 (Sex)	X_2 (Baseline)	A (Treatment)	Y (Outcome)
1	1	1	1	8
2	0	0	1	6
3	1	0	0	7
4	0	0	1	7
5	0	0	1	6
6	1	1	0	10
7	1	0	0	7
8	0	1	0	8
9	1	0	0	6
10	1	0	1	6
11	0	1	1	8
12	1	0	0	5
13	0	0	1	7
14	1	0	0	9
15	1	0	0	8
16	0	1	1	9
17	1	0	1	7
18	1	0	1	6
19	1	1	0	8
20	0	1	1	10

The variables X_1 and X_2 are considered measured confounders. Recall that the true value of the estimand is $\theta^* = 0.45$, while the unadjusted estimate is $\widehat{\theta}_{\text{unadj}} = -0.28$.

G-Computation Estimate

A linear model of the form $Y \sim X_1 + X_2 + A + X_1 \times X_2 + A \times X_1 + A \times X_2$ is fitted, and the following estimates of $Q(x_1, x_2, a)$ are obtained (see Exercise 9.3):

$$\widehat{Q}(0,0,1) = 6.50, \quad \widehat{Q}(0,0,0) = 5.17,$$
$$\widehat{Q}(1,0,1) = 6.33, \quad \widehat{Q}(1,0,0) = 7.00,$$
$$\widehat{Q}(0,1,1) = 9.00, \quad \widehat{Q}(0,1,0) = 8.00,$$
$$\widehat{Q}(1,1,1) = 8.00, \quad \widehat{Q}(1,1,0) = 9.00.$$

For each subject, the two potential outcomes under treatments 1 and 0, respectively, can be predicted by $\widehat{Q}(X_{1i}, X_{2i}, 1)$ and $\widehat{Q}(X_{1i}, X_{2i}, 0)$, as shown in Table 9.2. The g-computation estimate can then be obtained (see Exercise 9.4):

$$\widehat{\theta}_{\text{g-comp}} = \frac{1}{20}\sum_{i=1}^{20} \widehat{Q}(X_{1i}, X_{2i}, 1) - \frac{1}{20}\sum_{i=1}^{20} \widehat{Q}(X_{1i}, X_{2i}, 0) = 7.15 - 7.13 = 0.02.$$

TABLE 9.2
Computing the g-computation estimate

ID	X_1	X_2	$\widehat{Q}(X_1, X_2, 1)$	$\widehat{Q}(X_1, X_2, 0)$
1	1	1	8.00	9.00
2	0	0	6.50	5.17
3	1	0	6.33	7.00
4	0	0	6.50	5.17
5	0	0	6.50	5.17
6	1	1	8.00	9.00
7	1	0	6.33	7.00
8	0	1	9.00	8.00
9	1	0	6.33	7.00
10	1	0	6.33	7.00
11	0	1	9.00	8.00
12	1	0	6.33	7.00
13	0	0	6.50	5.17
14	1	0	6.33	7.00
15	1	0	6.33	7.00
16	0	1	9.00	8.00
17	1	0	6.33	7.00
18	1	0	6.33	7.00
19	1	1	8.00	9.00
20	0	1	9.00	8.00
mean			7.15	7.13

IPW Estimate

A logistic regression model of the form $A \sim X_1 + X_2$ is fitted, and the following estimates of the propensity score $g(a|x_1, x_2)$ are obtained (see Exercise 9.5):

$$\begin{aligned}
&\widehat{g}(1 \mid 0,0) = 0.63, \quad \widehat{g}(0 \mid 0,0) = 0.37,\\
&\widehat{g}(1 \mid 1,0) = 0.44, \quad \widehat{g}(0 \mid 1,0) = 0.56,\\
&\widehat{g}(1 \mid 0,1) = 0.34, \quad \widehat{g}(0 \mid 0,1) = 0.66,\\
&\widehat{g}(1 \mid 1,1) = 0.19, \quad \widehat{g}(0 \mid 1,1) = 0.81.
\end{aligned}$$

Based on these, the IPW estimate can be obtained (see Exercise 9.6):

$$\begin{aligned}
\widehat{\theta}_{\text{IPW}} &= \frac{1}{20}\sum_{i=1}^{20}\left[\frac{A_i Y_i}{\widehat{g}(1 \mid X_{1i}, X_{2i})} - \frac{(1-A_i)Y_i}{\widehat{g}(0 \mid X_{1i}, X_{2i})}\right]\\
&= \frac{1}{20}\sum_{i=1}^{20}\frac{(2A_i - 1)Y_i}{\widehat{g}(A_i|X_{1i}, X_{2i})} = 0.46.
\end{aligned}$$

TABLE 9.3
Computing the IPW estimate

ID	X_1	X_2	A	$2(A-1)$	Y	$\widehat{g}(A \mid X_1, X_2)$
1	1	1	1	1	8	0.19
2	0	0	1	1	6	0.63
3	1	0	0	−1	7	0.56
4	0	0	1	1	7	0.63
5	0	0	1	1	6	0.63
6	1	1	0	−1	10	0.81
7	1	0	0	−1	7	0.56
8	0	1	0	−1	8	0.66
9	1	0	0	−1	6	0.56
10	1	0	1	1	6	0.44
11	0	1	1	1	8	0.34
12	1	0	0	−1	5	0.56
13	0	0	1	1	7	0.63
14	1	0	0	−1	9	0.56
15	1	0	0	−1	8	0.56
16	0	1	1	1	9	0.34
17	1	0	1	1	7	0.44
18	1	0	1	1	6	0.44
19	1	1	0	−1	8	0.37
20	0	1	1	1	10	0.34

AIPW Estimate

By combining the data from Tables 9.2 and 9.3, the AIPW estimate is computed (see Exercise 9.7) as $\widehat{\theta}_{\text{AIPW}} = 0.02$.

Interestingly, $\widehat{\theta}_{\text{AIPW}}$ is equal to the g-computation estimate. This coincidence arises due to the small sample size ($n = 20$) and the sparsity of the data. All three adjusted estimates (0.02, 0.46, and 0.02) outperform the unadjusted estimate (−0.28), given that the true value is known to be $\theta^* = 0.45$ in this example. Furthermore, since the propensity score model is correctly specified while the outcome regression model is misspecified, the IPW estimate (0.46) outperforms the other two.

To assess the impact of sample size, the analysis is repeated for $n = 20$, 100, and 1000. The results are summarized in Table 9.4 (see Exercise 9.8).

TABLE 9.4
Results for different sample sizes

n	θ^*	$\widehat{\theta}_{\text{unadj}}$	$\widehat{\theta}_{\text{g-comp}}$	$\widehat{\theta}_{\text{IPW}}$	$\widehat{\theta}_{\text{AIPW}}$
20	0.45	−0.28	0.02	0.46	0.02
100	0.29	0.03	0.33	0.22	0.36
1000	0.275	0.002	0.311	0.295	0.305

9.4 Exercises

Exercise 9.1

Let $M_1(O;\theta,\beta,\gamma)$ be the following estimating equation:

$$\frac{AY}{g(1 \mid X;\gamma)} - \frac{(1-A)Y}{g(0 \mid X;\gamma)} - [A - g(1 \mid X;\gamma)]\left[\frac{Q(X,1;\beta)}{g(1 \mid X;\gamma)} + \frac{Q(X,0;\beta)}{g(0 \mid X;\gamma)}\right] - \theta.$$

Show that $\mathbb{E}[M_1(O;\theta_0,\beta,\gamma_0)] = 0$ for any β.

Exercise 9.2

Let $M_1(O;\theta,\beta,\gamma)$ be the following estimating equation:

$$\frac{AY}{g(1 \mid X;\gamma)} - \frac{(1-A)Y}{g(0 \mid X;\gamma)} - [A - g(1 \mid X;\gamma)]\left[\frac{Q(X,1;\beta)}{g(1 \mid X;\gamma)} + \frac{Q(X,0;\beta)}{g(0 \mid X;\gamma)}\right] - \theta.$$

Show that $\mathbb{E}[M_1(O;\theta_0,\beta_0,\gamma)] = 0$ for any γ.

Exercise 9.3

Using the data in Table 9.1, verify the estimates of $Q(x_1,x_2,a)$, where $x_1 = 0,1$, $x_2 = 0,1$, and $a = 0,1$.

Exercise 9.4

Using the data in Table 9.2, verify the g-computation estimate.

Exercise 9.5

Using the data in Table 9.1, verify the estimates of $g(a \mid x_1,x_2)$, where $x_1 = 0,1$, $x_2 = 0,1$, and $a = 0,1$.

Exercise 9.6

Using the data in Table 9.3, verify the IPW estimate.

Exercise 9.7

Using the data in Tables 9.2 and 9.3, verify the AIPW estimate.

Exercise 9.8

Verify the results in Table 9.4.

10

Sensitivity Analysis

10.1 Causal and Statistical Assumptions

A summary of the developments that have been achieved so far in Part II is provided now. In Chapter 6, the concept of potential outcomes was introduced. In Chapter 7, two primary classes of study designs—randomized controlled trials (RCTs) and non-interventional studies (NISs)—were described. In Chapter 8, the formulation of a causal estimand to reflect the research question was outlined, along with the translation of the causal estimand into a statistical estimand, under a specified set of identifiability assumptions. In Chapter 9, methods for estimating the statistical estimand were presented, assuming certain parametric models.

In Figure 10.1, the underlying assumptions are organized into two stages: causal assumptions, which include identifiability assumptions such as exchangeability and positivity, and statistical assumptions, which encompass model specifications such as the use of generalized linear models for the outcome regression function and the propensity score function.

Our ultimate estimation target is the causal estimand. To accomplish this, the statistical estimand must first be estimated, given that the equality between the statistical and causal estimands is ensured by the identifiability assumptions. Consequently, the validity of the estimator in estimating the causal estimand relies upon two conditions: (1) the validity of the causal assumptions employed during the identification stage, and (2) the validity of the statistical assumptions imposed during the estimation stage.

Gold Coin # 35. *A distinction must be made between causal assumptions assumed during identification and statistical assumptions assumed during estimation.*

Since assumptions do not constitute truths, the impact of their violations must be carefully evaluated. This underscores the importance of sensitivity analyses. The ICH E9(R1) guidance, *Addendum on Estimands and Sensitivity*

DOI: 10.1201/9781003635468-12

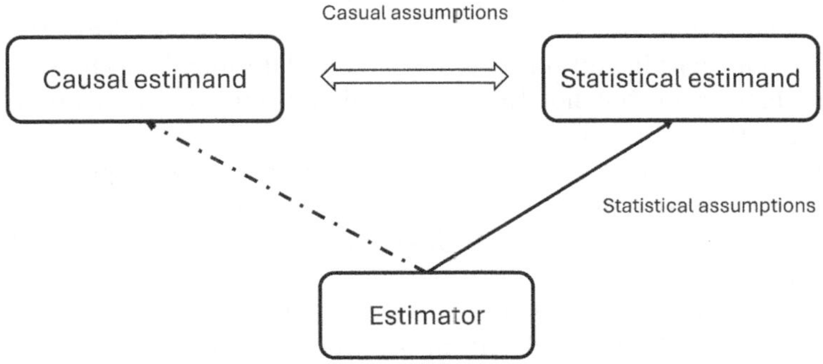

FIGURE 10.1
Underlying assumptions

Analysis in Clinical Trials, addresses two key topics: estimands and sensitivity analysis. Sensitivity analysis is defined therein as:

> "A series of analyses conducted with the intent to explore the robustness of inferences from the main estimator to deviations from its underlying modelling assumptions and limitations in the data."

Within this definition, two key components are emphasized: (1) the underlying modeling assumptions, and (2) the limitations in the data. These components often intersect, as data limitations frequently necessitate specific assumptions. For example, the existence of an unmeasured confounder—a limitation in the data—corresponds to a violation of the exchangeability assumption (a.k.a. the no unmeasured confounding assumption). Similarly, a detected violation of the positivity assumption may reflect data sparsity, such as the absence of observations for one treatment arm within certain covariate strata. In accordance with ICH E9(R1), sensitivity analyses should therefore be conducted for each relevant assumption or data limitation.

10.2 Two Causal Assumptions

In the identification process, two key assumptions are made: the exchangeability assumption and the positivity assumption. These two assumptions are often in tension—satisfying one may lead to the violation of the other.

This tension can be explained in more detail. Specifically, consider an alternative term for the exchangeability assumption: the *no unmeasured confounding* (NUC) assumption. To satisfy the NUC assumption, it is typically recommended that as many measured confounders as possible be included. However, when numerous suspected confounders are adjusted for, the chance increases that, within certain covariate strata, data may be available for only one treatment arm, but not the other. This phenomenon is referred to as a violation of the positivity assumption.

Conversely, to address violations of the positivity assumption, covariate levels that cause sparsity may be collapsed, or variables contributing to the violation may be removed. Yet, such actions increase the risk that the NUC assumption is violated.

Thus, variable selection is recognized as a critical component of causal inference and should be considered during the study design phase. While the topic of variable selection may conventionally belong in Chapter 7, it is introduced here, after the interplay between the NUC assumption and the positivity assumption has been established.

10.2.1 Variable Selection

The Odds in a Coin Flip Are Not Quite 50/50

In a well-known study,[45] Diaconis et al. (2007) demonstrated that "vigorously flipped coins tend to come up the same way they started." The limiting probability depends on a single parameter: the angle between the coin's normal vector and its angular momentum. Using high-speed photography, they found that for natural coin flips, the probability of landing in the same orientation as it started is approximately 0.51.

Variable Selection in an RCT

Figure 10.2 presents a directed acyclic graph (DAG) for a typical RCT. Here, X_Z represents covariates associated only with the treatment variable Z (e.g., variables influencing the outcome of a coin flip discovered by Diaconis), while X_Y includes covariates associated with the outcome variable Y.

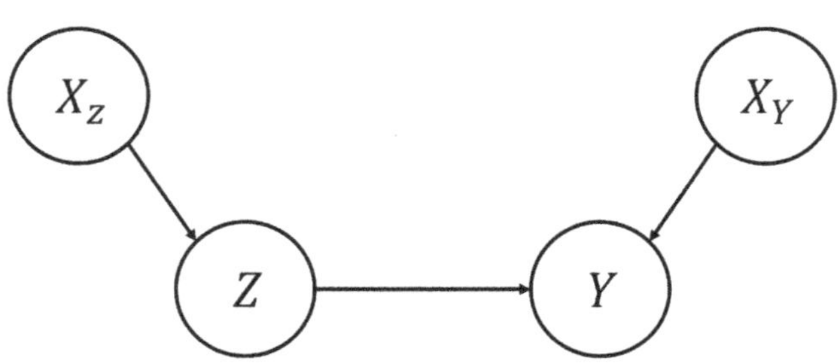

FIGURE 10.2
Covariate adjustment in an RCT

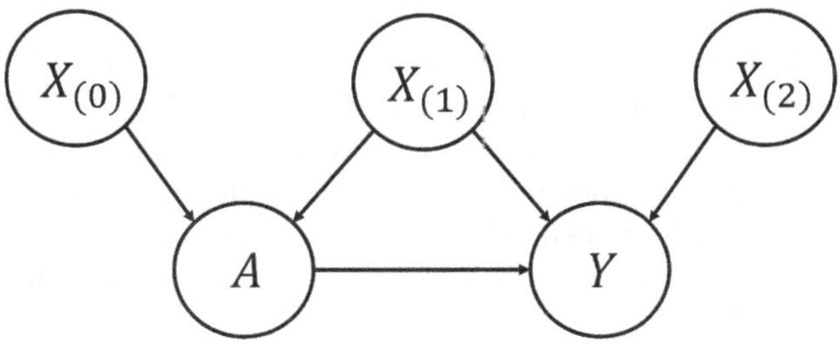

FIGURE 10.3
Variable selection in an NIS

According to FDA guidance titled *Adjusting for Covariates in Randomized Clinical Trials for Drugs and Biological Products*, the "main focus of the guidance is on the use of prognostic baseline covariates to improve statistical efficiency for estimating and testing treatment effects." Accordingly, variables in X_Y should be included in the treatment effect estimation model.

By contrast, variables in X_Z are not needed in estimating the treatment effect. For example, explicitly accounting for the variable that Diaconis identified as influencing coin-flip outcomes is not needed apparently. Their inclusion can degrade statistical efficiency and increase the risk of positivity violations.

Variable Selection in an NIS

Figure 10.3 depicts a DAG for a typical cohort study. In this setting:

- $X_{(0)}$ denotes variables associated only with A,
- $X_{(1)}$ denotes measured confounders associated with both A and Y,
- $X_{(2)}$ denotes variables associated only with Y.

In the estimation process, both $X_{(1)}$ and $X_{(2)}$ should be included—$X_{(1)}$ to control for confounding and $X_{(2)}$ to improve statistical efficiency. However, as in RCTs, variables in $X_{(0)}$ are not required.[46] Including such variables may reduce efficiency and exacerbate positivity violations.

Let $X = X_{(1)} \cup X_{(2)}$ represent the set of all covariates that influence the outcome Y. In summary, only variables that influence the outcome should be included in the treatment effect estimation process.[46]

10.2.2 Sensitivity Analysis for Unmeasured Confounding

The literature on sensitivity analysis for unmeasured confounding is extensive.[47,48,49,50] In this subsection, only two methods are briefly discussed: simulation-based sensitivity analysis and the E-value.[51]

The Tea Tasting Example Continued

The hypothetical tea-tasting example in Chapter 9 is revisited here. In that example, two confounders are measured: X_1 (i.e., sex) and X_2 (i.e., baseline taste rating). A plausible unmeasured confounder is U (e.g., regular coffee drinker), which may be associated with both how milk tea is prepared and how its taste is evaluated. More generally, the absence of randomization introduces the possibility of unmeasured confounding, though the specific confounders may not be identifiable.

To perform sensitivity analysis, the presence of a single unmeasured confounder U is assumed, as illustrated in Figure 10.4.

To explore the effect of the unmeasured confounder U on estimation, a binary U—with $U = 1$ indicating regular coffee drinker—can be simulated using a parametric model, such as logistic regression:

$$\text{logit}[\mathbb{P}(U = 1|A, Y)] = \beta_{u0} + \beta_{ua}A + \beta_{uy}Y,$$

where β_{ua} represents the assumed log odds ratio between U and A, and β_{uy} represents the assumed log odds ratio between U and Y. A range of values can be specified for these two parameters; e.g., $\beta_{ua} = 0, \pm 1, \pm 2, \dots$ and $\beta_{uy} = 0, \pm 0.5, \pm 1, \dots$.

For each parameter combination (β_{ua}, β_{uy}), a vector of U is simulated and added to the original dataset. Estimation is then repeated using the augmented dataset $\{(X_{1i}, X_{2i}, U_i, A_i, Y_i), i = 1, \dots, n\}$. The robustness of the main results is then examined across different values of (β_{ua}, β_{uy}). Of

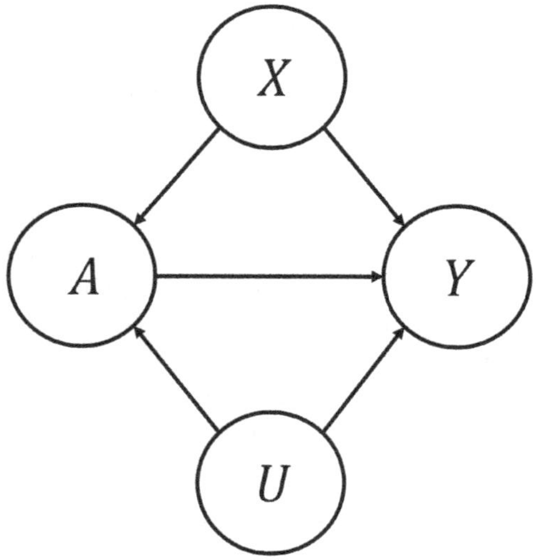

FIGURE 10.4
Measured and unmeasured confounders

particular interest is the parameter combination at which the result changes from statistically significant to non-significant.

E-Value

The E-value offers a model-free alternative to assess the influence of unmeasured confounding, without relying on simulation. VanderWeele and Ding[51] defined the E-value as follows:

> "The E-value is the minimum strength of association, on the risk ratio scale, that an unmeasured confounder would need to have with both the treatment and outcome, conditional on the measured covariates, to fully explain away a specific treatment–outcome association."

Let $RR_{AY|X}$ denote the estimated risk ratio of treatment A on outcome Y, conditional on measured confounders X, and let (LL, UL) be its 95% confidence interval. Without loss of generality, assume $RR_{AY|X} > 1$ and $LL > 1$, indicating that the result is statistically significant.

To evaluate the impact of unmeasured confounding, two parameters are conceptually defined: RR_{UY}, the maximum effect that U can have on Y conditional on $X = x$, and RR_{AU}, the maximum risk ratio relating A to any particular level of U conditional on $X = x$. These two values are used to compute the bias factor:[52]

$$B = \frac{RR_{UY} \times RR_{AU}}{RR_{UY} + RR_{AU} - 1}.$$

The estimated risk ratio can then be adjusted by dividing $RR_{AY|X}$ and its lower limit LL by the bias factor B.

The E-value is given by the following expression:

$$\text{E-value} = RR + \sqrt{RR \times (RR - 1)},$$

where RR may represent either $RR_{AY|X}$ or the lower bound LL.

For instance, if $RR_{AY|X} = 1.50$ and $LL = 1.33$, then the E-value for the point estimate is 2.37, and the E-value for the lower limit is 2.00. This implies that an unmeasured confounder would need to have associations of at least 2.00 with both treatment and outcome to explain away the observed effect.

Although defined on the risk ratio scale, the E-value can be applied to other effect measures—including odds ratios, hazard ratios, risk differences, and mean differences—using the R package `EValue`.[53]

Interpreting the E-Value

While the E-value is straightforward to compute, its interpretation can be challenging—another example of "it is easier done than said." Consider the following toy example. Let the causal estimand of interest be:

$$\theta^* = \frac{\mathbb{P}(Y^{a=1} = 1)}{\mathbb{P}(Y^{a=0} = 1)}.$$

Assume the unadjusted estimate is:

$$\widehat{\theta}_{\text{un}} = \frac{\sum_{i=1}^{n} A_i Y_i / \sum_{i=1}^{n} A_i}{\sum_{i=1}^{n} (1 - A_i) Y_i / \sum_{i=1}^{n} (1 - A_i)} = 1.87,$$

with 95% confidence interval $(1.64, 2.13)$.

Suppose the adjusted estimate obtained using AIPW is $\widehat{\theta} = 1.68$ with confidence interval $(1.55, 1.82)$. The corresponding E-values are 2.75 for the point estimate and 2.47 for the lower limit (see Exercise 10.1).

These values indicate the strength of association that an unmeasured confounder would need to have with both treatment and outcome to nullify the observed effect. RIn general, larger E-values indicate greater robustness of the finding to the unmeasured confounding. However, the question remains: "How large is large enough?"

Several approaches to interpreting the E-value have been proposed:[50]

Attempt 1: Fixed Threshold Based on Literature Review.

A threshold $\Delta = 2$ may be specified in the protocol, supported by literature suggesting that no known confounder has a risk ratio greater than 2. If the computed E-value exceeds this threshold (as in the toy example), robustness may be claimed.

Attempt 2: Fixed Threshold Based on Preliminary Data.

The bias from measured confounders can be estimated using the ratio:

$$B = \max \left\{ \frac{RR_{\text{un}}}{RR_{\text{adj}}}, \frac{RR_{\text{adj}}}{RR_{\text{un}}} \right\},$$

and the threshold computed as:

$$\Delta = B + \sqrt{B(B-1)}.$$

For example, if $B = 1.25$, then $\Delta = 1.81$ (see Exercise 10.2). An E-value above this threshold indicates robustness.

Attempt 3: Threshold Based on Current Study.

The bias factor B is calculated using the unadjusted and adjusted estimates from the current study. If $RR_{\text{un}} = 1.87$ and $RR_{\text{adj}} = 1.68$, then $B = 1.11$ and $\Delta = 1.46$ (see Exercise 10.3). An E-value exceeding this threshold supports robustness.

Attempt 4: Threshold Based on External Data.

The adjusted estimate from the current study is compared with that obtained by incorporating external data. If the estimate changes from 1.68 to 1.43, then $B = 1.17$ and $\Delta = 1.62$ (see Exercise 10.4). Again, an E-value exceeding this threshold supports the robustness of the findings.

10.2.3 Sensitivity Analysis for Positivity Violation

The positivity assumption requires sufficient variability in the treatment assignment across all levels of the confounders. Violations of this assumption can occur for two primary reasons.[54] First, a violation may arise at the population level when individuals with certain covariate patterns cannot, by design or reality, receive a particular treatment. In such cases, the causal estimand must be redefined by restricting the target population to a subpopulation in which positivity holds. Second, violations (or near-violations) may occur in finite samples due to random variation—particularly when sample sizes are small or the covariate space is high-dimensional. In this subsection, sensitivity to the second type of positivity violation is examined.

The Tea Tasting Example Continued

In the hypothetical tea-tasting example from Chapter 9, two confounders, X_1 and X_2, were measured. Table 10.1 presents the joint distribution of X_1, X_2, and A. It can be seen that, within the covariate level $(X_1, X_2) = (0, 0)$, all participants received treatment ($A = 1$), but no individuals received the control ($A = 0$). As a result, the outcome regression $\widehat{Q}(0, 0, 0) = \widehat{\mathbb{E}}(Y \mid x_1 = 0, x_2 = 0, A = 0)$ cannot be estimated from the data without extrapolation. Similarly, the propensity score $\widehat{g}(a \mid 0, 0)$ for $a = 0, 1$ cannot be reliably estimated within this stratum.

To perform a sensitivity analysis, four hypothetical participants with $(X_1, X_2, A) = (0, 0, 0)$ may be added to the dataset (see Exercise 10.5). Their outcomes can be constructed by subtracting a specified amount δ from the outcomes of the four existing participants with $(X_1, X_2, A) = (0, 0, 1)$. Based on Table 9.1, those outcomes are $(6, 7, 6, 7)$. Thus, the added outcomes would be $(6 - \delta, 7 - \delta, 6 - \delta, 7 - \delta)$.

With the augmented dataset of $24 = 20+4$ participants, adjusted estimates can be obtained using any of the three estimation methods. By varying δ (e.g.,

TABLE 10.1
Examining the positivity violation

X_1 (Sex)	X_2 (Baseline)	A (Treatment)	Frequency
0	0	1	4
0	0	0	0*
1	0	1	3
1	0	0	6
0	1	1	3
0	1	0	1
1	1	1	1
1	1	0	2

* This row indicates a violation of the positivity assumption

$\delta = \pm 0.5, \pm 1, \dots$), the robustness of the primary findings can be explored by identifying values of δ for which statistical significance is lost.

A Tipping-Point Method

The above naive strategy can be generalized to a more formal tipping-point method. Let $\mathcal{S}$ denote the set of individuals whose estimated propensity scores fall at the extremes, defined as:

$$\mathcal{S} = \{i : \widehat{g}(1 \mid X_{1i}, X_{2i}) > 0.95 \quad \text{or} \quad \widehat{g}(1 \mid X_{1i}, X_{2i}) < 0.05\},$$

where 0.95 and 0.05 (or any other appropriate values) are pre-specified thresholds used to classify suspected positivity violations.

Assume the g-computation estimator is used:

$$\widehat{\theta}_{\text{g-comp}} = \frac{1}{n}\sum_{i=1}^{n}\left[\widehat{Q}(X_{1i}, X_{2i}, 1) - \widehat{Q}(X_{1i}, X_{2i}, 0)\right].$$

For each individual in $\mathcal{S}$, one of the two components in the summation above, $\widehat{Q}(X_{1i}, X_{2i}, 1)$ or $\widehat{Q}(X_{1i}, X_{2i}, 0)$, is likely based on extrapolation due to positivity violation. Consequently, such extrapolation may introduce bias.

To evaluate the impact of this extrapolation, a disturbance term δ can be subtracted from the treatment contrast for each subject in $\mathcal{S}$. If higher values of Y indicate a more favorable treatment effect, then a positive value of $\widehat{\theta}_{\text{g-comp}}$ supports the treatment's effectiveness. Subtracting δ from each of the extrapolated contrasts creates a perturbed estimator:

$$\widehat{\theta}(\delta) = \widehat{\theta}_{\text{g-comp}} - \frac{m\delta}{n},$$

where $m = |\mathcal{S}|$ is the number of individuals in $\mathcal{S}$.

A tipping point δ_0 can be solved such that $\widehat{\theta}(\delta_0) = 0$. This value indicates the magnitude of bias required to nullify the estimated treatment effect. The larger the value of δ_0, the more robust the result is to potential positivity violations.

Gold Coin # 36. *Sensitivity analyses should be conducted to assess the robustness of inferences from the primary estimator to potential violations of the two causal assumptions in tension: exchangeability and positivity.*

10.3 On Statistical Assumptions

Causal assumptions are imposed during the identification stage, while statistical assumptions are imposed during the estimation stage, as illustrated in Figure 10.1. In the preceding section, sensitivity analysis methods for two causal assumptions were discussed.

A natural question then arises: Should the robustness of the main results with respect to violations of statistical assumptions (e.g., model misspecification) also be assessed? The answer is affirmative. However, rather than presenting specific sensitivity analysis methods for such violations, an alternative and more effective approach will be pursued.

As Albert Einstein famously stated, "Everything should be made as simple as possible, but not simpler." With this in mind, all underlying assumptions should be critically examined. Specifically, the necessity of each assumption should be questioned. If certain statistical assumptions cannot be avoided, then corresponding sensitivity analyses ought to be conducted. Conversely, when assumptions can be relaxed or removed, doing so is preferable, as it eliminates the need for additional sensitivity analysis.

In each of the three estimation methods discussed in Chapter 9, parametric models were assumed either for the outcome regression function or for the propensity score function. It should be emphasized that each parametric model specification constitutes a statistical assumption. Therefore, sensitivity to each modeling assumption should, in principle, be evaluated.

Replacing parametric models with nonparametric alternatives allows the estimation framework to be situated within the broader and more flexible domain of **semiparametric statistics**—a subject that will be explored in depth in the second half of this book.

Gold Coin # 37. *Semiparametric statistics focuses on studying the properties of estimators for the target parameter (i.e., the parametric component), while allowing for flexible, nonparametric modeling of functions such as the outcome regression function and the propensity score function (i.e., the nonparametric components).*

10.4 Exercises

Exercise 10.1

Find the E-value for the point estimate 1.68 and the lower level of its CI $= (1.55, 1.82)$.

Exercise 10.2

Calculate the threshold Δ corresponding to the bounding factor $B = 1.25$.

Exercise 10.3

Based on the unadjusted estimate 1.87 and the adjusted estimate 1.68, calculate the bounding factor B and its corresponding threshold Δ.

Exercise 10.4

Based on the adjusted estimate 1.68 and the new adjusted estimate 1.43, calculate the bounding factor B and its corresponding threshold Δ.

Exercise 10.5

Create an augmented dataset for sensitivity analysis of positivity violation using the data in Table 9.1 and four hypothetical participants with $(X_1, X_2, A) = (0, 0, 0)$.

Part III

An Introduction to Semiparametric Statistics

11

Regular and Asymptotically Linear Estimator

We now move into the second half of this book, embarking on a new and exciting journey: the study of advanced statistics, with a focus on *semiparametrics* and *targeted learning*. Introducing semiparametrics is not a simple task, demanding both intuition and solid mathematics. This chapter offers a concise introduction to semiparametrics within the framework of causal inference.

For readers seeking a comprehensive and rigorous learning of semiparametric theory in its full generality, a highly recommended reference is Bickel, Klaassen, Ritov, and Wellner (1993), *Efficient and Adaptive Estimation for Semiparametric Models*. For those interested in a deeper exploration of semiparametrics specifically within the fields of causal inference and missing data, Tsiatis (2006), *Semiparametric Theory and Missing Data*, provides a practical but insightful foundation.

But where should we begin if our goal is to capture the essence of semiparametrics in just a single chapter? Perhaps the best starting point is a short yet profound paper by Charles Stein (1920–2016), with many important findings named after him, including *Stein's paradox*, *Stein's lemma*, and *Stein's method*. In his 1956 paper presented at the Third Berkeley Symposium on Mathematical Statistics and Probability,[55] Stein states:

> "When Θ is infinite-dimensional, that is, in nonparametric problems, the maximum likelihood method often breaks down. Frequently the maximum likelihood estimate is undefined, and it is not clear that it is good when it exists. However, the existence of a one-dimensional subproblem asymptotically as difficult as the original problem (of estimating a single real-valued function) often persists, at least formally."

This paragraph captures the main idea of what would later evolve into the concept of *least favorable parametric submodels*. It also reflects Stein's remarkable humility. In the same paper, he refers to one of his own results as "somewhat trivial" in the one of his statements: "The somewhat trivial lemma obtained in this section is often useful in proving that an estimation problem is not made more difficult asymptotically by introducing additional unknown parameters in a certain way."

DOI: 10.1201/9781003635468-14

To me, this so-called "trivial lemma" is anything but trivial—it is truly phenomenal. Let us take inspiration from Stein and explore how the idea of *least favorable parametric submodels* plays a central and influential role in semiparametric theory and targeted learning. Indeed, it will guide us throughout the remainder of this book.

11.1 Regularity

11.1.1 Revisiting MLE

Assume $X_1, \ldots, X_n$ are independent and identically distributed (i.i.d.) with $X \sim p(x;\theta)$. The maximum likelihood estimator (MLE) of the true value of θ, θ_0, as discussed in Chapter 4, can be expressed as

$$\widehat{\theta}_n = \theta_0 + \frac{1}{n}\sum_{i=1}^{n} \mathcal{I}_{\theta_0}^{-1} s_{\theta_0}(X_i) + o_p(n^{-1/2}),$$

where the following definitions are involved:

$$\begin{aligned} \ell_\theta(X) &= \log p(X;\theta), \\ s_\theta(X) &= \frac{\partial \ell_\theta(X)}{\partial \theta}, \\ \mathcal{I}_\theta &= \mathrm{Cov}_\theta\left[s_\theta(X)\right], \end{aligned}$$

along with the identity

$$\mathrm{Cov}_\theta\left[s_\theta(X)\right] = \mathbb{E}_\theta\left[-\frac{\partial^2 \ell_\theta(X)}{\partial\theta\partial\theta^\top}\right].$$

In asymptotic statistics, an estimator of the form

$$\widetilde{\theta}_n - \theta_0 = \frac{1}{n}\sum_{i=1}^{n}\phi(X_i) + o_p(n^{-1/2}),$$

is referred to as an *asymptotically linear estimator*, where $\mathbb{E}_\theta[\phi(X_i)] = 0$ and $\mathrm{Cov}_\theta[\phi(X_i)] < \infty$. The function $\phi(X_i)$ is known as the *influence function (IF)*, which quantifies the impact of each observation X_i on the estimator.

By multiplying both sides by $\sqrt{n}$, it follows that

$$\sqrt{n}(\widetilde{\theta}_n - \theta_0) = \frac{1}{\sqrt{n}}\sum_{i=1}^{n}\phi(X_i) + o_p(1),$$

and hence, consistency and asymptotic normality are attained as $n \to \infty$:

$$\widetilde{\theta}_n \xrightarrow{p} \theta_0,$$
$$\sqrt{n}(\widetilde{\theta}_n - \theta_0) \xrightarrow{d} \mathcal{N}\left(0, \mathrm{Cov}_{\theta_0}(\phi(X))\right),$$

where $\mathrm{Cov}_{\theta_0}(\phi(X))$ is the asymptotic covariance.

In particular, due to its asymptotic linearity, the MLE $\widehat{\theta}_n$ satisfies

$$\widehat{\theta}_n \xrightarrow{p} \theta_0,$$
$$\sqrt{n}(\widehat{\theta}_n - \theta_0) \xrightarrow{d} \mathcal{N}\left(0, \mathcal{I}_{\theta_0}^{-1}\right),$$

implying that it is both consistent and asymptotically normal.

Moreover, the asymptotic covariance of the MLE $\widehat{\theta}_n$ achieves the Cramér–Rao lower bound (CRLB), $\mathcal{I}_{\theta_0}^{-1}$. The CRLB is defined over the class of all unbiased estimators:

$$\mathcal{C}_{\text{UB}} = \left\{\widetilde{\theta}_n : \mathbb{E}_\theta(\widetilde{\theta}_n) = \theta\right\},$$

and represents a lower bound on their covariances:

$$\inf_{\widetilde{\theta}_n \in \mathcal{C}_{\text{UB}}} \mathrm{Cov}_\theta(\widetilde{\theta}_n) \geq \mathcal{I}_\theta^{-1}.$$

It can therefore be concluded that, if the MLE $\widehat{\theta}_n$ is unbiased, it is efficient, since its asymptotic covariance attains the CRLB.

However, the MLE $\widehat{\theta}_n$ is generally not unbiased. Given that $\widehat{\theta}_n$ is asymptotically linear, it is natural to desire that the CRLB also serves as the lower bound over the class of all asymptotically linear estimators:

$$\mathcal{C}_{\text{AL}} = \left\{\widetilde{\theta}_n : \widetilde{\theta}_n \text{ is asymptotically linear}\right\},$$

and to establish that

$$\inf_{\widetilde{\theta}_n \in \mathcal{C}_{\text{AL}}} \mathrm{Cov}_\theta(\widetilde{\theta}_n) \geq \mathcal{I}_\theta^{-1}.$$

If such a result were to hold, it could be concluded that the MLE $\widehat{\theta}_n$, being asymptotically linear and achieving the CRLB, is asymptotically efficient. Unfortunately, due to the phenomenon of *super-efficiency*, such a conclusion cannot be universally sustained.

11.1.2 Super-Efficiency

To understand the issue of super-efficiency, consider the following hypothetical asymptotically linear estimator:

$$\widetilde{\theta}_n - \theta_0 = \frac{1}{n}\sum_{i=1}^{n} 0 + o_p(n^{-1/2}),$$

in which the influence function is given by $\phi(X_i) = 0$. If such an asymptotically linear estimator can be constructed, its asymptotic variance becomes zero. This provides a counterexample, demonstrating that the CRLB does not serve as a universal lower bound for the asymptotic variances of asymptotically linear estimators in the class $\mathcal{C}_{\text{AL}}$.

A concrete example of such a phenomenon is given by Hodges' estimator, named for Joseph Hodges (1922–2000). Assume that $X_1, \ldots, X_n$ are i.i.d. random variables with $X \sim \mathcal{N}(\theta, 1)$, and let $\overline{X}_n$ denote the sample mean. Hodges' estimator is then defined as:

$$\widetilde{\theta}_n = \begin{cases} \overline{X}_n & \text{if } \overline{X}_n \geq n^{-1/4}, \\ 0 & \text{if } \overline{X}_n < n^{-1/4}. \end{cases}$$

Since $\overline{X}_n \sim \mathcal{N}(\theta_0, n^{-1})$, the probability that $\overline{X}_n$ falls within the interval $(\theta_0 - Cn^{-1/2}, \theta_0 + Cn^{-1/2})$ can be made arbitrarily close to one for sufficiently large values of C and n. When $\theta_0 \neq 0$, the intervals $(\theta_0 - Cn^{-1/2}, \theta_0 + Cn^{-1/2})$ and $(-n^{-1/4}, n^{-1/4})$ become disjoint for large n. Therefore, in this case, the following holds:

$$\widetilde{\theta}_n - \theta_0 = \frac{1}{n} \sum_{i=1}^{n} (X_i - \theta_0) + o_p(n^{-1/2}), \quad \text{if } \theta_0 \neq 0.$$

In contrast, when $\theta_0 = 0$, the interval $(-Cn^{-1/2}, Cn^{-1/2})$ is entirely contained within $(-n^{-1/4}, n^{-1/4})$ for large n, and hence:

$$\widetilde{\theta}_n - \theta_0 = \frac{1}{n} \sum_{i=1}^{n} 0 + o_p(n^{-1/2}), \quad \text{if } \theta_0 = 0.$$

At first glance, super-efficiency may appear to be a desirable property. However, such a "desirable" gain at $\theta_0 = 0$ comes at a cost: performance in neighborhoods around $\theta_0 = 0$ is degraded. This trade-off reflects the principle that "there is no free lunch" in estimation theory.

To visualize this behavior, a simulation is conducted to examine the mean squared error (MSE) of Hodges' estimator when $n = 50$ (see Exercise 11.1), motivated by a simulation in var der Vaart (2000).[12] As shown in Figure 11.1, the MSE is near zero when θ_0 is close to zero, remains below 0.02 near the origin, then rises sharply as θ_0 deviates slightly from zero, before eventually stabilizing at approximately 0.02—the MSE of the sample mean $\overline{X}_n$.

11.1.3 Regularity

To exclude the possibility of super-efficiency at θ_0, a conceptual framework known as the *local data-generating process* (LDGP) is considered. It is assumed that $X_1, \ldots, X_n$ are generated from a distribution $p(x; \theta_n)$, where $\theta_n = \theta_0 + c/\sqrt{n}$, for a given constant c. Estimators whose asymptotic behaviors are

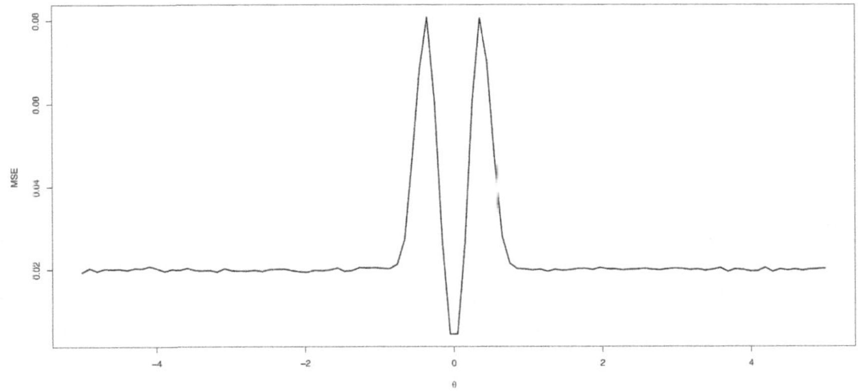

FIGURE 11.1
MSE of Hodges' estimator ($n = 50$)

independent of the specific value of c (i.e., independent of how θ_n approaches θ_0) are preferred. Such estimators are referred to as *regular estimators.*

An estimator $\widetilde{\theta}_n$ is said to be *regular* if the distribution of $\sqrt{n}(\widetilde{\theta}_n - \theta_n)$ converges to a limit that is invariant with respect to the LDGP. Specifically, an asymptotically linear estimator $\widetilde{\theta}_n$ is regular if, for any constant c and for i.i.d. observations $X_1, \ldots, X_n \sim p(x; \theta_n)$, with $\theta_n = \theta_0 + c/\sqrt{n}$, the following convergence holds:

$$\sqrt{n}(\widetilde{\theta}_n - \theta_n) \xrightarrow{d} \mathcal{N}(0, \Sigma_{\theta_0}),$$

where Σ_{θ_0} denotes the asymptotic variance, which is independent of c.

To evaluate whether Hodges' estimator satisfies regularity, consider the setting where $X_1, \ldots, X_n$ are i.i.d. from $\mathcal{N}(\theta_n, 1)$, with $\theta_n = 0 + c/\sqrt{n}$. Since $\overline{X}_n \sim \mathcal{N}(\theta_n, n^{-1})$, the probability that $\overline{X}_n$ falls within the interval

$$\begin{aligned} &(\theta_n - Cn^{-1/2}, \theta_n + Cn^{-1/2}) \\ =&(cn^{-1/2} - Cn^{-1/2}, cn^{-1/2} + Cn^{-1/2}) \\ =&\left(-(c+C)n^{-1/2}, (c+C)n^{-1/2}\right) \end{aligned}$$

can be arbitrarily close to one for sufficiently large C and n. Note that the interval $(-(c+C)n^{-1/2}, (c+C)n^{-1/2})$ is a subset of $(-n^{-1/4}, n^{-1/4})$ when n is large. Under this LDGP, the Hodges estimator satisfies $\widetilde{\theta}_n = o_p(n^{-1/2})$, and therefore:

$$\sqrt{n}(\widetilde{\theta}_n - \theta_n) = \sqrt{n}\left[o_p(n^{-1/2}) - cn^{-1/2}\right] = -c + o_p(1).$$

Since the limiting distribution is degenerate and concentrated at $-c$, which depends on c, it follows that Hodges' estimator is not regular.

Because regularity effectively excludes super-efficient estimators, the focus is henceforth placed on the class of *regular and asymptotically linear (RAL)* estimators:

$$\mathcal{C}_{\text{RAL}} = \left\{\widetilde{\theta}_n : \widetilde{\theta}_n \text{ is regular and asymptotically linear}\right\}.$$

The question then arises: How can it be shown that the MLE $\widehat{\theta}_n$ is regular? Since the definition of regularity is somewhat abstract, the *Fundamental Theorem of Regularity* (FTR)* will be invoked in the next section to verify the regularity of asymptotically linear estimators. For now, it is assumed that it has been shown that the MLE $\widehat{\theta}_n$ is regular, i.e., $\widehat{\theta}_n \in \mathcal{C}_{\text{RAL}}$.

Thanks to this regularity, the CRLB is well-defined within the class $\mathcal{C}_{\text{RAL}}$. In summary, the following results are obtained:

$$\begin{aligned}
\widehat{\theta}_n &\in \mathcal{C}_{\text{RAL}}, \\
\sqrt{n}(\widehat{\theta}_n - \theta_0) &\xrightarrow{d} \mathcal{N}\left(0, \mathcal{I}_{\theta_0}^{-1}\right), \\
\mathcal{I}_{\theta}^{-1} &= \inf_{\widetilde{\theta}_n \in \mathcal{C}_{\text{RAL}}} \left\{\text{asymptotic variance of } \widetilde{\theta}_n\right\},
\end{aligned}$$

and thus it can be concluded that the MLE $\widehat{\theta}_n$ is efficient.

Gold Coin # 38. *To study efficiency, neither the class of all unbiased estimators nor the class of all asymptotically linear estimators is considered appropriate. Instead, the focus is placed on the class of regular and asymptotically linear (RAL) estimators. Each RAL estimator satisfies two essential conditions: regularity and asymptotic linearity.*

11.2 Parametric Modeling

11.2.1 Estimand and Estimator

A general research question is posed within the context of parametric modeling. Suppose $X_1, \ldots, X_n$ are i.i.d. from the distribution $X \sim p(x; \nu)$, where ν

*As far as I know, the existing literature—expect for my previous book[22]—does not refer to it as the *fundamental theorem of regularity.* I use this term here to highlight its importance in the development of semiparametric theory in this book.

is a d-dimensional parameter. The objective is to estimate a specific function of ν, referred to as the *target estimand*, which is defined as

$$\theta = \theta(\nu).$$

If the true value of ν is denoted by ν_0, then the corresponding true value of the estimand is $\theta_0 = \theta(\nu_0)$.

Note that the parameter ν takes the same role as the parameter θ in the previous section. However, to reserve the symbol θ for the target estimand, a new notation ν is adopted here for the distributional parameter.

Within the parametric setting, the MLE of v, $\widehat{\nu}_n$, takes the form

$$\widehat{\nu}_n = \nu_0 + \frac{1}{n}\sum_{i=1}^{n} \mathcal{I}_{\nu_0}^{-1} s_{\nu_0}(X_i) + o_p(n^{-1/2}),$$

where the following quantities are defined:

$$\begin{aligned} \ell_\nu(X) &= \log p(X;\nu), \\ s_\nu(X) &= \frac{\partial \ell_\nu(X)}{\partial \nu}, \\ \mathcal{I}_\nu &= \mathrm{Cov}_\nu[s_\nu(X)], \end{aligned}$$

representing the log-likelihood, score function, and Fisher information matrix, respectively.

Using the plug-in principle, the MLE for θ is given by

$$\widehat{\theta}_n = \theta(\widehat{\nu}_n).$$

A first-order Taylor expansion yields the approximation:

$$\widehat{\theta}_n - \theta_0 = \theta(\widehat{\nu}_n) - \theta(\nu_0) = \frac{\partial \theta(\nu_0)}{\partial \nu^\top}(\widehat{\nu}_n - \nu_0) + o_p(n^{-1/2}),$$

where the first derivative is defined as

$$\frac{\partial \theta(\nu_0)}{\partial \nu^\top} = \left.\frac{\partial \theta(\nu)}{\partial \nu^\top}\right|_{\nu=\nu_0}.$$

Hence, the MLE $\widehat{\theta}_n$ is seen to be an asymptotically linear estimator:

$$\widehat{\theta}_n = \theta_0 + \frac{1}{n}\sum_{i=1}^{n} \frac{\partial \theta(\nu_0)}{\partial \nu^\top}\mathcal{I}_{\nu_0}^{-1} s_{\nu_0}(X_i) + o_p(n^{-1/2}).$$

Thus, the corresponding influence function of $\widehat{\theta}_n$ is

$$\phi(X) = \frac{\partial \theta(\nu_0)}{\partial \nu^\top}\mathcal{I}_{\nu_0}^{-1} s_{\nu_0}(X),$$

whose variance is given by

$$\mathbb{V}_{\nu_0}[\phi(X)] = \frac{\partial\theta(\nu_0)}{\partial\nu^\top}\mathcal{I}_{\nu_0}^{-1}\text{Cov}_{\nu_0}(s_{\nu_0}(X))\mathcal{I}_{\nu_0}^{-1}\frac{\partial\theta(\nu_0)}{\partial\nu} = \frac{\partial\theta(\nu_0)}{\partial\nu^\top}\mathcal{I}_{\nu_0}^{-1}\frac{\partial\theta(\nu_0)}{\partial\nu}.$$

Consequently, as an asymptotically linear estimator, the MLE $\widehat{\theta}_n$ satisfies both consistency and asymptotic normality:

$$\widehat{\theta}_n \xrightarrow{p} \theta_0,$$
$$\sqrt{n}(\widehat{\theta}_n - \theta_0) \xrightarrow{d} \mathcal{N}\left(0, \frac{\partial\theta(\nu_0)}{\partial\nu^\top}\mathcal{I}_{\nu_0}^{-1}\frac{\partial\theta(\nu_0)}{\partial\nu}\right).$$

11.2.2 Regular and Asymptotically Linear Estimator

Recall that the MLE $\widehat{\theta}_n$ is asymptotically linear. To show that it is also regular, a theorem will be utilized. Since it is a special case of the FTR, it is referred to here as the *lemma of regularity*. Although it formally follows as a corollary of the FTR, it is stated as a lemma for immediate use.

Lemma of Regularity

Let $\widetilde{\theta}_n$ be an asymptotically linear estimator with influence function $\varphi(X)$, such that $\mathbb{E}_\nu[\varphi^2(X)]$ exists and is continuous in ν in a neighborhood of ν_0. Then $\widetilde{\theta}_n$ is regular if and only if

$$\mathbb{E}_{\nu_0}\left[\varphi(X)s_{\nu_0}^\top(X)\right] = \frac{\partial\theta(\nu_0)}{\partial\nu^\top}.$$

The proof of the lemma is omitted here.†

MLE is Regular

Applying the lemma, the regularity of the MLE $\widehat{\theta}_n$, whose influence function is $\phi(X) = \frac{\partial\theta(\nu_0)}{\partial\nu^\top}\mathcal{I}_{\nu_0}^{-1}s_{\nu_0}(X)$, can be easily verified. In fact,

$$\begin{aligned}\mathbb{E}_{\nu_0}\left[\phi(X)s_{\nu_0}^\top(X)\right] &= \mathbb{E}_{\nu_0}\left[\frac{\partial\theta(\nu_0)}{\partial\nu^\top}\mathcal{I}_{\nu_0}^{-1}s_{\nu_0}(X)s_{\nu_0}^\top(X)\right]\\ &= \frac{\partial\theta(\nu_0)}{\partial\nu^\top}\mathcal{I}_{\nu_0}^{-1}\mathbb{E}_{\nu_0}\left[s_{\nu_0}(X)s_{\nu_0}^\top(X)\right]\\ &= \frac{\partial\theta(\nu_0)}{\partial\nu^\top}\mathcal{I}_{\nu_0}^{-1}\mathcal{I}_{\nu_0}\\ &= \frac{\partial\theta(\nu_0)}{\partial\nu^\top}.\end{aligned}$$

†The proof of the *Fundamental Theorem of Regularity* relies on the pioneering work of Lucien Le Cam (1924–2000), particularly his theory of *contiguity* and the associated *fundamental lemmas*—notably, Le Cam's First and Third Lemmas. Le Cam's contributions are profound enough to merit an independent chapter devoted entirely to his asymptotic theory. However, to keep this introduction focused on causal inference, we will not undertake a detailed exposition of Le Cam's theory in this book.

11.2.3 Efficient Estimator

Consider any RAL estimator $\widetilde{\theta}_n$ with influence function $\varphi(X)$ satisfying the regularity condition:

$$\mathbb{E}_{\nu_0}\left[\varphi(X)s_{\nu_0}^\top(X)\right] = \frac{\partial\theta(\nu_0)}{\partial\nu^\top}.$$

Define the following non-negative definite matrix:

$$\begin{pmatrix}\varphi(X)\\ s_{\nu_0}(X)\end{pmatrix}\begin{pmatrix}\varphi(X)\\ s_{\nu_0}(X)\end{pmatrix}^\top = \begin{pmatrix}\varphi^2(X) & \varphi(X)s_{\nu_0}^\top(X)\\ s_{\nu_0}(X)\varphi(X) & s_{\nu_0}(X)s_{\nu_0}^\top(X)\end{pmatrix} \geq 0.$$

Taking expectations, the following matrix inequality is obtained:

$$\begin{pmatrix}\mathbb{V}_{\nu_0}[\varphi(X)] & \frac{\partial\theta(\nu_0)}{\partial\nu^\top}\\ \frac{\partial\theta(\nu_0)}{\partial\nu} & \mathcal{I}_{\nu_0}\end{pmatrix} \geq 0.$$

This implies the CRLB on the asymptotic variance of all RAL estimators (see Exercise 11.2):

$$\mathbb{V}_{\nu_0}[\varphi(X)] \geq \frac{\partial\theta(\nu_0)}{\partial\nu^\top}\mathcal{I}_{\nu_0}^{-1}\frac{\partial\theta(\nu_0)}{\partial\nu}.$$

In summary, the MLE $\widehat{\theta}_n$ belongs to the class of RAL estimators and achieves the CRLB. Therefore, $\widehat{\theta}_n$ is an efficient estimator of θ_0.

Gold Coin # 39. *In the parametric setting where $X_1,\ldots,X_n \overset{i.i.d.}{\sim} p(x;\nu)$, $\theta = \theta(\nu)$ is the target estimand, and $\widehat{\nu}_n$ the MLE of ν, the plug-in estimator $\widehat{\theta}_n = \theta(\widehat{\nu}_n)$ is regular and asymptotically linear (RAL):*

$$\widehat{\theta}_n = \theta_0 + \frac{1}{n}\sum_{i=1}^n \frac{\partial\theta(\nu_0)}{\partial\nu^\top}\mathcal{I}_{\nu_0}^{-1}s_{\nu_0}(X_i) + o_p(n^{-1/2}),$$

with asymptotic variance attaining the Cramér–Rao lower bound (CRLB):

$$\frac{\partial\theta(\nu_0)}{\partial\nu^\top}\mathcal{I}_{\nu_0}^{-1}\frac{\partial\theta(\nu_0)}{\partial\nu}.$$

An Example

As an illustration, consider the setting in which $X_1,\ldots,X_n$ are i.i.d. from the normal distribution $\mathcal{N}(\mu,\sigma^2)$, and the coefficient of variation $\theta = \sigma/\mu$ is identified as the target estimand. Based on the preceding discussion, the asymptotic variance of the MLE of θ is derived as follows.

Define the parameter vector as $\nu = (\mu, \sigma)^\top$. The MLE of ν is given by $\widehat{\nu}_n = (\widehat{\mu}_n, \widehat{\sigma}_n)^\top$, where

$$\widehat{\mu}_n = \overline{X}_n, \quad \widehat{\sigma}_n = \sqrt{\frac{1}{n}\sum_{i=1}^{n}(X_i - \overline{X}_n)^2}.$$

The Fisher information matrix corresponding to ν is expressed as

$$\mathcal{I}_\nu = \begin{pmatrix} \frac{n}{\sigma^2} & 0 \\ 0 & \frac{n}{2\sigma^2} \end{pmatrix}.$$

The gradient of the target estimand θ with respect to ν is computed as

$$\frac{\partial \theta}{\partial \nu^\top} = \left(-\frac{\sigma}{\mu^2}, \frac{1}{\mu}\right).$$

Given the above quantities, the asymptotic variance of the MLE $\widehat{\theta}_n$ can be obtained by applying the CRLB formula. The explicit calculation is provided in Exercise 11.3.

11.3 Nonparametric Modeling

11.3.1 Estimand and Estimator

A general research question is now posed in the context of nonparametric modeling. It is assumed that $X_1, \ldots, X_n$ are i.i.d. with $X \sim p(x)$, where the density function $p(x)$ is completely unknown. The goal is to estimate a functional of the unknown distribution $p(x)$, referred to as the *target estimand.* This target estimand is defined as

$$\theta = \theta(p(\cdot)).$$

If the true density is denoted by $p_0(x)$, then the true value of the estimand is given by $\theta_0 = \theta(p_0(\cdot))$. The function $p_0(\cdot)$ can be interpreted as an unknown, infinite-dimensional parameter.

If an estimator of $p_0(\cdot)$, denoted by $\widehat{p}_n(\cdot)$, is obtained—e.g., through the empirical distribution function or a spline-based method[56]—then the corresponding estimator of θ_0 is constructed as

$$\widehat{\theta}_n = \theta(\widehat{p}_n(\cdot)).$$

Although this estimator is straightforward to construct, the derivation of its asymptotic properties is nontrivial. Suppose a first-order Taylor expansion were possible:

$$\widehat{\theta}_n - \theta_0 = \theta(\widehat{p}_n(\cdot)) - \theta(p_0(\cdot)) = \left[\frac{\partial\theta(p_0(\cdot))}{\partial p(\cdot)}\right]^\top [\widehat{p}_n(\cdot) - p_0(\cdot)] + o_p(n^{-1/2}).$$

However, the interpretation of the first derivative with respect to the infinite-dimensional function $p(\cdot)$ evaluated at $p_0(\cdot)$,

$$\left.\frac{\partial\theta(p(\cdot))}{\partial p(\cdot)}\right|_{p(\cdot)=p_0(\cdot)},$$

poses significant analytical challenges. To address this, the framework of parametric submodels is introduced.[55,41]

11.3.2 Parametric Submodels

Let $\mathcal{M}_\infty$ denote the class of all data-generating processes considered sufficiently rich to contain the true distribution $p_0(\cdot)$.

As illustrated in Figure 11.2, this class can be expanded via a collection of one-dimensional parametric submodels. Each point in the figure represents a distribution $p(\cdot)$, and each curve corresponds to a one-dimensional submodel $p(\cdot;\nu)$ that intersects $p_0(\cdot)$, i.e., $p(\cdot;0) = p_0(\cdot)$. Although only two such submodels are shown in the illustration, the inclusion of any number of them allows the full space $\mathcal{M}_\infty$ to be effectively covered.

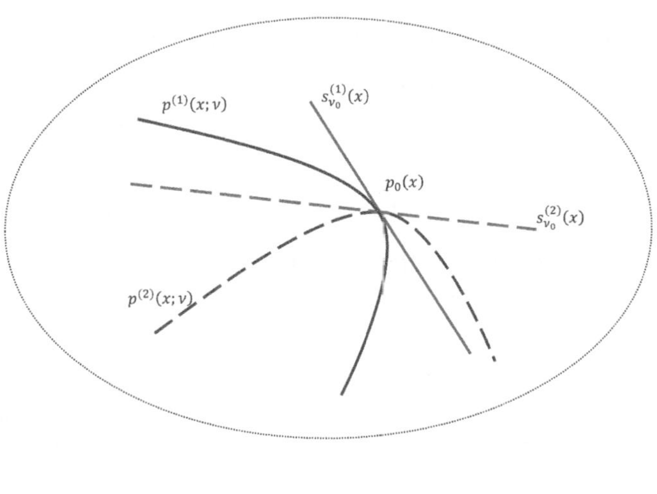

FIGURE 11.2
One-dimensional parametric submodels

Such parametric submodels can be explicitly constructed. Let $s(X)$ be a mean-zero function satisfying

$$\mathbb{E}_0[s(X)] = 0 \quad \text{and} \quad \mathbb{V}_0[s(X)] < \infty,$$

where the subscript 0 indicates the true distribution $p_0(x)$.

Two methods for constructing one-dimensional submodels based on $s(X)$ are considered. The first method is analytically simple but less robust; the second is more complex but more robust.

Method 1

$$p(x;\nu) = p_0(x)\left[1 + \nu\, s(x)\right].$$

It can be verified that $\int p(x;\nu)\,dx = 1$, $p(x;0) = p_0(x)$, and the score function evaluated at $\nu = 0$ is $s(X)$ (see Exercise 11.4).

Method 2

$$p(x;\nu) = \frac{p_0(x)\exp\left[\nu\, s(x)\right]}{\int p_0(t)\exp[\nu\, s(t)]\,dt}.$$

It can also be shown that $\int p(x;\nu)\,dx = 1$, $p(x;0) = p_0(x)$, and the score function at $\nu = 0$ equals $s(X)$ (see Exercise 11.5).

Because $s(X)$ can be chosen as any mean-zero function, the union of such submodels generated by either method is sufficient to span the entire model class $\mathcal{M}_\infty$.

11.3.3 Fundamental Theorem of Regularity

With the introduction of parametric submodels, it becomes feasible to determine whether an asymptotically linear estimator is regular. Let a family of submodels $p(x;\nu)$ be constructed via Method 1 or 2, such that

$$\mathcal{M}_\infty = \bigcup \{p(x;\nu) : \text{for a given mean-zero function } S_{0;\nu}(X)\}.$$

For each submodel $p(x;\nu)$, the condition $p(x;0) = p_0(x)$ holds, and the score function at $\nu = 0$ is denoted by

$$\left.\frac{\partial \log p(x;\nu)}{\partial \nu}\right|_{\nu=0} = S_{0;\nu}(x).$$

Here, ν in the subscript of $S_{0;\nu}(x)$ indicates that the score function corresponds to the parametric submodel indexed by ν, and 0 in the subscript of $S_{0;\nu}(x)$ indicates that the score function is evaluated at the true value of ν, $\nu = 0$.

For each submodel, the target estimand becomes a function of ν,

$$\theta = \theta(\nu) = \theta(p(\cdot;\nu)).$$

The true value θ_0 is given by $\theta_0 = \theta(0) = \theta(p_0(\cdot))$.

The following theorem is adapted from Newey (1990).[57] For brevity, we present only the main statement, omitting the technical conditions on the parametric submodels—the so-called *regular parametric submodel*—and also omitting the mathematical proof. Our goal here is solely to highlight how the theorem is applied to verify regularity.

The Fundamental Theorem of Regularity (FTR)

Assume that $\widetilde{\theta}_n$ is an asymptotically linear estimator:

$$\widetilde{\theta}_n = \theta_0 + \frac{1}{n}\sum_{i=1}^{n} \varphi(X_i) + o_p(n^{-1/2}),$$

where the influence function $\varphi(X)$ satisfies $\mathbb{E}_0[\varphi(X)] = 0$ and $\mathbb{E}_0[\varphi^2(X)] < \infty$. Then, $\widetilde{\theta}_n$ is regular if and only if, for any regular parametric submodel $p(x;\nu)$ whose score function at $v = 0$ is $S_{0;\nu}(X)$, the following condition holds:

$$\left.\frac{\partial\theta(\nu)}{\partial\nu}\right|_{\nu=0} = \mathbb{E}_0\left[\varphi(X)S_{0;\nu}(X)\right].$$

Gold Coin # 40. *The Fundamental Theorem of Regularity (FTR) establishes necessary and sufficient conditions for the regularity of an asymptotically linear estimator. It plays a pivotal role in the derivation of the efficient influence function (EIF), which is central to the semiparametric efficiency theory.*

11.3.4 First Application of the FTR

Let $U_1, \ldots, U_n$ be i.i.d. random variables with distribution $U \sim f_0(u)$. The estimand of interest is defined as $\theta_0 = \mathbb{E}_0(U)$. The distribution of U is assumed to be unrestricted, except for the requirement that $\mathbb{V}_0(U) < \infty$.

The sample mean,

$$\widehat{\theta}_n = \frac{1}{n}\sum_{i=1}^{n} U_i,$$

is an asymptotically linear estimator of θ_0, with influence function $\varphi(U) = U - \mathbb{E}_0(U)$, as shown by:

$$\sqrt{n}(\widehat{\theta}_n - \theta_0) = \frac{1}{\sqrt{n}}\sum_{i=1}^{n}[U_i - \mathbb{E}_0(U)].$$

Consider any regular parametric submodel $f(u;\nu)$, where $f(u;0) = f_0(u)$. The corresponding estimand is $\theta(\nu) = \int u f(u;\nu)\,du$. Then,

$$\begin{aligned}\frac{\partial\theta(\nu_0)}{\partial\nu} &= \int u \frac{\partial f(u;\nu_0)}{\partial\nu}\,du = \int u \frac{\partial \log f(u;\nu_0)}{\partial\nu} f(u;\nu_0)\,du \\ &= \mathbb{E}_0[US_{0;\nu}] = \mathbb{E}_0[(U - \mathbb{E}_0(U))S_{0;\nu}] = \mathbb{E}_0[\varphi(U)S_{0;\nu}],\end{aligned}$$

where $S_{0;\nu} = \partial \log f(u;\nu_0)/\partial\nu$ is the score function.

Therefore, by the FTR, $\widehat{\theta}_n$ is regular and hence is an RAL estimator.

11.4 Fundamentals of Semiparametric Theory

This section outlines the foundational results of semiparametric theory, upon which the estimation of treatment effects will be developed in the subsequent four chapters.

11.4.1 Generalized Cramér–Rao Lower Bound

Continue the nonparametric setting, in which the target estimand is denoted by $\theta_0 = \theta(p_0(\cdot))$. For any RAL estimator $\widetilde{\theta}_n$ with influence function $\varphi(X)$, and any regular parametric submodel $p(x;\nu)$, the FTR implies

$$\mathbb{E}_0\left[\varphi(X)S_{0;\nu}(X)\right] = \frac{\partial\theta(\nu_0)}{\partial\nu}.$$

The following non-negative definite matrix can then be constructed:

$$\begin{pmatrix}\varphi(X)\\ S_{0;\nu}(X)\end{pmatrix}\begin{pmatrix}\varphi(X)\\ S_{0;\nu}(X)\end{pmatrix}^{\top} = \begin{pmatrix}\varphi^2(X) & \varphi(X)S_{0;\nu}^{\top}(X)\\ S_{0;\nu}(X)\varphi(X) & S_{0;\nu}(X)S_{0;\nu}^{\top}(X)\end{pmatrix} \geq 0.$$

Taking expectations on both sides yields

$$\begin{pmatrix}\mathbb{V}_0[\varphi(X)] & \mathbb{E}_0[\varphi(X)S_{0;\nu}^{\top}(X)]\\ \mathbb{E}_0[S_{0;\nu}(X)\varphi(X)] & \mathrm{Cov}_0[S_{0;\nu}(X)]\end{pmatrix} \geq 0.$$

By regularity, the matrix simplifies to

$$\begin{pmatrix}\mathbb{V}_0[\varphi(X)] & \frac{\partial\theta(\nu_0)}{\partial\nu^{\top}}\\ \frac{\partial\theta(\nu_0)}{\partial\nu} & \mathcal{I}_{0;\nu}\end{pmatrix} \geq 0,$$

where $\mathcal{I}_{0;\nu} = \mathrm{Cov}_0[S_{0;\nu}(X)]$ is the Fisher information matrix of the regular parametric submodel $p(x;\nu)$ at $\nu_0 = 0$.

By the Schur complement, the following inequality holds:

$$\mathbb{V}_0[\varphi(X)] \geq \frac{\partial\theta(\nu_0)}{\partial\nu^{\top}}\mathcal{I}_{0;\nu}^{-1}\frac{\partial\theta(\nu_0)}{\partial\nu}.$$

Since this inequality holds for any regular parametric submodel, a tighter lower bound is obtained by taking the supremum:

$$\mathbb{V}_0[\varphi(X)] \geq \sup_{\text{all regular submodels}} \frac{\partial\theta(\nu_0)}{\partial\nu^\top}\mathcal{I}_{0;\nu}^{-1}\frac{\partial\theta(\nu_0)}{\partial\nu}.$$

This supremum defines the *generalized Cramér–Rao lower bound (CRLB)* for the asymptotic variance of any RAL estimator:

$$\sup_{\text{all regular submodels}} \frac{\partial\theta(\nu_0)}{\partial\nu^\top}\mathcal{I}_{0;\nu}^{-1}\frac{\partial\theta(\nu_0)}{\partial\nu}.$$

11.4.2 Least Favorable Submodel

Although the generalized CRLB is well-defined, interpreting the supremum over all regular submodels can be challenging. If a submodel can be identified such that its CRLB exactly equals the generalized CRLB, then that submodel is referred to as a *least favorable submodel*.[55,41] Specifically, a submodel $p(x;\nu^*)$ is least favorable if

$$\frac{\partial\theta(\nu_0^*)}{\partial\nu^{*\top}}\mathcal{I}_{0;\nu^*}^{-1}\frac{\partial\theta(\nu_0^*)}{\partial\nu^*} = \sup_\nu \frac{\partial\theta(\nu_0)}{\partial\nu^\top}\mathcal{I}_{0;\nu}^{-1}\frac{\partial\theta(\nu_0)}{\partial\nu},$$

where $\sup_\nu$ is over all the regular submodels indicated by $\{p(x;\nu)\}$.

11.4.3 Efficient Influence Function

The generalized CRLB and the least favorable submodel can be used to derive the *efficient influence function (EIF)*.

Based on the least favorable submodel $p(x;\nu^*)$, define

$$\varphi^*(X) = \frac{\partial\theta(\nu_0^*)}{\partial\nu^{*\top}}\mathcal{I}_{0;\nu^*}^{-1}S_{0;\nu^*}(X).$$

Then,

$$\mathbb{V}_0[\varphi^*(X)] = \frac{\partial\theta(\nu_0^*)}{\partial\nu^{*\top}}\mathcal{I}_{\nu^*}^{-1}\frac{\partial\theta(\nu_0^*)}{\partial\nu^*} = \sup_\nu \frac{\partial\theta(\nu_0)}{\partial\nu^\top}\mathcal{I}_{\nu}^{-1}\frac{\partial\theta(\nu_0)}{\partial\nu}.$$

Since every regular influence function $\varphi(X)$ satisfies

$$\mathbb{V}_0[\varphi(X)] \geq \sup_\nu \frac{\partial\theta(\nu_0)}{\partial\nu^\top}\mathcal{I}_{\nu}^{-1}\frac{\partial\theta(\nu_0)}{\partial\nu},$$

it follows that

$$\mathbb{V}_0[\varphi(X)] \geq \mathbb{V}_0[\varphi^*(X)].$$

Thus, $\varphi^*(X)$ attains the lower bound and is therefore the EIF.

Gold Coin # 41. *The definition of the efficient influence function (EIF) involves both an infimum and a supremum:*

$$\mathbb{V}_0[\varphi^*(X)] = \inf_{\varphi} \mathbb{V}_0[\varphi(X)],$$

$$\mathbb{V}_0[\varphi^*(X)] = \sup_{\nu} \frac{\partial \theta(\nu_0)}{\partial \nu^\top} \mathcal{I}_{0;\nu}^{-1} \frac{\partial \theta(\nu_0)}{\partial \nu}.$$

11.4.4 Efficient Estimator

The remaining chapters in Part III are devoted to deriving EIFs for treatment effect estimation. In Part IV, the construction of *efficient estimators* is discussed. In short, an efficient estimator is an RAL estimator whose influence function is the EIF $\varphi^*(X)$. That is, such an estimator satisfies

$$\widehat{\theta}_n = \theta_0 + \frac{1}{n}\sum_{i=1}^{n} \varphi^*(X_i) + o_p(n^{-1/2}).$$

11.5 Exercises

Exercise 11.1

Generate Figure 11.1, a plot of MSE of Hodges' estimator for $n = 50$.

Exercise 11.2

Show that if the following matrix is positive definite,

$$\begin{pmatrix} A, & B \\ B^\top, & C \end{pmatrix} > 0,$$

then the Schur complement of C, $S = C - B^\top A^{-1} B$, is also positive definite.

Exercise 11.3

Calculate the asymptotic variance of the MLE of the coefficient of variation, $\theta = \sigma/\mu$, based on data $X_1, \ldots, X_n$ which are i.i.d. with $X \sim \mathcal{N}(\mu, \sigma^2)$.

Exercise 11.4

Consider the parametric submodel

$$p(x;\nu) = p_0(x)\left[1 + \nu s(X)\right],$$

where $\mathbb{E}_0[s(X)] = 0$. Find the score function at $\nu_0 = 0$.

Exercise 11.5

Consider the parametric submodel

$$p(x;\nu) = p_0(x)\exp\left[\nu s(x)\right] / \int p_0(t)\exp[\nu s(t)]dt,$$

where $\mathbb{E}_0[s(X)] = 0$. Find the score function at $\nu_0 = 0$.

12

Efficient Influence Function

12.1 Average Treatment Effect Revisited

12.1.1 Two Versions of a Statistical Estimand

The central research question in pharmaceutical statistics is revisited. Consider the task of estimating the following causal estimand:

$$\theta^*_{\text{ATE}} = \mathbb{E}(Y^{a=1}) - \mathbb{E}(Y^{a=0}).$$

Consider a cohort study of sample size n that generates the observed data $\{O_i = (X_i, A_i, Y_i)\}_{i=1}^n$. Under the standard identifiability assumptions, two strategies can be employed to identify this estimand.

Using the standardization strategy, θ^*_{ATE} is expressed as:

$$\theta_{\text{ATE}} = \mathbb{E}[\mathbb{E}(Y \mid A = 1, X)] - \mathbb{E}[\mathbb{E}(Y \mid A = 0, X)].$$

Alternatively, through the weighting strategy, θ^*_{ATE} is expressed as:

$$\theta_{\text{ATE}} = \mathbb{E}\left[\frac{I(A=1)}{\mathbb{P}(A=1 \mid X)}Y\right] - \mathbb{E}\left[\frac{I(A=0)}{\mathbb{P}(A=0 \mid X)}Y\right] = \mathbb{E}\left[\frac{(2A-1)Y}{\mathbb{P}(A \mid X)}\right].$$

12.1.2 Two Approaches to Deriving the EIF

Based on the principles of semiparametric theory, the efficient influence function (EIF) plays a key role in the construction of efficient estimators for θ_{ATE}. Therefore, approaches to deriving the EIF to facilitate the estimation of θ_{ATE} are discussed in this chapter.

Given the two equivalent representations of θ_{ATE}, this chapter presents two approaches to deriving the EIF, using simplified mathematical arguments. These approaches correspond to those treated with greater mathematical rigor in the existing semiparametric literature.[41, 58, 42]

Gold Coin # 42. *Two approaches can be used to derive the efficient influence function (EIF) for the estimation of* θ_{ATE}.

DOI: 10.1201/9781003635468-15

The first approach relies on the following form of the estimand:

$$\theta_{ATE} = \mathbb{E}[\mathbb{E}(Y \mid A = 1, X)] - \mathbb{E}[\mathbb{E}(Y \mid A = 0, X)].$$

The second approach relies on the alternative representation:

$$\theta_{ATE} = \mathbb{E}\left[\frac{(2A-1)Y}{\mathbb{P}(A \mid X)}\right].$$

12.2 Approach One to Deriving the EIF

The derivation of the EIF corresponding to the following statistical estimand will now be considered:

$$\theta_{\text{ATE}} = \mathbb{E}[\mathbb{E}(Y \mid A = 1, X)] - \mathbb{E}[\mathbb{E}(Y \mid A = 0, X)].$$

12.2.1 Saturated Model

Assume that the distribution of Y is discrete, with sample space $\{y_1, \ldots, y_M\}$, and that the distribution of X is also discrete, with sample space $\{x_1, \ldots, x_K\}$. Given that A takes values in $\{0, 1\}$, the joint distribution of (X, A, Y) is discrete and is denoted by:

$$\nu_{k,a,m} = \mathbb{P}(X = x_k, A = a, Y = y_m),$$

for $k = 1, \ldots, K$, $a = 1, 0$, and $m = 1, \ldots, M$. Let ν denote the vector containing all such joint probabilities:

$$\nu = (\nu_{1,1,1}, \cdots, \nu_{1,1,M}, \ldots, \nu_{K,0,1}, \ldots, \nu_{K,0,M})^\top.$$

The true value of ν is denoted as

$$\nu_0 = (\nu_{1,1,1;0}, \cdots, \nu_{1,1,M;0}, \ldots, \nu_{K,0,1;0}, \ldots, \nu_{K,0,M;0})^\top.$$

The maximum likelihood estimator (MLE) of ν is given by:

$$\widehat{\nu}_n = (\widehat{\nu}_{1,1,1;n}, \cdots, \widehat{\nu}_{1,1,M;n}, \ldots, \widehat{\nu}_{K,0,1;n}, \ldots, \widehat{\nu}_{K,0,M;n})^\top,$$

where

$$\widehat{\nu}_{k,a,m;n} = \frac{1}{n}\sum_{i=1}^{n} I(X_i = x_k, A_i = a, Y_i = y_m).$$

The MLE $\widehat{\nu}_n$ can be expressed in asymptotically linear form as:

$$\widehat{\nu}_{k,a,m;n} = \nu_{k,a,m;0} + \frac{1}{n}\sum_{i=1}^{n}\left[I(X_i = x_k, A_i = a, Y_i = y_m) - \nu_{k,a,m;0}\right].$$

Next, θ_{ATE} can be seen as a function of ν. To see this, define

$$\theta_a = \mathbb{E}[\mathbb{E}(Y \mid A = a, X)], \quad a = 1, 0.$$

Hence,

$$\theta_{\text{ATE}} = \theta_{a=1} - \theta_{a=0}.$$

It follows that the EIF of θ_{ATE} is equal to the difference between the EIFs of $\theta_{a=1}$ and $\theta_{a=0}$. Each θ_a can be written as a function of ν, as shown below:

$$\begin{aligned}
\theta_a &= \mathbb{E}(Y \mid A = a, X) \\
&= \sum_{k=1}^{K}\left[\mathbb{E}(Y \mid A = a, X = x_k)\right]\mathbb{P}(X = x_k) \\
&= \sum_{k=1}^{K}\left[\sum_{m=1}^{M} y_m\,\mathbb{P}(Y = y_m \mid A = a, X = x_k)\right]\mathbb{P}(X = x_k) \\
&= \sum_{k=1}^{K}\left[\frac{\sum_{m=1}^{M} y_m\,\mathbb{P}(X = x_k, A = a, Y = y_m)}{\mathbb{P}(X = x_k, A = a)}\right]\mathbb{P}(X = x_k) \\
&= \sum_{k=1}^{K}\left[\frac{\sum_{m=1}^{M} y_m \nu_{k,a,m}}{\sum_{m=1}^{M} \nu_{k,a,m}}\right]\sum_{m=1}^{M}\sum_{a'=0}^{1} \nu_{k,a',m} \triangleq \theta_a(\nu).
\end{aligned}$$

Therefore,

$$\theta_{\text{ATE}} = \theta_{a=1} - \theta_{a=0} = \theta_{a=1}(\nu) - \theta_{a=0}(\nu) \triangleq \theta_{\text{ATE}}(\nu),$$

and the MLE of θ_{ATE} is given by:

$$\widehat{\theta}_{\text{ATE},n} = \theta_{\text{ATE}}(\widehat{\nu}_n) = \theta_{a=1}(\widehat{\nu}_n) - \theta_{a=0}(\widehat{\nu}_n).$$

12.2.2 MLE's Influence Function

The same Taylor expansion as considered in Section 11.2 is applied:

$$\begin{aligned}
\widehat{\theta}_{\text{ATE},n} - \theta_{\text{ATE},0} &= \theta_{\text{ATE}}(\widehat{\nu}_n) - \theta_{\text{ATE}}(\nu_0) \\
&= \frac{\partial\theta_{\text{ATE}}(\nu_0)}{\partial\nu^{\top}}(\widehat{\nu}_n - \nu_0) + o_p(n^{-1/2}) \\
&= \left[\frac{\partial\theta_{a=1}(\nu_0)}{\partial\nu} - \frac{\partial\theta_{a=0}(\nu_0)}{\partial\nu}\right]^{\top}(\widehat{\nu}_n - \nu_0) + o_p(n^{-1/2}).
\end{aligned}$$

The partial derivatives are given by: for $a = 1, 0$ (see Exercise 12.1),

$$\frac{\partial \theta_a(\nu_0)}{\partial \nu_{k,a,m}} = \frac{y_m - \mathbb{E}(Y \mid A = a, X = x_k)}{\mathbb{P}(A = a \mid X = x_k)} + \mathbb{E}(Y \mid A = a, X = x_k),$$

and for $a \neq a'$ (see Exercise 12.2),

$$\frac{\partial \theta_a(\nu_0)}{\partial \nu_{k,a',m}} = \mathbb{E}(Y \mid A = a, X = x_k).$$

The following inner product is written as:

$$\begin{aligned}\frac{\partial \theta_a(\nu_0)}{\partial \nu^\top}(\widehat{\nu}_n - \nu_0) = &\sum_{k=1}^{K}\sum_{m=1}^{M} \frac{\partial \theta_a(\nu_0)}{\partial \nu_{k,a,m}}(\widehat{\nu}_{k,a,m;n} - \nu_{k,a,m;0}) \\ &+ \sum_{k=1}^{K}\sum_{m=1}^{M} \frac{\partial \theta_a(\nu_0)}{\partial \nu_{k,a',m}}(\widehat{\nu}_{k,a',m;n} - \nu_{k,a',m;0}).\end{aligned}$$

Note that:

$$\begin{aligned}&\sum_{k=1}^{K}\sum_{m=1}^{M} \frac{\partial \theta_a(\nu_0)}{\partial \nu_{k,a,m}}(\widehat{\nu}_{k,a,m;n} - \nu_{k,a,m;0}) \\ =&\frac{1}{n}\sum_{i=1}^{n}\left\{\sum_{k=1}^{K}\sum_{m=1}^{M} \frac{\partial \theta_a(\nu_0)}{\partial \nu_{k,a,m}}\left[I(X_i = x_k, A_i = a, Y_i = y_m) - \nu_{k,a,m;0}\right]\right\} \\ =&\frac{1}{n}\sum_{i=1}^{n}\left\{\sum_{k=1}^{K}\sum_{m=1}^{M} \frac{y_m - \mathbb{E}(Y \mid A = a, X = x_k)}{\mathbb{P}(A = a \mid X = x_k)} I(X_i = x_k, A_i = a, Y_i = y_m)\right\} \\ &+ \frac{1}{n}\sum_{i=1}^{n}\left\{\sum_{k=1}^{K}\sum_{m=1}^{M} \mathbb{E}(Y \mid A = a, X = x_k) I(X_i = x_k, A_i = a, Y_i = y_m)\right\} \\ &- \frac{1}{n}\sum_{i=1}^{n}\left\{\sum_{k=1}^{K}\sum_{m=1}^{M} \frac{y_m - \mathbb{E}(Y \mid A = a, X = x_k)}{\mathbb{P}(A = a \mid X = x_k)} \mathbb{P}(Y = y_m, A = a, X = x_k)\right\} \\ &- \frac{1}{n}\sum_{i=1}^{n}\left\{\sum_{k=1}^{K}\sum_{m=1}^{M} \mathbb{E}(Y \mid A = a, X = x_k)\mathbb{P}(Y = y_m, A = a, X = x_k)\right\} \\ =&\frac{1}{n}\sum_{i=1}^{n} \frac{Y_i - \mathbb{E}(Y \mid A = a, X = X_i)}{\mathbb{P}(A = a \mid X = X_i)} I(A_i = a) \\ &+ \frac{1}{n}\sum_{i=1}^{n} \mathbb{E}(Y \mid A = a, X = X_i) I(A_i = a) \\ &- \frac{1}{n}\sum_{i=1}^{n} 0 \\ &- \frac{1}{n}\sum_{i=1}^{n}\left\{\sum_{k=1}^{K} \mathbb{E}(Y \mid A = a, X = x_k)\mathbb{P}(A = a, X = x_k)\right\}.\end{aligned}$$

Similarly,

$$
\begin{aligned}
&\sum_{k=1}^{K}\sum_{m=1}^{M}\frac{\partial\theta_a(\nu_0)}{\partial\nu_{k,a',m}}(\widehat{\nu}_{k,a',m;n}-\nu_{k,a',m;0})\\
=&\frac{1}{n}\sum_{i=1}^{n}\left\{\sum_{k=1}^{K}\sum_{m=1}^{M}\mathbb{E}(Y\mid A=a,X=x_k)I(X_i=x_k,A_i=a',Y_i=y_m)\right\}\\
&-\frac{1}{n}\sum_{i=1}^{n}\left\{\sum_{k=1}^{K}\sum_{m=1}^{M}\mathbb{E}(Y\mid A=a,X=x_k)\mathbb{P}(Y=y_m,A=a',X=x_k)\right\}\\
=&\frac{1}{n}\sum_{i=1}^{n}\mathbb{E}(Y\mid A=a,X=X_i)I(A_i=a')\\
&-\frac{1}{n}\sum_{i=1}^{n}\left\{\sum_{k=1}^{K}\mathbb{E}(Y\mid A=a,X=x_k)\mathbb{P}(A=a',X=x_k)\right\}.
\end{aligned}
$$

Combining both expressions, the following is obtained:

$$
\begin{aligned}
\frac{\partial\theta_a(\nu_0)}{\partial\nu^\top}(\widehat{\nu}_n-\nu_0)=&\frac{1}{n}\sum_{i=1}^{n}\frac{Y_i-\mathbb{E}(Y\mid A=a,X=X_i)}{\mathbb{P}(A=a\mid X=X_i)}I(A_i=a)\\
&+\frac{1}{n}\sum_{i=1}^{n}\{\mathbb{E}(Y\mid A=a,X=X_i)-\mathbb{E}\left[\mathbb{E}(Y\mid A=a,X)\right]\}.
\end{aligned}
$$

Asymptotic linearity for the MLE $\widehat{\theta}_{\text{ATE},n}$ is thus established:

$$
\begin{aligned}
\widehat{\theta}_{\text{ATE},n}-\theta_{\text{ATE},0}=&\frac{1}{n}\sum_{i=1}^{n}\frac{Y_i-\mathbb{E}(Y\mid A=1,X=X_i)}{\mathbb{P}(A=1\mid X=X_i)}I(A_i=1)\\
&+\frac{1}{n}\sum_{i=1}^{n}\{\mathbb{E}(Y\mid A=1,X=X_i)-\mathbb{E}\left[\mathbb{E}(Y\mid A=1,X)\right]\}\\
&-\frac{1}{n}\sum_{i=1}^{n}\frac{Y_i-\mathbb{E}(Y\mid A=0,X=X_i)}{\mathbb{P}(A=0\mid X=X_i)}I(A_i=0)\\
&-\frac{1}{n}\sum_{i=1}^{n}\{\mathbb{E}(Y\mid A=0,X=X_i)-\mathbb{E}\left[\mathbb{E}(Y\mid A=0,X)\right]\}\\
&+o_p(n^{-1/2}).
\end{aligned}
$$

Thus, the influence function of the MLE $\widehat{\theta}_{\text{ATE},n}$ has been derived:

$$
\begin{aligned}
\phi(X,A,Y)=&\frac{Y-\mathbb{E}(Y\mid A=1,X)}{\mathbb{P}(A=1\mid X)}I(A=1)-\frac{Y-\mathbb{E}(Y\mid A=0,X)}{\mathbb{P}(A=0\mid X)}I(A=0)\\
&+\left[\mathbb{E}(Y\mid A=1,X)-\mathbb{E}(Y\mid A=0,X)\right]-\theta_{\text{ATE},0}.
\end{aligned}
$$

Since the MLE is efficient, the above influence function is the EIF for the estimation of θ_{ATE}.

An equivalent expression of the EIF is (see Exercise 12.3):

$$\begin{aligned}\phi(X, A, Y) = &\frac{(2A-1)}{\mathbb{P}(A \mid X)}[Y - \mathbb{E}(Y \mid A, X)] \\ &+ [\mathbb{E}(Y \mid A = 1, X) - \mathbb{E}(Y \mid A = 0, X)] - \theta_{\text{ATE},0}.\end{aligned}$$

Gold Coin # 43. *The validity of Approach One for deriving the EIF is attributed to the multinomial distribution's dual nature—being both parametric and nonparametric. As a result, the MLE of the estimand is regarded as both a parametric MLE and a nonparametric MLE. Consequently, the EIF derived within the parametric framework is also valid in the nonparametric context.*

12.3 Approach Two to Deriving the EIF

The EIF for the estimation of the following statistical estimand is now derived:

$$\theta_{\text{ATE}} = \mathbb{E}\left[\frac{(2A-1)Y}{\mathbb{P}(A \mid X)}\right].$$

12.3.1 Propensity Score Function is Known

Assume that the propensity score function, $g_0(a \mid X) = \mathbb{P}(A = a \mid X)$, is known. Under this assumption, and by applying the first application of the fundamental theorem of regularity (FTR) discussed in Subsection 11.3.4, the following estimator is obtained:

$$\widetilde{\theta}_{\text{ATE},n} = \frac{1}{n}\sum_{i=1}^{n} \frac{2(A_i - 1)Y_i}{\mathbb{P}(A_i \mid X_i)},$$

which serves as a regular and asymptotically linear (RAL) estimator for θ_{ATE}. The corresponding influence function is given by:

$$\widetilde{\varphi}(X, A, Y) = \frac{(2A-1)Y}{\mathbb{P}(A \mid X)} - \theta_{\text{ATE},0}.$$

Let $p_0(x)$ denote the true marginal distribution of X, and let $f_0(y \mid A, X)$ denote the true conditional distribution of Y given A and X. Consider the parametric submodels $p(x;\nu_1)$ and $f(y \mid A, X; \nu_3)$ such that $p(x;\nu_1)$ passes through $p_0(x)$ at $\nu_1 = 0$, and $f(y \mid A, X; \nu_3)$ passes through $f_0(y \mid A, X)$ at $\nu_3 = 0$.

Under these submodels, the estimand can be expressed as a function of the parameters ν_1 and ν_3:

$$\theta_{\text{ATE}} = \theta_{\text{ATE}}(\nu_1, \nu_3).$$

The corresponding log-likelihood function under these submodels is:

$$\ell(\nu_1, \nu_3) = \log p(X; \nu_1) + \log g_0(A \mid X) + \log f(Y \mid A, X; \nu_3).$$

Consequently, the score functions evaluated at the true parameter values are:

$$S_{0;\nu_1}(X) = \left.\frac{\partial \log p(X; \nu_1)}{\partial \nu_1}\right|_{\nu_1=0},$$

$$S_{0;\nu_3}(X, A, Y) = \left.\frac{\partial \log f(Y \mid A, X; \nu_3)}{\partial \nu_3}\right|_{\nu_3=0}.$$

According to the FTR, since $\widetilde{\theta}_{\text{ATE},n}$ is RAL, the following equations hold:

$$\left.\frac{\partial \theta_{\text{ATE}}(\nu_1, \nu_3)}{\partial \nu_1}\right|_{\nu_1=0,\nu_3=0} = \mathbb{E}[\widetilde{\varphi}(X, A, Y) S_{0;\nu_1}(X)],$$

$$\left.\frac{\partial \theta_{\text{ATE}}(\nu_1, \nu_3)}{\partial \nu_3}\right|_{\nu_1=0,\nu_3=0} = \mathbb{E}[\widetilde{\varphi}(X, A, Y) S_{0;\nu_3}(X, A, Y)].$$

These two equations will be used in the next subsection.

12.3.2 Propensity Score Function is Unknown

In this subsection, the following lemma will be used multiple times.

A Lemma

Any function $S(A, X)$ satisfying $\mathbb{E}[S(A, X) \mid X] = 0$ can be expressed as

$$S(A, X) = T(X)[A - \mathbb{P}(A = 1 \mid X)],$$

for some function $T(X)$.

Proof: Given that $\mathbb{E}[S(A, X) \mid X] = 0$, it follows that

$$S(1, X)\mathbb{P}(A = 1 \mid X) + S(0, X)\mathbb{P}(A = 0 \mid X) = 0.$$

Then, the following identities can be derived:

$$S(1, X) = -\frac{S(0, X)}{\mathbb{P}(A = 1 \mid X)}\left[1 - \mathbb{P}(A = 1 \mid X)\right],$$

$$S(0, X) = -\frac{S(0, X)}{\mathbb{P}(A = 1 \mid X)}\left[0 - \mathbb{P}(A = 1 \mid X)\right].$$

By defining $T(X) = -S(0, X)/\mathbb{P}(A = 1 \mid X)$, the result follows. □

From Known PS to Unknown PS

Let $p_0(x)$ denote the probability distribution of X, let $g_0(a \mid X)$ denote the unknown propensity score function, and let $f_0(y \mid A, X)$ denote the conditional distribution of Y given A and X. Let $p(x; \nu_1)$ be a parametric submodel that coincides with $p_0(x)$ when $\nu_1 = 0$, $g(a \mid X; \nu_2)$ a parametric submodel passing through $g_0(a \mid X)$ at $\nu_2 = 0$, and $f(y \mid A, X; \nu_3)$ a submodel passing through $f_0(y \mid A, X)$ at $\nu_3 = 0$.

The estimand θ_{ATE} under these submodels is denoted as

$$\theta_{\text{ATE}} = \theta_{\text{ATE}}(\nu_1, \nu_2, \nu_3).$$

Actually, θ_{ATE} is independent of ν_2 (see Exercise 12.4), that is,

$$\theta_{\text{ATE}} = \theta_{\text{ATE}}(\nu_1, \nu_2, \nu_3) = \theta_{\text{ATE}}(\nu_1, \nu_3).$$

The corresponding log-likelihood is given by

$$\ell(\nu_1, \nu_2, \nu_3) = \log p(X; \nu_1) + \log g(A \mid X; \nu_2) + \log f(y \mid A, X; \nu_3),$$

and the associated score functions evaluated at the true parameter values are

$$\begin{aligned} S_{0;\nu_1}(X) &= \left. \frac{\partial \log p(X; \nu_1)}{\partial \nu_1} \right|_{\nu_1 = 0}, \\ S_{0;\nu_2}(X, A) &= \left. \frac{\partial \log g(A \mid X; \nu_2)}{\partial \nu_2} \right|_{\nu_2 = 0}, \\ S_{0;\nu_3}(X, A, Y) &= \left. \frac{\partial \log f(Y \mid A, X; \nu_3)}{\partial \nu_3} \right|_{\nu_3 = 0}. \end{aligned}$$

From likelihood theory, the following expectations hold:

$$\begin{aligned} \mathbb{E}[S_{0;\nu_1}(X)] &= 0, \\ \mathbb{E}[S_{0;\nu_2}(X, A) \mid X] &= 0, \\ \mathbb{E}[S_{0;\nu_3}(X, A, Y) \mid A, X] &= 0. \end{aligned}$$

A regular influence function satisfying the conditions in the FTR is to be constructed by augmenting the influence function derived under the assumption of a known propensity score function. That is, a regular influence function of the form

$$\begin{aligned} \varphi(X, A, Y) &= \widetilde{\varphi}(X, A, Y) + S(X, A) \\ &= \frac{(2A - 1)Y}{\mathbb{P}(A \mid X)} - \theta_{\text{ATE},0} + S(X, A), \end{aligned}$$

is considered, and it is required that the following equalities are satisfied:

$$\left.\frac{\partial \theta_{\text{ATE}}(\nu_1, \nu_2, \nu_3)}{\partial \nu_1}\right|_0 \overset{(a)}{=} \mathbb{E}[\varphi(X, A, Y) S_{0;\nu_1}(X)],$$

$$\left.\frac{\partial \theta_{\text{ATE}}(\nu_1, \nu_2, \nu_3)}{\partial \nu_2}\right|_0 \overset{(b)}{=} \mathbb{E}[\varphi(X, A, Y) S_{0;\nu_2}(X, A)],$$

$$\left.\frac{\partial \theta_{\text{ATE}}(\nu_1, \nu_2, \nu_3)}{\partial \nu_3}\right|_0 \overset{(c)}{=} \mathbb{E}[\varphi(X, A, Y) S_{0;\nu_3}(X, A, Y)].$$

Given that (i.e. the last two equations in the preceding subsection)

$$\left.\frac{\partial \theta_{\text{ATE}}(\nu_1, \nu_3, \nu_3)}{\partial \nu_1}\right|_0 = \mathbb{E}[\widetilde{\varphi}(X, A, Y) S_{0;\nu_1}(X)],$$

$$\left.\frac{\partial \theta_{\text{ATE}}(\nu_1, \nu_2, \nu_3)}{\partial \nu_3}\right|_0 = \mathbb{E}[\widetilde{\varphi}(X, A, Y) S_{0;\nu_3}(X, A, Y)],$$

to satisfy the above conditions (a) and (c), the function $S(X, A)$ must satisfy

$$\mathbb{E}[S(X, A) S_{0;\nu_1}(X)] \overset{(1)}{=} 0,$$

$$\mathbb{E}[S(X, A) S_{0;\nu_3}(X, A, Y)] \overset{(2)}{=} 0,$$

for all parametric submodels. The second condition (2) is automatically satisfied because

$$\begin{aligned}\mathbb{E}[S(X, A) S_{0;\nu_3}(X, A, Y)] &= \mathbb{E}\left\{\mathbb{E}[S(X, A) S_{0;\nu_3}(X, A, Y) \mid A, X]\right\} \\ &= \mathbb{E}\{S(X, A) \mathbb{E}[S_{0;\nu_3}(X, A, Y) \mid A, X]\} = 0.\end{aligned}$$

To satisfy the first condition (1), it is sufficient that

$$\mathbb{E}[S(X, A) \mid X] = 0.$$

By the earlier lemma, this implies

$$S(X, A) = T(X)[A - \mathbb{P}(A = 1 \mid X)],$$

for some function $T(X)$. Consequently, a candidate influence function of the form

$$\varphi(X, A, Y) = \frac{(2A - 1)Y}{\mathbb{P}(A \mid X)} - \theta_{\text{ATE},0} + T(X)[A - \mathbb{P}(A = 1 \mid X)],$$

satisfies

$$\left.\frac{\partial \theta_{\text{ATE}}(\nu_1, \nu_2, \nu_3)}{\partial \nu_1}\right|_0 = \mathbb{E}[\varphi(X, A, Y) S_{0;\nu_1}(X)],$$

$$\left.\frac{\partial \theta_{\text{ATE}}(\nu_1, \nu_2, \nu_3)}{\partial \nu_3}\right|_0 = \mathbb{E}[\varphi(X, A, Y) S_{0;\nu_3}(X, A, Y)].$$

The remaining requirement is to determine $T(X)$ that satisfies the condition (b) for all submodels $g(a \mid X; \nu_2)$. Because

$$\theta_{\text{ATE}}(\nu_1, \nu_2, \nu_3) = \mathbb{E}_{\nu_1}[\mathbb{E}_{\nu_3}(Y \mid A = 1, X)] - \mathbb{E}_{\nu_1}[\mathbb{E}_{\nu_3}(Y \mid A = 0, X)],$$

is independent of ν_2,

$$\left.\frac{\partial \theta_{\text{ATE}}(\nu_1, \nu_2, \nu_3)}{\partial \nu_2}\right|_0 = 0.$$

By the earlier lemma again, the score function $S_{0;\nu_2}(X, A)$ can be written as

$$S_{0;\nu_2}(X, A) = T'(X)[A - \mathbb{P}(A = 1 \mid X)].$$

Therefore, $T(X)$ must be chosen such that

$$\mathbb{E}\left(\left\{\frac{(2A-1)Y}{\mathbb{P}(A \mid X)} + T(X)[A - \mathbb{P}(A = 1 \mid X)]\right\}[A - \mathbb{P}(A = 1 \mid X)] \;\middle|\; X\right) = 0.$$

Solving the above equation (see Exercise 12.5), the solution is

$$T(X) = -\left[\frac{\mathbb{E}(Y \mid A = 1, X)}{\mathbb{P}(A = 1 \mid X)} + \frac{\mathbb{E}(Y \mid A = 0, X)}{\mathbb{P}(A = 0 \mid X)}\right].$$

In summary, the following regular influence function for the estimation of θ_{ATE} when the PS is unknown is obtained:

$$\begin{aligned}\varphi(X, A, Y) = &\frac{(2A-1)Y}{\mathbb{P}(A \mid X)} - \theta_{\text{ATE},0}\\ &- \left[\frac{\mathbb{E}(Y \mid A = 1, X)}{\mathbb{P}(A = 1 \mid X)} + \frac{\mathbb{E}(Y \mid A = 0, X)}{\mathbb{P}(A = 0 \mid X)}\right][A - \mathbb{P}(A = 1 \mid X)].\end{aligned}$$

Gold Coin # 44. *Approach Two for deriving the EIF proceeds in two stages: first, an influence function is derived under the assumption of a known propensity score; then, this function is adjusted to obtain a regular influence function using the fundamental theorem of regularity (FTR).*

However, the mission of deriving the EIF is not yet accomplished. The fact that a function is regular does not imply that it is EIF. In the next chapter, it will be demonstrated that the regular influence function $\varphi(X, A, Y)$ is in fact unique in the estimation of θ_{ATE}, and hence qualifies as the EIF.

12.4 Exercises

Exercise 12.1

Verify that

$$\frac{\partial \theta_a(\nu_0)}{\partial \nu_{k,a,m}} = \frac{y_m - \mathbb{E}(Y \mid A = a, X = x_k)}{\mathbb{P}(A = a \mid X = x_k)} + \mathbb{E}(Y \mid A = a, X = x_k),$$

Exercise 12.2

Verify that

$$\frac{\partial \theta_a(\nu_0)}{\partial \nu_{k,a',m}} = \mathbb{E}(Y \mid A = a, X = x_k).$$

Exercise 12.3

Show that the following two are equivalent:

$$\phi(X, A, Y) = \frac{Y - \mathbb{E}(Y \mid A = 1, X)}{\mathbb{P}(A = 1 \mid X)} I(A = 1) - \frac{Y - \mathbb{E}(Y \mid A = 0, X)}{\mathbb{P}(A = 0 \mid X)} I(A = 0)$$
$$+ [\mathbb{E}(Y \mid A = 1, X) - \mathbb{E}(Y \mid A = 0, X)] - \theta_{\text{ATE},0},$$

and

$$\phi(X, A, Y) = \frac{(2A - 1)}{\mathbb{P}(A \mid X)}[Y - \mathbb{E}(Y \mid A, X)]$$
$$+ [\mathbb{E}(Y \mid A = 1, X) - \mathbb{E}(Y \mid A = 0, X)] - \theta_{\text{ATE},0}.$$

Exercise 12.4

Show that

$$\theta_{\text{ATE}}(\nu_1, \nu_2, \nu_3) = \theta_{\text{ATE}}(\nu_1, \nu_3).$$

Exercise 12.5

Solve the following equation for $T(X)$,

$$\mathbb{E}\left(\left\{\frac{(2A - 1)Y}{\mathbb{P}(A \mid X)} + T(X)[A - \mathbb{P}(A = 1 \mid X)]\right\}[A - \mathbb{P}(A = 1 \mid X)] \;\middle|\; X\right) = 0.$$

13

A Convenient Approach

13.1 Geometric Viewpoint

In the preceding chapter, two approaches were discussed for deriving the efficient influence function (EIF) for the average treatment effect (ATE). The first approach yields the following EIF:

$$\begin{aligned}\phi(X, A, Y) =& \frac{(2A-1)}{\mathbb{P}(A \mid X)}[Y - \mathbb{E}(Y \mid A, X)] \\ &+ [\mathbb{E}(Y \mid A=1, X) - \mathbb{E}(Y \mid A=0, X)] - \theta_{\text{ATE},0},\end{aligned}$$

while the second approach results in the following regular influence function:

$$\begin{aligned}\varphi(X, A, Y) =& \frac{(2A-1)Y}{\mathbb{P}(A \mid X)} - \theta_{\text{ATE},0} \\ &- \left[\frac{\mathbb{E}(Y \mid A=1, X)}{\mathbb{P}(A=1 \mid X)} + \frac{\mathbb{E}(Y \mid A=0, X)}{\mathbb{P}(A=0 \mid X)}\right][A - \mathbb{P}(A=1 \mid X)].\end{aligned}$$

Fortunately, these two expressions are equivalent (see Exercise 13.1):

$$\phi(X, A, Y) = \varphi(X, A, Y).$$

To establish this equivalence, it suffices to verify that both functions coincide when $A = 1$ and $A = 0$, respectively.

The question is remaining as to whether the regular influence function—via the second approach—satisfying the following conditions in the fundamental theorem of regularity (FTR) is unique in the estimation of the ATE:

$$\left.\frac{\partial \theta_{\text{ATE}}(\nu_1, \nu_2, \nu_3)}{\partial \nu_1}\right|_{\nu_1=\nu_2=\nu_3=0} = \mathbb{E}\varphi(X, A, Y) S_{0;\nu_1}(X),$$

$$\left.\frac{\partial \theta_{\text{ATE}}(\nu_1, \nu_2, \nu_3)}{\partial \nu_2}\right|_{\nu_1=\nu_2=\nu_3=0} = \mathbb{E}\varphi(X, A, Y) S_{0;\nu_2}(X, A),$$

$$\left.\frac{\partial \theta_{\text{ATE}}(\nu_1, \nu_2, \nu_3)}{\partial \nu_3}\right|_{\nu_1=\nu_2=\nu_3=0} = \mathbb{E}\varphi(X, A, Y) S_{0;\nu_3}(X, A, Y),$$

DOI: 10.1201/9781003635468-16

for any parametric submodel characterized by the following log-likelihood:

$$\ell(\nu_1, \nu_2, \nu_3) = \log p(X; \nu_1) + \log g(A \mid X; \nu_2) + \log f(y \mid A, X; \nu_3),$$

and the corresponding score functions evaluated at the truth:

$$\begin{aligned} S_{0;\nu_1}(X) &= \frac{\partial \log p(X; \nu_1)}{\partial \nu_1}\Big|_{\nu_1=0}, \\ S_{0;\nu_2}(X, A) &= \frac{\partial \log g(A \mid X; \nu_2)}{\partial \nu_2}\Big|_{\nu_2=0}, \\ S_{0;\nu_3}(X, A, Y) &= \frac{\partial \log f(Y \mid A, X; \nu_3)}{\partial \nu_3}\Big|_{\nu_3=0}. \end{aligned}$$

Spoiler alert: The regular influence function derived via the second approach is indeed unique in the estimation of the ATE. This uniqueness can be established by viewing influence functions from a geometric perspective.[41, 42]

13.1.1 Hilbert Space and Tangent Space

Define the set of all random variables with zero mean and finite variance as:

$$\mathcal{H} = \{h(X, A, Y) : \mathbb{E}[h(X, A, Y)] = 0,\ \mathbb{V}[h(X, A, Y)] < \infty\},$$

which constitutes a Hilbert space equipped with the inner product

$$\begin{aligned} \langle h_1(X, A, Y), h_2(X, A, Y)\rangle &= \mathrm{Cov}(h_1(X, A, Y), h_2(X, A, Y)) \\ &= \mathbb{E}[h_1(X, A, Y)h_2(X, A, Y)], \end{aligned}$$

for any $h_1(X, A, Y), h_2(X, A, Y) \in \mathcal{H}$.

According to the FTR, an influence function is said to be regular if and only if the FTR conditions are satisfied. The goal is to identify a regular influence function $\varphi_{\text{ATE-EIF}}(X, A, Y)$ from $\mathcal{H}$ that minimizes the variance among all such regular influence functions.

The following three subspaces are defined:

$$\begin{aligned} \mathcal{T}_1 &= \{S_{0;\nu_1}(X) : \text{for any regular parametric submodel } p(x; \nu_1)\} \\ &= \{S_1(X) : \mathbb{E}[S_1(X)] = 0,\ \mathbb{V}[S_1(X)] < \infty\}, \\ \mathcal{T}_2 &= \{S_{0;\nu_2}(X, A) : \text{for any regular parametric submodel } g(a \mid X; \nu_2)\} \\ &= \{S_2(X, A) : \mathbb{E}[S_2(X, A) \mid X] = 0,\ \mathbb{V}[S_2(X, A)] < \infty\}, \\ \mathcal{T}_3 &= \{S_{0;\nu_3}(X, A, Y) : \text{for any regular parametric submodel } f(y \mid A, X; \nu_3)\} \\ &= \{S_3(X, A, Y) : \mathbb{E}[S_3(X, A, Y) \mid A, X] = 0,\ \mathbb{V}[S_3(X, A, Y)] < \infty\}. \end{aligned}$$

The *tangent space* is then defined as

$$\begin{aligned} \mathcal{T} &= \mathcal{T}_1 \oplus \mathcal{T}_2 \oplus \mathcal{T}_3 \\ &= \{S(X, A, Y) = S_1(X) + S_2(X, A) + S_3(X, A, Y) : S_j \in \mathcal{T}_j,\ j = 1, 2, 3\}. \end{aligned}$$

13.1.2 Existence and Uniqueness

Based on the definitions of the Hilbert space $\mathcal{H}$ and the tangent space $\mathcal{T}$ in the context of estimating the ATE, it follows that

$$\mathcal{T} \subset \mathcal{H},$$

since for any $S_j \in \mathcal{T}_j$, the sum satisfies $\mathbb{E}(S_1 + S_2 + S_3) = 0$. Conversely,

$$\mathcal{H} \subset \mathcal{T},$$

as any $h(X, A, Y) \in \mathcal{H}$ can be decomposed as

$$h(X, A, Y) = \mathbb{E}[h \mid X] + \{\mathbb{E}[h \mid X, A] - \mathbb{E}[h \mid X]\} + \{h - \mathbb{E}[h \mid X, A]\},$$

where $\mathbb{E}[h \mid X] \in \mathcal{T}_1$, $\mathbb{E}[h \mid X, A] - \mathbb{E}[h \mid X] \in \mathcal{T}_2$, and $h - \mathbb{E}[h \mid X, A] \in \mathcal{T}_3$.

Therefore, in the estimation of the ATE,

$$\mathcal{T} = \mathcal{H}.$$

As a result, if a regular influence function can be constructed, it must be unique and hence is the EIF. In fact, suppose that two regular influence functions $\varphi_1, \varphi_2 \in \mathcal{H}$ both satisfy the FTR conditions:

$$\left.\frac{\partial \theta_{\text{ATE}}(\nu_1, \nu_2, \nu_3)}{\partial \nu_1}\right|_0 = \mathbb{E}[\varphi_k(X, A, Y) S_{\nu_1}(X)],$$

$$\left.\frac{\partial \theta_{\text{ATE}}(\nu_1, \nu_2, \nu_3)}{\partial \nu_2}\right|_0 = \mathbb{E}[\varphi_k(X, A, Y) S_{\nu_2}(X, A)],$$

$$\left.\frac{\partial \theta_{\text{ATE}}(\nu_1, \nu_2, \nu_3)}{\partial \nu_3}\right|_0 = \mathbb{E}[\varphi_k(X, A, Y) S_{\nu_3}(X, A, Y)],$$

for $k = 1, 2$, where the subscript 0 means $\nu_1 = \nu_2 = \nu_3 = 0$. Then, for any $S_j \in \mathcal{T}_j$,

$$\mathbb{E}[(\varphi_1 - \varphi_2) S_j] = 0.$$

Thus, for all $S \in \mathcal{T}$,

$$\mathbb{E}[(\varphi_1 - \varphi_2) S] = 0.$$

In particular, since $\mathcal{T} = \mathcal{H}$, by setting $S = \varphi_1 - \varphi_2$, it follows that

$$\mathbb{E}[(\varphi_1 - \varphi_2)^2] = 0,$$

which implies $\varphi_1 = \varphi_2$ with probability one (that is, almost surely). □

Given the uniqueness of the regular influence function in the estimation of the ATE, the incomplete task from the preceding chapter can now be completed. In Chapter 12, it was shown—via the second approach—that a regular influence function can be constructed. Based on the uniqueness just shown here, it can now be concluded that this function is indeed the EIF.

In summary, the EIF in the estimation of the ATE is given by:

$$\begin{aligned}\varphi_{\text{ATE-EIF}}(X, A, Y) = \;& \frac{(2A-1)}{\mathbb{P}(A \mid X)}[Y - \mathbb{E}(Y \mid A, X)] \\ &+ [\mathbb{E}(Y \mid A = 1, X) - \mathbb{E}(Y \mid A = 0, X)] - \theta_{\text{ATE},0}.\end{aligned}$$

Gold Coin # 45. *In estimating the ATE, the existence and uniqueness of the regular influence function are established, thereby identifying it as the efficient influence function (EIF).*

13.1.3 Projection

Although in all treatment effect estimation settings discussed in this book the equality $\mathcal{T} = \mathcal{H}$ holds, situations do exist in which $\mathcal{T} \subsetneq \mathcal{H}$.

In such settings, the concept of projection may be employed to construct the EIF. Suppose a regular influence function $\varphi \in \mathcal{H}$ has already been obtained. The projection of φ onto the tangent space $\mathcal{T}$ is defined as

$$\varphi^* = \Pi(\varphi \mid \mathcal{T}).$$

That is, for any $\widetilde{\varphi} \in \mathcal{T}$,

$$\langle \varphi - \varphi^*, \widetilde{\varphi} \rangle = 0.$$

This implies that φ^* also satisfies the FTR and is therefore regular. By the Pythagorean theorem,

$$\mathbb{E}[(\varphi^*)^2] + \mathbb{E}[(\varphi - \varphi^*)^2] = \mathbb{E}[\varphi^2],$$

and consequently,

$$\mathbb{V}(\varphi^*) \leq \mathbb{V}(\varphi).$$

Therefore, φ^* is the EIF that is constructed given a regular influence function.

13.2 A Convenient Approach to Deriving EIFs

13.2.1 Approach

A convenient approach is now presented for constructing a regular influence function that satisfies the FTR conditions. Given the established existence and uniqueness of the regular influence function, any such function that satisfies the FTR conditions must be the EIF.

This approach relies on the fact that two equivalent representations exist for the ATE estimand:

$$\theta_{\text{ATE},0} = \mathbb{E}\left[\mathbb{E}(Y \mid A = 1, X) - \mathbb{E}(Y \mid A = 0, X)\right],$$
$$\theta_{\text{ATE},0} = \mathbb{E}\left\{\frac{(2A-1)}{\mathbb{P}(A \mid X)}Y\right\}.$$

Let the integrands of the above expectations be denoted as:

$$\eta_1(X) = \mathbb{E}(Y \mid A = 1, X) - \mathbb{E}(Y \mid A = 0, X),$$
$$\eta_2(X, A, Y) = \frac{(2A-1)}{\mathbb{P}(A \mid X)}Y,$$

Therefore, the following holds:

$$\theta_{\text{ATE},0} = \mathbb{E}[\eta_1(X)] = \mathbb{E}[\eta_2(X, A, Y)].$$

The procedure for identifying a regular influence function proceeds in two steps: subtraction and addition.

- **Step 1 (Subtraction):** The mean of $\eta_1(X)$ is subtracted, and the conditional expectation of $\eta_2(X, A, Y)$ given (X, A) is also subtracted, yielding two mean-zero functions:

$$\phi_1(X) = \mathbb{E}(Y \mid A = 1, X) - \mathbb{E}(Y \mid A = 0, X) - \theta_{\text{ATE},0},$$
$$\phi_2(X, A, Y) = \frac{(2A-1)}{\mathbb{P}(A \mid X)}\left[Y - \mathbb{E}(Y \mid A, X)\right].$$

  It can be verified that $\phi_1(X) \in \mathcal{T}_1$ and $\phi_2(X, A, Y) \in \mathcal{T}_3$.

- **Step 2 (Addition):** The two mean-zero functions are then added to yield the desired regular influence function:

$$\phi_{\text{ATE}}(X, A, Y) = \phi_1(X) + \phi_2(X, A, Y).$$

13.2.2 Proof

To establish the validity of the above method for constructing a regular influence function, it is sufficient to verify the following:

$$\left.\frac{\partial\theta_{\text{ATE}}(\nu_1, \nu_2, \nu_3)}{\partial\nu_1}\right|_0 = \mathbb{E}\left\{[\phi_1(X) + \phi_2(X, A, Y)]S_{0;\nu_1}(X)\right\},$$
$$\left.\frac{\partial\theta_{\text{ATE}}(\nu_1, \nu_2, \nu_3)}{\partial\nu_2}\right|_0 = \mathbb{E}\left\{[\phi_1(X) + \phi_2(X, A, Y)]S_{0;\nu_2}(X, A)\right\},$$
$$\left.\frac{\partial\theta_{\text{ATE}}(\nu_1, \nu_2, \nu_3)}{\partial\nu_3}\right|_0 = \mathbb{E}\left\{[\phi_1(X) + \phi_2(X, A, Y)]S_{0;\nu_3}(X, A, Y)\right\},$$

for any parametric submodel with log-likelihood

$$\ell(\nu_1, \nu_2, \nu_3) = \log p(X; \nu_1) + \log g(A \mid X; \nu_2) + \log f(Y \mid A, X; \nu_3),$$

and corresponding score functions evaluated at the truth:

$$\begin{aligned} S_{0;\nu_1}(X) &= \left.\frac{\partial \log p(X; \nu_1)}{\partial \nu_1}\right|_{\nu_1=0}, \\ S_{0;\nu_2}(X, A) &= \left.\frac{\partial \log g(A \mid X; \nu_2)}{\partial \nu_2}\right|_{\nu_2=0}, \\ S_{0;\nu_3}(X, A, Y) &= \left.\frac{\partial \log f(Y \mid A, X; \nu_3)}{\partial \nu_3}\right|_{\nu_3=0}. \end{aligned}$$

Using the law of iterated expectations (LIE), the following identities hold (see Exercise 13.2):

$$\begin{aligned} \mathbb{E}\left\{\phi_2(X, A, Y) S_{0;\nu_1}(X)\right\} &= 0, \\ \mathbb{E}\left\{\phi_1(X) S_{0;\nu_2}(X, A)\right\} &= 0, \\ \mathbb{E}\left\{\phi_2(X, A, Y) S_{0;\nu_2}(X, A)\right\} &= 0, \\ \mathbb{E}\left\{\phi_1(X) S_{0;\nu_3}(X, A, Y)\right\} &= 0. \end{aligned}$$

Therefore, it is sufficient to confirm that:

$$\begin{aligned} &\text{(a):} \quad \left.\frac{\partial \theta_{\text{ATE}}(\nu)}{\partial \nu_1}\right|_0 = \mathbb{E}\left\{\phi_1(X) S_{0;\nu_1}(X)\right\}, \\ &\text{(b):} \quad \left.\frac{\partial \theta_{\text{ATE}}(\nu)}{\partial \nu_2}\right|_0 = 0, \\ &\text{(c):} \quad \left.\frac{\partial \theta_{\text{ATE}}(\nu)}{\partial \nu_3}\right|_0 = \mathbb{E}\left\{\phi_2(X, A, Y) S_{0;\nu_3}(X, A, Y)\right\}. \end{aligned}$$

Verification of (a):

$$\begin{aligned} &\left.\frac{\partial \theta_{\text{ATE}}(\nu)}{\partial \nu_1}\right|_0 \\ &= \left.\frac{\partial}{\partial \nu_1} \int \left[\mathbb{E}_{\nu_3}(Y \mid A = 1, x) - \mathbb{E}_{\nu_3}(Y \mid A = 0, x)\right] p(x; \nu_1)\, dx\right|_0 \\ &= \left.\frac{\partial}{\partial \nu_1} \int \left[\mathbb{E}(Y \mid A = 1, x) - \mathbb{E}(Y \mid A = 0, x)\right] p(x; \nu_1)\, dx\right|_{\nu_1=0} \\ &= \left.\frac{\partial}{\partial \nu_1} \int \left[\mathbb{E}(Y \mid A = 1, x) - \mathbb{E}(Y \mid A = 0, x) - \theta_{\text{ATE},0}\right] p(x; \nu_1)\, dx\right|_{\nu_1=0} \\ &= \left.\frac{\partial}{\partial \nu_1} \int \phi_1(x) p(x; \nu_1) dx\right|_{\nu_1=0} = \left.\int \phi_1(x) \frac{\partial p(x; \nu_1)/\partial \nu_1}{p(x; \nu_1)} p(x; \nu_1)\, dx\right|_{\nu_1=0} \\ &= \int \phi_1(x) S_{0;\nu_1}(x) p(x; 0)\, dx = \mathbb{E}[\phi_1(X) S_{0;\nu_1}(X)]. \end{aligned}$$

Verification of (b):

$$\begin{aligned}
&\left.\frac{\partial \theta_{\text{ATE}}(\nu)}{\partial \nu_2}\right|_0 \\
=&\frac{\partial}{\partial \nu_2}\mathbb{E}_{\nu_1}\left[\mathbb{E}_{\nu_3}(Y \mid A=1, X) - \mathbb{E}_{\nu_3}(Y \mid A=0, X)\right]\Big|_0 = 0.
\end{aligned}$$

Verification of (c):

$$\begin{aligned}
&\left.\frac{\partial \theta_{\text{ATE}}(\nu)}{\partial \nu_3}\right|_0 \\
=&\frac{\partial}{\partial \nu_3}\mathbb{E}_{\nu_1}\left[\frac{(2A-1)\mathbb{E}_{\nu_3}(Y \mid A, X)}{\mathbb{P}_{\nu_2}(A \mid X)}\right]\Bigg|_0 \\
=&\frac{\partial}{\partial \nu_3}\mathbb{E}\left[\frac{(2A-1)\mathbb{E}_{\nu_3}(Y \mid A, X)}{\mathbb{P}(A \mid X)}\right]\Bigg|_{\nu_3=0} \\
=&\frac{\partial}{\partial \nu_3}\mathbb{E}\left[\frac{(2A-1)\int y f(y \mid A, X; \nu_3)\, dy}{\mathbb{P}(A \mid X)}\right]\Bigg|_{\nu_3=0} \\
=&\mathbb{E}\left[\frac{(2A-1)\int y \partial f(y \mid A, X; \nu_3)/\partial \nu_3|_{\nu_3=0}\, dy}{\mathbb{P}(A \mid X)}\right] \\
=&\mathbb{E}\left[\frac{(2A-1)\int y S_{0;\nu_3}(X, A, y) f_0(y|A, X)\, dy}{\mathbb{P}(A \mid X)}\right] \\
=&\mathbb{E}\left[\frac{(2A-1)\mathbb{E}\{Y S_{0;\nu_3}(X, A, Y) \mid A, X\}}{\mathbb{P}(A \mid X)}\right] \\
=&\mathbb{E}\left[\frac{(2A-1)\mathbb{E}\{[Y - \mathbb{E}(Y \mid A, X)] S_{0;\nu_3}(X, A, Y) \mid A, X\}}{\mathbb{P}(A \mid X)}\right] \\
=&\mathbb{E}\left[\frac{(2A-1)[Y - \mathbb{E}(Y \mid A, X)] S_{0;\nu_3}(X, A, Y)}{\mathbb{P}(A \mid X)}\right] \\
=&\mathbb{E}\left\{\phi_2(X, A, Y) S_{0;\nu_3}(X, A, Y)\right\}.
\end{aligned}$$

Thus, the constructed influence function $\phi_{\text{ATE}}(X, A, Y)$ is confirmed to be regular. Given the aforementioned uniqueness, it is therefore the EIF. □

Gold Coin # 46. *The convenient approach consists of two steps: subtraction and addition. Its validity hinges on the fact that the statistical estimand has two equivalent representations.*

13.2.3 Notation

For notational simplicity, define the following:

$$Q_0(X, A) = \mathbb{E}(Y \mid A, X), \quad \text{(outcome regression function)}$$
$$g_0(A \mid X) = \mathbb{P}(A \mid X), \quad \text{(propensity score function)}$$

which allow the influence function components to be rewritten as:

$$\phi_1(X) = Q_0(X, 1) - Q_0(X, 0) - \theta_{\text{ATE},0},$$
$$\phi_2(X, A, Y) = \frac{(2A-1)}{g_0(A \mid X)} \left[Y - Q_0(X, A)\right],$$

and hence the EIF for estimating the ATE estimand becomes:

$$\begin{aligned}\phi_{\text{ATE}}(X, A, Y) &= \phi_1(X) + \phi_2(X, A, Y) \\ &= \frac{(2A-1)}{g_0(A \mid X)} \left[Y - Q_0(X, A)\right] + Q_0(X, 1) - Q_0(X, 0) - \theta_{\text{ATE},0}.\end{aligned}$$

13.3 EIFs for ATT and ATC

13.3.1 ATT

Suppose the causal estimand of interest is the average treatment effect on the treated (ATT), defined as

$$\theta^*_{\text{ATT}} = \mathbb{E}(Y^{a=1} - Y^{a=0} \mid A = 1),$$

and assume that observed data are given by $\{O_i = (X_i, A_i, Y_i)\}_{i=1}^n$.

Recall that the causal estimand can be translated into a statistical estimand using two strategies. By the standardization strategy,

$$\theta_{\text{ATT},0} = \mathbb{E}\left\{\frac{I(A=1)}{\mathbb{P}(A=1)} \left[Q_0(X, 1) - Q_0(X, 0)\right]\right\}.$$

By the weighting strategy,

$$\theta_{\text{ATT},0} = \mathbb{E}\left\{\left[\frac{I(A=1)}{\mathbb{P}(A=1)} - \frac{I(A=0)g_0(1 \mid X)}{\mathbb{P}(A=1)g_0(0 \mid X)}\right] Y\right\}.$$

The EIF for estimating θ_{ATT} can be derived using the convenient approach. After applying the subtraction step, two mean-zero functions are obtained:

$$\phi_1(X, A) = \frac{I(A=1)}{\mathbb{P}(A=1)} \left[Q_0(X, 1) - Q_0(X, 0) - \theta_{\text{ATT},0}\right],$$
$$\phi_2(X, A, Y) = \left[\frac{I(A=1)}{\mathbb{P}(A=1)} - \frac{I(A=0)g_0(1 \mid X)}{\mathbb{P}(A=1)g_0(0 \mid X)}\right] \left[Y - Q_0(X, A)\right].$$

After the addition step, the regular influence function is obtained:

$$\begin{aligned}
\phi_{\text{ATT}}(X,A,Y) &= \phi_1(X,A) + \phi_2(X,A,Y) \\
&= \frac{I(A=1)}{\mathbb{P}(A=1)}\left[Q_0(X,1) - Q_0(X,0) - \theta_{\text{ATT},0}\right] \\
&\quad + \left[\frac{I(A=1)}{\mathbb{P}(A=1)} - \frac{I(A=0)g_0(1 \mid X)}{\mathbb{P}(A=1)g_0(0 \mid X)}\right]\left[Y - Q_0(X,A)\right].
\end{aligned}$$

To confirm that $\phi_{\text{ATT}}(X,A,Y)$ is a valid regular influence function, it is sufficient to verify the following equalities:

$$\begin{aligned}
\left.\frac{\partial \theta_{\text{ATT}}(\nu_1,\nu_2,\nu_3)}{\partial \nu_1}\right|_0 &= \mathbb{E}\left\{[\phi_1(X,A) + \phi_2(X,A,Y)]S_{0;\nu_1}(A)\right\}, \\
\left.\frac{\partial \theta_{\text{ATT}}(\nu_1,\nu_2,\nu_3)}{\partial \nu_2}\right|_0 &= \mathbb{E}\left\{[\phi_1(X,A) + \phi_2(X,A,Y)]S_{0;\nu_2}(X,A)\right\}, \\
\left.\frac{\partial \theta_{\text{ATT}}(\nu_1,\nu_2,\nu_3)}{\partial \nu_3}\right|_0 &= \mathbb{E}\left\{[\phi_1(X,A) + \phi_2(X,A,Y)]S_{0;\nu_3}(X,A,Y)\right\},
\end{aligned}$$

for any parametric submodels with likelihood

$$L(\nu_1,\nu_2,\nu_3) = g(1;\nu_1)^A[1 - g(1;\nu_1)]^{1-A}p(X \mid A;\nu_2)f(Y \mid A,X;\nu_3),$$

where the parametric submodel $g(A=1;\nu_1)$ passes $g_0(1) = \mathbb{P}(A=1)$ at $\nu_1 = 0$, the parametric submodel $p(x \mid A;\nu_2)$ passes $p_0(x \mid A)$—the true distribution of X conditional on A—at $\nu_2 = 0$, the parametric submodel $f(y \mid A,X;\nu_3)$ passes $f_0(y \mid A,X)$—the true distribution of Y conditional on A and X—at $\nu_3 = 0$,

$$\begin{aligned}
&\theta_{\text{ATT}}(\nu_1,\nu_2,\nu_3) \\
=&\mathbb{E}_{\nu_2}\left[\mathbb{E}_{\nu_3}(Y \mid A=1,X) - \mathbb{E}_{\nu_3}(Y \mid A=0,X)\right] \\
=&\int\left[\int yf(y \mid 1,x;\nu_3)\,dy - \int yf(y \mid 0,x;\nu_3)\,dy\right]p(x \mid 1;\nu_2)\,dx,
\end{aligned}$$

and the score functions at 0 are given by (see Exercise 13.3)

$$\begin{aligned}
S_{0;\nu_1}(A) &= \frac{A}{g_0(1)} - \frac{1-A}{1-g_0(1)}, \\
S_{0;\nu_2}(X,A) &= \left.\frac{\partial \log p(X \mid A;\nu_2)}{\partial \nu_2}\right|_{\nu_2=0}, \\
S_{0;\nu_3}(X,A,Y) &= \left.\frac{\partial \log f(Y \mid A,X;\nu_3)}{\partial \nu_3}\right|_{\nu_3=0}.
\end{aligned}$$

By the LIE, the following orthogonality conditions hold:

$$\begin{aligned}
\mathbb{E}\left\{\phi_1(X,A)S_{0;\nu_1}(A)\right\} &= 0,\\
\mathbb{E}\left\{\phi_2(X,A,Y)S_{0;\nu_1}(A)\right\} &= 0,\\
\mathbb{E}\left\{\phi_2(X,A,Y)S_{0;\nu_2}(X,A)\right\} &= 0,\\
\mathbb{E}\left\{\phi_1(X,A)S_{0;\nu_3}(X,A,Y)\right\} &= 0.
\end{aligned}$$

Hence, it suffices to verify that:

$$\begin{aligned}
&\text{(a):} \quad \left.\frac{\partial\theta_{\text{ATT}}(\nu)}{\partial\nu_1}\right|_0 = 0,\\
&\text{(b):} \quad \left.\frac{\partial\theta_{\text{ATT}}(\nu)}{\partial\nu_2}\right|_0 = \mathbb{E}\left\{\phi_1(X,A)S_{0;\nu_2}(X,A)\right\},\\
&\text{(c):} \quad \left.\frac{\partial\theta_{\text{ATT}}(\nu)}{\partial\nu_3}\right|_0 = \mathbb{E}\left\{\phi_2(X,A,Y)S_{0;\nu_3}(X,A,Y)\right\}.
\end{aligned}$$

To verify (a), note that

$$\begin{aligned}
&\left.\frac{\partial\theta_{\text{ATT}}(\nu)}{\partial\nu_1}\right|_0\\
=&\left.\frac{\partial}{\partial\nu_1}\mathbb{E}_{\nu_2}\left[\mathbb{E}_{\nu_3}(Y \mid A=1,X) - \mathbb{E}_{\nu_3}(Y \mid A=0,X)\right]\right|_0 = 0.
\end{aligned}$$

To verify (b), note that (see Exercise 13.4)

$$\begin{aligned}
&\left.\frac{\partial\theta_{\text{ATT}}(\nu)}{\partial\nu_2}\right|_0\\
=&\left.\frac{\partial}{\partial\nu_2}\int\left[Q_0(x,1)-Q_0(x,0)\right]p(x \mid A=1;\nu_2)dx\right|_{\nu_2=0}\\
=&\mathbb{E}\left\{\phi_1(X,A)S_{0;\nu_2}(X,A)\right\}.
\end{aligned}$$

To verify (c), note that (see Exercise 13.5)

$$\begin{aligned}
&\left.\frac{\partial\theta_{\text{ATT}}(\nu)}{\partial\nu_3}\right|_0\\
=&\left.\frac{\partial}{\partial\nu_3}\mathbb{E}_{\nu_3}\left\{\left[\frac{I(A=1)}{\mathbb{P}(A=1)}-\frac{I(A=0)\mathbb{P}(A=1\mid X)}{\mathbb{P}(A=1)\mathbb{P}(A=0\mid X)}\right]Y\right\}\right|_{\nu_3=0}\\
=&\mathbb{E}\left\{\phi_2(X,A,Y)S_{0;\nu_3}(X,A,Y)\right\}.
\end{aligned}$$

13.3.2 ATC

To obtain the EIF for the average treatment effect on the controls (ATC), the values of the treatment indicator A in the ATT expressions can be reversed—replacing $A=1$ with $A=0$ and vice versa. Accordingly, the EIF for

ATC is given by:

$$\begin{aligned}\phi_{\text{ATC}}(X, A, Y) = &\frac{I(A=0)}{\mathbb{P}(A=0)}\left[Q_0(X,1) - Q_0(X,0) - \theta_{\text{ATC},0}\right] \\ &+ \left[\frac{I(A=0)}{\mathbb{P}(A=0)} - \frac{I(A=1)g_0(0 \mid X)}{\mathbb{P}(A=0)g_0(1 \mid X)}\right]\left[Y - Q_0(X, A)\right].\end{aligned}$$

Gold Coin # 47. *The same convenient approach proposed for deriving the EIF for the ATE can also be applied to obtain the EIFs for the ATT and ATC.*

13.4 Exercises

Exercise 13.1

Show that the following two influence functions are the same:

$$\begin{aligned}\phi(X, A, Y) =& \frac{(2A-1)}{\mathbb{P}(A \mid X)}[Y - \mathbb{E}(Y \mid A, X)] \\ &+ [\mathbb{E}(Y \mid A=1, X) - \mathbb{E}(Y \mid A=0, X)] - \theta_{\text{ATE},0}, \\ \varphi(X, A, Y) =& \frac{(2A-1)Y}{\mathbb{P}(A \mid X)} - \theta_{\text{ATE},0} \\ &- \left[\frac{\mathbb{E}(Y \mid A=1, X)}{\mathbb{P}(A=1 \mid X)} + \frac{\mathbb{E}(Y \mid A=0, X)}{\mathbb{P}(A=0 \mid X)}\right][A - \mathbb{P}(A=1 \mid X)].\end{aligned}$$

Exercise 13.2

Show that

$$\begin{aligned}\mathbb{E}\left\{\phi_2(X, A, Y)S_{0;\nu_1}(X)\right\} &= 0, \\ \mathbb{E}\left\{\phi_1(X)S_{0;\nu_2}(X, A)\right\} &= 0, \\ \mathbb{E}\left\{\phi_2(X, A, Y)S_{0;\nu_2}(X, A)\right\} &= 0, \\ \mathbb{E}\left\{\phi_1(X)S_{0;\nu_3}(X, A, Y)\right\} &= 0.\end{aligned}$$

Exercise 13.3

Consider the following likelihood,

$$L(\nu_1, \nu_2, \nu_3) = g(1;\nu_1)^A[1 - g(1;\nu_1)]^{1-A}p(X \mid A;\nu_2)f(Y \mid A, X;\nu_3).$$

Show that

$$S_{0;\nu_1}(A) = \frac{A}{g_0(1)} - \frac{1-A}{1-g_0(1)}.$$

Exercise 13.4

Verify that

$$\begin{aligned}&\partial \int \left[Q_0(x,1) - Q_0(x,0)\right] p(x|A=1;\nu_2)dx / \partial \nu_2\Big|_{\nu_2=0}\\ =&\mathbb{E}\left\{\phi_1(X,A)S_{0;\nu_2}(X,A)\right\}.\end{aligned}$$

Exercise 13.5

Verify that

$$\begin{aligned}&\partial \mathbb{E}_{\nu_3}\left\{\left[\frac{I(A=1)}{\mathbb{P}(A=1)} - \frac{I(A=0)\mathbb{P}(A=1\mid X)}{\mathbb{P}(A=1)\mathbb{P}(A=0\mid X)}\right]Y\right\} / \partial \nu_3\Big|_{\nu_3=0}\\ =&\mathbb{E}\left\{\phi_2(X,A,Y)S_{0;\nu_3}(X,A,Y)\right\}.\end{aligned}$$

14

Missing Data

14.1 Two Missing Mechanisms

A study is considered with data $(X_i, A_i, \Delta_i, \Delta_i Y_i)$, $i = 1, \ldots, n$, where $\Delta_i = 1$ indicates that the outcome of the ith subject is observed, and $\Delta_i = 0$ indicates that the outcome is missing.

A taxonomy for missing data was introduced by Donald Rubin,[59] classifying mechanisms as missing completely at random (MCAR), missing at random (MAR), and missing not at random (MNAR). Since MCAR is rarely assumed in clinical studies and is a special case of MAR, in this chapter the focus will be restricted to only two missing mechanisms: MAR and MNAR.

Under MAR, the probability of missingness may depend on the observed variables of the study. Under MNAR, the probability of missingness may depend on the values of the unobserved (missing) data.

In Figure 14.1, a directed acyclic graph (DAG) representing a typical study is shown. An assumption implied by this DAG is that Δ is conditionally independent of Y given X and A, as no arrow from Y to Δ is present.

If it is suspected that Δ depends on Y (e.g., if patients with higher values of Y are more likely to drop out), then an arrow should be added from Y to Δ, resulting in the DAG shown in Figure 14.2. The missing data mechanism implied by this DAG is MNAR.

It is common practice for analyses to be performed under the MAR assumption, followed by sensitivity analyses to assess robustness of results under the MNAR assumption.[60]

Gold Coin # 48. *In pharmaceutical statistics, attention is often limited to two missing mechanisms: missing at random (MAR) and missing not at random (MNAR).*

When there are no missing data, besides the stable unit treatment value assumption (SUTVA) and consistency assumption, the following two identifiability assumptions are made:

DOI: 10.1201/9781003635468-17

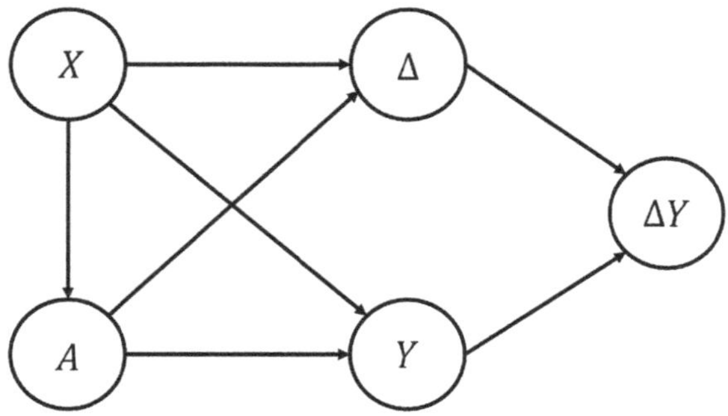

FIGURE 14.1
An example of MAR

Exchangeability:

$$(Y^{a=1}, Y^{a=0}) \perp\!\!\!\perp A \mid X,$$

Positivity:

$$\mathbb{P}(A = a \mid X = x) > 0, \quad \text{for } a = 1, 0;\; x \in \text{supp}(X).$$

When missing data are present, as in the setting illustrated in Figure 14.1, two additional identifiability assumptions are invoked:

Exchangeability with respect to missingness:

$$(Y^{a=1}, Y^{a=0}) \perp\!\!\!\perp \Delta \mid (A, X),$$

Positivity with respect to missingness:

$$\mathbb{P}(\Delta = 1 \mid A = a, X = x) > 0, \quad \text{for } a = 1, 0;\; x \in \text{supp}(X).$$

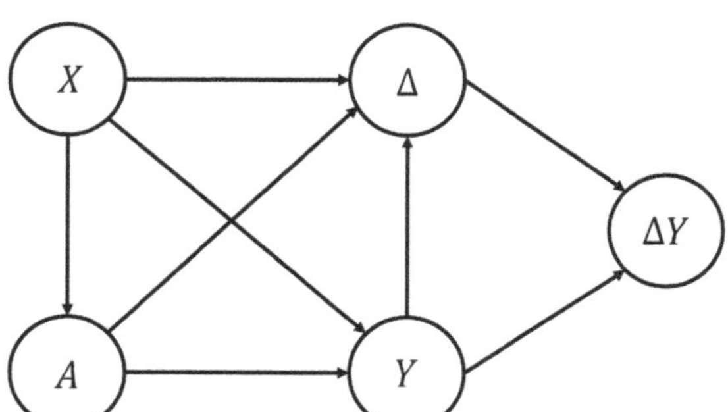

FIGURE 14.2
An example of MNAR

The need for the positivity assumption is intuitive: some data must be observed in order to perform analyses. If, for a given level (a, x), $\mathbb{P}(\Delta = 1 \mid A = a, X = x) = 0$, then no data would be available within that stratum.

The DAG in Figure 14.1 implies the following exchangeability assumptions:

$$(Y^{a=1}, Y^{a=0}) \perp\!\!\!\perp A \mid X,$$
$$(Y^{a=1}, Y^{a=0}) \perp\!\!\!\perp \Delta \mid (A, X).$$

These two assumptions can be combined into a single exchangeability condition (see Exercise 14.1):

$$(Y^{a=1}, Y^{a=0}) \perp\!\!\!\perp (A, \Delta) \mid X.$$

To demonstrate how the DAG implies these assumptions, the following structural causal model (SCM), equivalent to the DAG, is considered:

$$X = f_X(U_X),$$
$$A = f_A(X, U_A),$$
$$\Delta = f_\Delta(X, A, U_\Delta),$$
$$Y = f_Y(X, A, U_Y),$$

where $U_.$'s denote exogenous variables, which are mutually independent, and $f_.$'s are deterministic functions.

The potential outcomes $Y^{a=1}$ and $Y^{a=0}$ can be expressed as:

$$Y^{a=1} = f_Y(X, 1, U_Y),$$
$$Y^{a=0} = f_Y(X, 0, U_Y).$$

Given the independence of U_Y and (U_Δ, U_A), the following conditional independencies hold:

$$(f_Y(X, 1, U_Y), f_Y(X, 0, U_Y)) \perp\!\!\!\perp f_A(X, U_A) \mid X,$$
$$(f_Y(X, 1, U_Y), f_Y(X, 0, U_Y)) \perp\!\!\!\perp f_\Delta(X, A, U_\Delta) \mid (A, X).$$

That is, $(Y^{a=1}, Y^{a=0}) \perp\!\!\!\perp A \mid X$ and $(Y^{a=1}, Y^{a=0}) \perp\!\!\!\perp \Delta \mid (A, X)$.

Moreover, the mutual independence of U_Y and (U_Δ, U_A) implies:

$$(f_Y(X, 1, U_Y), f_Y(X, 0, U_Y)) \perp\!\!\!\perp (f_A(X, U_A), f_\Delta(X, f_A(X, U_A), U_\Delta)) \mid X,$$

which is equivalent to:

$$(Y^{a=1}, Y^{a=0}) \perp\!\!\!\perp (A, \Delta) \mid X. \quad \square$$

14.2 Two Versions of a Statistical Estimand

Let the causal estimand of interest be defined as

$$\theta^*_{\text{ATE}} = \mathbb{E}(Y^{a=1} - Y^{a=0}).$$

For the purpose of identification, two assumptions are imposed: the exchangeability assumption and the positivity assumption. The exchangeability assumption assumes that:

$$(Y^{a=0}, Y^{a=1}) \perp\!\!\!\perp A \mid X,$$
$$(Y^{a=0}, Y^{a=1}) \perp\!\!\!\perp \Delta \mid (A, X).$$

The positivity assumption assumes that:

$$\mathbb{P}(A = a \mid X = x) > 0, \quad \text{for } a = 1, 0;\ x \in \text{supp}(X),$$
$$\mathbb{P}(\Delta = 1 \mid A = a, X = x) > 0, \quad \text{for } a = 1, 0;\ x \in \text{supp}(X).$$

The above assumptions can be expressed more concisely:

$$(Y^{a=0}, Y^{a=1}) \perp\!\!\!\perp (A, \Delta) \mid X,$$
$$\mathbb{P}(A = a, \Delta = 1 \mid X = x) > 0, \quad \text{for } a = 1, 0;\ x \in \text{supp}(X).$$

14.2.1 Standardization Strategy

Identification via the standardization strategy can be established through the following sequence of equalities:

$$\begin{aligned}
\theta^*_{\text{ATE}} &= \mathbb{E}(Y^{a=1}) - \mathbb{E}(Y^{a=0}) \\
&\overset{(a)}{=} \mathbb{E}[\mathbb{E}(Y^{a=1} \mid X)] - \mathbb{E}[\mathbb{E}(Y^{a=0} \mid X)] \\
&\overset{(b)}{=} \mathbb{E}[\mathbb{E}(Y^{a=1} \mid A = 1, X)] - \mathbb{E}[\mathbb{E}(Y^{a=0} \mid A = 0, X)] \\
&\overset{(c)}{=} \mathbb{E}[\mathbb{E}(Y^{a=1} \mid \Delta = 1, A = 1, X) - \mathbb{E}(Y^{a=0} \mid \Delta = 1, A = 0, X)] \\
&\overset{(d)}{=} \mathbb{E}[\mathbb{E}(Y \mid \Delta = 1, A = 1, X) - \mathbb{E}(Y \mid \Delta = 1, A = 0, X)],
\end{aligned}$$

where step (a) follows from the law of iterated expectations (LIE), steps (b) and (c) rely on the exchangeability and positivity assumptions, and step (d) follows from the consistency assumption.

Thus, one version of the statistical estimand can be expressed as

$$\theta_{\text{ATE}} = \mathbb{E}[\mathbb{E}(Y \mid \Delta = 1, A = 1, X) - \mathbb{E}(Y \mid \Delta = 1, A = 0, X)].$$

By defining

$$\widetilde{Q}_0(a, X) = \mathbb{E}(Y \mid \Delta = 1, A = a, X), \quad a = 1, 0,$$

the statistical estimand can be written more compactly as

$$\theta_{\text{ATE}} = \mathbb{E}[\widetilde{Q}_0(1, X) - \widetilde{Q}_0(0, X)].$$

14.2.2 Weighting Strategy

Using the weighting strategy, the following identity can be established:

$$\begin{aligned}
&\mathbb{E}\left\{\frac{I(A = a, \Delta = 1)}{\mathbb{P}(A = a, \Delta = 1 \mid X)} Y\right\} \\
&\overset{(a)}{=} \mathbb{E}\left\{\frac{I(A = a, \Delta = 1)}{\mathbb{P}(A = a, \Delta = 1 \mid X)} Y^a\right\} \\
&\overset{(b)}{=} \mathbb{E}\left[\mathbb{E}\left\{\frac{I(A = a, \Delta = 1)}{\mathbb{P}(A = a, \Delta = 1 \mid X)} Y^a \,\middle|\, X\right\}\right] \\
&\overset{(c)}{=} \mathbb{E}\left[\mathbb{E}\left\{\frac{I(A = a, \Delta = 1)}{\mathbb{P}(A = a, \Delta = 1 \mid X)} \,\middle|\, X\right\} \mathbb{E}\{Y^a \mid X\}\right] \\
&\overset{(d)}{=} \mathbb{E}[\mathbb{E}\{Y^a \mid X\}] \\
&\overset{(e)}{=} \mathbb{E}(Y^a),
\end{aligned}$$

where step (a) is well-defined under the positivity assumption and holds under the consistency assumption, step (b) follows from the LIE, step (c) uses the exchangeability assumption, step (d) follows from the identity $\mathbb{E}\{I(A = a \mid X)\} = \mathbb{P}(A = a \mid X)$, and step (e) again uses the LIE.

Thus, another version of the statistical estimand can be expressed as

$$\theta_{\text{ATE}} = \mathbb{E}\left[\frac{I(A = 1, \Delta = 1)}{\mathbb{P}(A = 1, \Delta = 1 \mid X)} Y\right] - \mathbb{E}\left[\frac{I(A = 0, \Delta = 1)}{\mathbb{P}(A = 0, \Delta = 1 \mid X)} Y\right].$$

By introducing the functions

$$\begin{aligned}
g_0(a \mid X) &= \mathbb{P}(A = a \mid X), \\
h_0(X, A) &= \mathbb{P}(\Delta = 1 \mid A, X),
\end{aligned}$$

the statistical estimand can be re-expressed as

$$\theta_{\text{ATE}} = \mathbb{E}\left[\frac{I(A = 1, \Delta = 1)}{g_0(1 \mid X) h_0(X, 1)} Y\right] - \mathbb{E}\left[\frac{I(A = 0, \Delta = 1)}{g_0(0 \mid X) h_0(X, 0)} Y\right].$$

Gold Coin # 49. *Under the exchangeability and positivity assumptions, two equivalent forms of the statistical estimand are given by*

$$\theta_{ATE} = \mathbb{E}[\mathbb{E}(Y \mid \Delta = 1, A = 1, X) - \mathbb{E}(Y \mid \Delta = 1, A = 0, X)],$$

and

$$\theta_{ATE} = \mathbb{E}\left[\frac{I(A=1,\Delta=1)}{\mathbb{P}(A=1,\Delta=1 \mid X)}Y\right] - \mathbb{E}\left[\frac{I(A=0,\Delta=1)}{\mathbb{P}(A=0,\Delta=1 \mid X)}Y\right].$$

14.3 EIF for Missing Data Handling

Since the statistical estimand has two equivalent representations, the convenient approach developed in the preceding chapter can be applied to derive the efficient influence function (EIF) under the MAR assumption. This approach consists of two main steps: subtraction and addition.

- **Subtraction:** The following two mean-zero functions are obtained:

$$\phi_1(X) = \mathbb{E}(Y \mid \Delta = 1, A = 1, X) - \mathbb{E}(Y \mid \Delta = 1, A = 0, X) - \theta_{\mathrm{ATE},0},$$
$$\phi_2(X, A, \Delta, Y) = \frac{(2A-1)\Delta}{\mathbb{P}(A, \Delta = 1 \mid X)}\left[Y - \mathbb{E}(Y \mid \Delta = 1, A, X)\right].$$

- **Addition:** The EIF is constructed as the sum of the above components:

$$\phi_{\mathrm{ATE}}(X, A, \Delta, Y) = \phi_1(X) + \phi_2(X, A, \Delta, Y).$$

To validate this approach for constructing the regular influence function, consider the following parametric submodel:

- $p(x; \nu_1)$, with $p(x; 0) = p_0(x)$,
- $g(a \mid X; \nu_2)$, with $g(a \mid X; 0) = g_0(a \mid X)$,
- $h(X, A; \nu_3)$, with $h(X, A; 0) = h_0(X, A)$,
- $f(y \mid X, A, \Delta = 1; \nu_4)$, with $f(y \mid X, A, \Delta = 1; 0) = f_0(y \mid X, A, \Delta = 1)$.

Let $\boldsymbol{\nu} = (\nu_1, \nu_2, \nu_3, \nu_4)^\top$ and, to abuse the notation, $0 = (0, 0, 0, 0)^\top$ in $\boldsymbol{\nu} = 0$. The likelihood under the parametric submodel is given by

$$\begin{aligned} L(\boldsymbol{\nu}) = p(x; \nu_1) &\times [g(1 \mid X; \nu_2)]^A [1 - g(1 \mid X; \nu_2)]^{1-A} \\ &\times [h(X, A; \nu_3)]^\Delta [1 - h(X, A; \nu_3)]^{1-\Delta} \\ &\times [f(y \mid X, A, \Delta = 1; \nu_4)]^\Delta. \end{aligned}$$

The score functions evaluated at the truth are given by (see Exercise 14.2):

$$
\begin{aligned}
S_{0;\nu_1}(X) &= S(X),\\
S_{0;\nu_2}(X,A) &= S'(X)[A - g_0(1 \mid X)],\\
S_{0;\nu_3}(X,A,\Delta) &= S''(X,A)[\Delta - h_0(X,A)],\\
S_{0;\nu_4}(X,A,\Delta,Y) &= \Delta\, S'''(X,A,\Delta,Y),
\end{aligned}
$$

where $S(X)$ is a function of X such that $\mathbb{E}[S(X)] = 0$, $S'(X)$ is a function of X, $S''(X,A)$ is a function of (X,A), and $S'''(X,A,Y)$ is a function of (X,A,Y) such that $\mathbb{E}[S'''(X,A,Y) \mid X, A, \Delta = 1] = 0$.

To establish that $\phi_{\text{ATE}}(X,A,\Delta,Y)$ is a regular influence function, it is sufficient to verify the following equalities:

$$
\begin{aligned}
\left.\frac{\partial \theta_{\text{ATE}}(\boldsymbol{\nu})}{\partial \nu_1}\right|_{\boldsymbol{\nu}=0} &= \mathbb{E}\left\{[\phi_1(X) + \phi_2(X,A,\Delta,Y)]S_{0;\nu_1}(X)\right\},\\
\left.\frac{\partial \theta_{\text{ATE}}(\boldsymbol{\nu})}{\partial \nu_2}\right|_{\boldsymbol{\nu}=0} &= \mathbb{E}\left\{[\phi_1(X) + \phi_2(X,A,\Delta,Y)]S_{0;\nu_2}(X,A)\right\},\\
\left.\frac{\partial \theta_{\text{ATE}}(\boldsymbol{\nu})}{\partial \nu_3}\right|_{\boldsymbol{\nu}=0} &= \mathbb{E}\left\{[\phi_1(X) + \phi_2(X,A,\Delta,Y)]S_{0;\nu_3}(X,A,\Delta)\right\},\\
\left.\frac{\partial \theta_{\text{ATE}}(\boldsymbol{\nu})}{\partial \nu_4}\right|_{\boldsymbol{\nu}=0} &= \mathbb{E}\left\{[\phi_1(X) + \phi_2(X,A,\Delta,Y)]S_{0;\nu_4}(X,A,\Delta,Y)\right\}.
\end{aligned}
$$

By the LIE, the following identities hold (see Exercise 14.3):

$$
\begin{aligned}
\mathbb{E}[\phi_2(X,A,\Delta,Y)S_{0;\nu_1}(X)] &= 0,\\
\mathbb{E}[\phi_1(X)S_{0;\nu_2}(X,A)] &= 0,\\
\mathbb{E}[\phi_2(X,A,\Delta,Y)S_{0;\nu_2}(X,A)] &= 0,\\
\mathbb{E}[\phi_1(X)S_{0;\nu_3}(X,A,\Delta)] &= 0,\\
\mathbb{E}[\phi_2(X,A,\Delta,Y)S_{0;\nu_3}(X,A,\Delta)] &= 0,\\
\mathbb{E}[\phi_1(X)S_{0;\nu_4}(X,A,\Delta,Y)] &= 0.
\end{aligned}
$$

Therefore, it suffices to verify:

$$
\begin{aligned}
(a):&\quad \left.\frac{\partial \theta_{\text{ATE}}(\boldsymbol{\nu})}{\partial \nu_1}\right|_{\boldsymbol{\nu}=0} = \mathbb{E}[\phi_1(X)S_{0;\nu_1}(X)],\\
(b):&\quad \left.\frac{\partial \theta_{\text{ATE}}(\boldsymbol{\nu})}{\partial \nu_2}\right|_{\boldsymbol{\nu}=0} = 0,\\
(c):&\quad \left.\frac{\partial \theta_{\text{ATE}}(\boldsymbol{\nu})}{\partial \nu_3}\right|_{\boldsymbol{\nu}=0} = 0,\\
(d):&\quad \left.\frac{\partial \theta_{\text{ATE}}(\boldsymbol{\nu})}{\partial \nu_4}\right|_{\boldsymbol{\nu}=0} = \mathbb{E}[\phi_2(X,A,\Delta,Y)S_{0;\nu_4}(X,A,\Delta,Y)].
\end{aligned}
$$

To verify (a), note that (see Exercise 14.4)

$$\begin{aligned}
&\left.\frac{\partial\theta_{\text{ATE}}(\boldsymbol{\nu})}{\partial\nu_1}\right|_{\boldsymbol{\nu}=\mathbf{0}}\\
=&\left.\frac{\partial\mathbb{E}_{\nu_1}\left[\mathbb{E}\left(Y \mid \Delta=1, A=1, X\right)-\mathbb{E}(Y \mid \Delta=1, A=0, X)\right]}{\partial\nu_1}\right|_{\nu_1=0}\\
=&\int\left[\mathbb{E}\left(Y \mid \Delta=1 A=1, x\right)-\mathbb{E}(Y \mid \Delta=1, A=0, x)\right] S_{0;\nu_1}(x) p_0(x) dx\\
=&\mathbb{E}\left\{\phi_1(X) S_{0;\nu_1}(X)\right\}.
\end{aligned}$$

To verify (b), note that

$$\begin{aligned}
&\left.\frac{\partial\theta_{\text{ATE}}(\boldsymbol{\nu})}{\partial\nu_2}\right|_{\boldsymbol{\nu}=\mathbf{0}}\\
=&\left.\frac{\partial\mathbb{E}_{\nu_1}\left[\mathbb{E}_{\nu_4}\left(Y \mid \Delta=1, A=1, X\right)-\mathbb{E}_{\nu_3}(Y \mid \Delta=1, A=0, X)\right]}{\partial\nu_2}\right|_{\boldsymbol{\nu}=\mathbf{0}}\\
=&0.
\end{aligned}$$

To verify (c), note that

$$\begin{aligned}
&\left.\frac{\partial\theta_{\text{ATE}}(\boldsymbol{\nu})}{\partial\nu_3}\right|_{\boldsymbol{\nu}=\mathbf{0}}\\
=&\left.\frac{\partial\mathbb{E}_{\nu_1}\left[\mathbb{E}_{\nu_4}\left(Y \mid \Delta=1, A=1, X\right)-\mathbb{E}_{\nu_3}(Y \mid \Delta=1, A=0, X)\right]}{\partial\nu_3}\right|_{\boldsymbol{\nu}=\mathbf{0}}\\
=&0.
\end{aligned}$$

To verify (d), note that (see Exercise 14.5)

$$\begin{aligned}
&\left.\frac{\partial\theta_{\text{ATE}}(\boldsymbol{\nu})}{\partial\nu_4}\right|_{\boldsymbol{\nu}=\mathbf{0}}\\
=&\,\partial\left\{\mathbb{E}_{\boldsymbol{\nu}}\frac{(2A-1)\Delta Y}{\mathbb{P}(A, \Delta=1 \mid X)}\right\}\Big/\partial\nu_4\Big|_{\boldsymbol{\nu}=\mathbf{0}}\\
=&\,\partial\left\{\mathbb{E}\frac{(2A-1)\Delta\mathbb{E}_{\nu_4}(Y \mid \Delta=1, A, X)}{\mathbb{P}(A, \Delta=1 \mid X)}\right\}\Big/\partial\nu_4\Big|_{\nu_4=0}\\
=&\mathbb{E}\frac{(2A-1)\Delta\partial\mathbb{E}_{\nu_4}(Y \mid \Delta=1, A, X)/\partial\nu_4|_{\nu_4=0}}{\mathbb{P}(A, \Delta=1 \mid X)}\\
=&\mathbb{E}\left\{\frac{(2A-1)\Delta}{\mathbb{P}(A, \Delta=1 \mid X)}\left.\frac{\partial\int y f(y \mid \Delta=1, A, X; \nu_4)}{\partial\nu_4}\right|_{\nu_4=0}\right\}\\
=&\mathbb{E}\left\{\frac{(2A-1)\Delta}{\mathbb{P}(A, \Delta=1 \mid X)}\mathbb{E}[Y S_{0;\nu_4}(X, A, \Delta, Y) \mid \Delta, A, X]\right\}\\
=&\mathbb{E}\left\{\frac{(2A-1)\Delta}{\mathbb{P}(A, \Delta=1 \mid X)}[Y-\mathbb{E}(Y \mid \Delta=1, A, X)] S_{0;\nu_4}(X, A, \Delta, Y)\right\}\\
=&\mathbb{E}\left\{\phi_2(X, A, \Delta, Y) S_{0;\nu_4}(X, A, \Delta, Y)\right\}.
\end{aligned}$$

Combining the above results, it is concluded that $\phi_{\text{ATE}}(X, A, \Delta, Y)$ represents the regular influence function. □

Gold Coin # 50. *The same convenient approach that has been proposed for deriving the EIF for the ATE estimand in the absence of missing data can be applied to the case with missing data under the MAR assumption.*

14.4 What If the Missingness Were Viewed Differently?

There is an alternative way—more "active" in nature—by which missingness can be interpreted, as illustrated in Figure 14.3, where

$$\widetilde{Y} = \Delta Y$$

denotes the observed outcome, and

$$\widetilde{A} = (A, \Delta)$$

represents a newly defined intervention variable.

The two DAGs shown in Figure 14.3 are implied by the original DAG in Figure 14.1. To demonstrate this, consider the following SCM, which

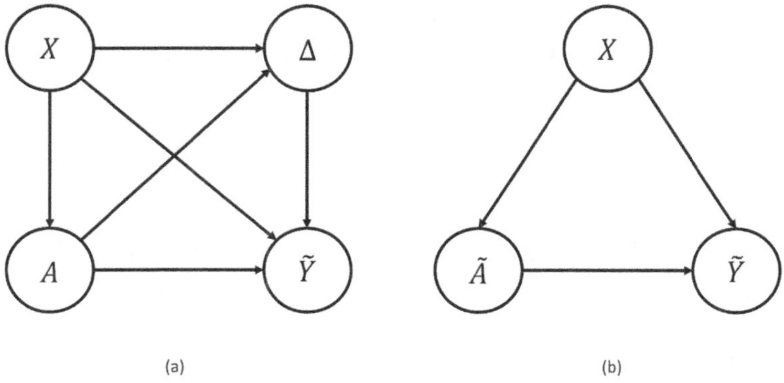

FIGURE 14.3
Two alternative DAGs for missingness

corresponds to Figure 14.1:

$$\begin{aligned} X &= f_X(U_X), \\ A &= f_A(X, U_A), \\ \Delta &= f_\Delta(X, A, U_\Delta), \\ Y &= f_Y(X, A, U_Y), \\ \widetilde{Y} &= \Delta Y. \end{aligned}$$

From this SCM, the following two SCMs can be derived. The first is:

$$\begin{aligned} X &= f_X(U_X), \\ A &= f_A(X, U_A), \\ \Delta &= f_\Delta(X, A, U_\Delta), \\ \widetilde{Y} &= f_{\widetilde{Y}}(X, A, \Delta, U_Y) = \Delta f_Y(X, A, U_Y). \end{aligned}$$

This SCM corresponds to the DAG in Figure 14.1(a). Consider the following two potential outcomes:

$$\begin{aligned} \widetilde{Y}^{a=1,\delta=1} &= f_{\widetilde{Y}}(X, A = 1, \Delta = 1, U_Y) = f_Y(X, 1, U_Y) = Y^{a=1}, \\ \widetilde{Y}^{a=0,\delta=1} &= f_{\widetilde{Y}}(X, A = 0, \Delta = 1, U_Y) = f_Y(X, 0, U_Y) = Y^{a=0}. \end{aligned}$$

Thus, the condition "$\Delta = 1$" can be interpreted as an intervention: the participant is intervened upon to attend the follow-up visit, and that the data are intervened upon to be collected.

Furthermore, the second implied SCM is:

$$\begin{aligned} X &= f_X(U_X), \\ \widetilde{A} &= (A, \Delta) = (f_A(X, U_A), f_\Delta(X, A, U_\Delta)), \\ \widetilde{Y} &= f_{\widetilde{Y}}(X, \widetilde{A}, U_Y) = \Delta f_Y(X, A, U_Y). \end{aligned}$$

This SCM corresponds to the DAG in Figure 14.1(b). Consider the following two potential outcomes:

$$\begin{aligned} \widetilde{Y}^{(a,\delta)=(1,1)} &= f_{\widetilde{Y}}(X, \widetilde{A} = (1,1), U_Y) = f_Y(X, 1, U_Y) = Y^{a=1}, \\ \widetilde{Y}^{(a,\delta)=(0,1)} &= f_{\widetilde{Y}}(X, \widetilde{A} = (0,1), U_Y) = f_Y(X, 0, U_Y) = Y^{a=0}. \end{aligned}$$

Thus, the condition "$A = a, \Delta = 1$" can be interpreted as a joint intervention: the participant is intervened upon to receive treatment $A = a$, to attend the follow-up visit, and the data are intervened upon to be observed.

When viewed through this new lens, where non-missingness is interpreted as an intervention, the EIF for for the setting with missing data can be derived directly based on the EIF previously obtained for the setting without missingness.

Consider the following estimand:

$$\theta^*_{\text{ATE}} = \mathbb{E}(\widetilde{Y}^{\widetilde{a}=(1,1)} - \widetilde{Y}^{\widetilde{a}=(0,1)}) = \mathbb{E}(Y^{a=1} - Y^{a=0}).$$

Assume the following assumptions (exchangeability and positivity):

$$(\widetilde{Y}^{\widetilde{a}=(1,1)}, \widetilde{Y}^{\widetilde{a}=(0,1)}) \perp\!\!\!\perp \widetilde{A} \mid X,$$
$$\mathbb{P}(\widetilde{A} = (a,1) \mid X = x) > 0, \quad \text{for } a = 1, 0;\ x \in \text{supp}(X).$$

These are equivalent to:

$$(Y^{a=0}, Y^{a=1}) \perp\!\!\!\perp (A, \Delta) \mid X,$$
$$\mathbb{P}(A = a, \Delta = 1 \mid X = x) > 0, \quad \text{for } a = 1, 0;\ x \in \text{supp}(X).$$

Accordingly, the two equivalent expressions for the statistical estimand are (by the standardization strategy and the weighting strategy, respectively):

$$\begin{aligned}\theta_{\text{ATE}} &= \mathbb{E}[\mathbb{E}(\widetilde{Y} \mid \widetilde{A} = (1,1), X) - \mathbb{E}(\widetilde{Y} \mid \widetilde{A} = (0,1), X)] \\ &= \mathbb{E}[\mathbb{E}(Y \mid A = 1, \Delta = 1, X) - \mathbb{E}(Y \mid A = 0, \Delta = 1, X)],\end{aligned}$$

and

$$\begin{aligned}\theta_{\text{ATE}} &= \mathbb{E}\left[\frac{I(\widetilde{A} = (1,1))}{\mathbb{P}(\widetilde{A} = (1,1) \mid X)}\widetilde{Y}\right] - \mathbb{E}\left[\frac{I(\widetilde{A} = (0,1))}{\mathbb{P}(\widetilde{A} = (0,1) \mid X)}\widetilde{Y}\right] \\ &= \mathbb{E}\left[\frac{I(A = 1, \Delta = 1)}{\mathbb{P}(A = 1, \Delta = 1 \mid X)}Y\right] - \mathbb{E}\left[\frac{I(A = 0, \Delta = 1)}{\mathbb{P}(A = 0, \Delta = 1 \mid X)}Y\right].\end{aligned}$$

Consequently, using the same approach for the setting without missing data, the EIF for θ_{ATE} for the setting with missing data is given by:

$$\begin{aligned}&\phi_{\text{ATE}}(X, \widetilde{A}, \widetilde{Y}) \\ &= \mathbb{E}(\widetilde{Y} \mid \widetilde{A} = (1,1), X) - \mathbb{E}(\widetilde{Y} \mid \widetilde{A} = (0,1), X) - \theta_{\text{ATE},0} \\ &\quad + \frac{I(\widetilde{A} = (1,1))}{\mathbb{P}(\widetilde{A} = (1,1) \mid X)}[\widetilde{Y} - \mathbb{E}(\widetilde{Y} \mid \widetilde{A} = (1,1), X)] \\ &\quad - \frac{I(\widetilde{A} = (0,1))}{\mathbb{P}(\widetilde{A} = (0,1) \mid X)}[\widetilde{Y} - \mathbb{E}(\widetilde{Y} \mid \widetilde{A} = (0,1), X)],\end{aligned}$$

which is the same as the EIF derived in the preceding section:

$$\begin{aligned}&\phi_{\text{ATE}}(X, A, \Delta, Y) \\ &= \mathbb{E}(Y \mid A = 1, \Delta = 1, X) - \mathbb{E}(Y \mid A = 0, \Delta = 1, X) - \theta_{\text{ATE},0} \\ &\quad + \frac{I(A = 1, \Delta = 1)}{\mathbb{P}(A = 1, \Delta = 1 \mid X)}[Y - \mathbb{E}(Y \mid A = 1, \Delta = 1, X)] \\ &\quad - \frac{I(A = 0, \Delta = 1)}{\mathbb{P}(A = 0, \Delta = 1 \mid X)}[Y - \mathbb{E}(Y \mid A = 0, \Delta = 1, X)].\end{aligned}$$

Gold Coin # 51. *Two equivalent perspectives on missingness lead to two equivalent expressions of the EIF:*

$$\begin{aligned}
&\phi_{ATE}(X, \widetilde{A}, \widetilde{Y}) \\
&= \mathbb{E}(\widetilde{Y} \mid \widetilde{A} = (1,1), X) - \mathbb{E}(\widetilde{Y} \mid \widetilde{A} = (0,1), X) - \theta_{ATE,0} \\
&\quad + \frac{I(\widetilde{A} = (1,1))}{\mathbb{P}(\widetilde{A} = (1,1) \mid X)}[\widetilde{Y} - \mathbb{E}(\widetilde{Y} \mid \widetilde{A} = (1,1), X)] \\
&\quad - \frac{I(\widetilde{A} = (0,1))}{\mathbb{P}(\widetilde{A} = (0,1) \mid X)}[\widetilde{Y} - \mathbb{E}(\widetilde{Y} \mid \widetilde{A} = (0,1), X)],
\end{aligned}$$

and

$$\begin{aligned}
&\phi_{ATE}(X, A, \Delta, Y) \\
&= \mathbb{E}(Y \mid A = 1, \Delta = 1, X) - \mathbb{E}(Y \mid A = 0, \Delta = 1, X) - \theta_{ATE,0} \\
&\quad + \frac{I(A = 1, \Delta = 1)}{\mathbb{P}(A = 1, \Delta = 1 \mid X)}[Y - \mathbb{E}(Y \mid A = 1, \Delta = 1, X)] \\
&\quad - \frac{I(A = 0, \Delta = 1)}{\mathbb{P}(A = 0, \Delta = 1 \mid X)}[Y - \mathbb{E}(Y \mid A = 0, \Delta = 1, X)].
\end{aligned}$$

14.5 Exercises

Exercise 14.1

Assume that X, Y, Z are three random variables. Show that the following two statements are equivalent:

$$X \perp\!\!\!\perp Y \text{ and } X \perp\!\!\!\perp Z \mid Y,$$

and

$$X \perp\!\!\!\perp (Y, Z).$$

Exercise 14.2

Consider the following likelihood:

$$\begin{aligned}
L(\boldsymbol{\nu}) = p(x; \nu_1) &\times [g(1 \mid X; \nu_2)]^A [1 - g(1 \mid X; \nu_2)]^{1-A} \\
&\times [h(X, A; \nu_3)]^{\Delta} [1 - h(X, A; \nu_3)]^{1-\Delta} \\
&\times [f(y \mid X, A, \Delta = 1; \nu_4)]^{\Delta}.
\end{aligned}$$

Show that the score functions evaluated at $\boldsymbol{\nu} = 0$ are

$$\begin{aligned}
S_{0;\nu_1}(X) &= S(X), \\
S_{0;\nu_2}(X, A) &= S'(X)[A - g_0(1 \mid X)], \\
S_{0;\nu_3}(X, A, \Delta) &= S''(X, A)[\Delta - h_0(X, A)], \\
S_{0;\nu_4}(X, A, \Delta, Y) &= \Delta\, S'''(X, A, \Delta, Y),
\end{aligned}$$

Exercise 14.3

Given that $\mathbb{E}[\phi_1(X)] = 0$ and $\mathbb{E}[\phi_2(X, A, \Delta, Y) \mid X, A, \Delta = 1] = 0$, show that

$$\begin{aligned}
\mathbb{E}[\phi_2(X, A, \Delta, Y) S_{0;\nu_1}(X)] &= 0, \\
\mathbb{E}[\phi_1(X) S_{0;\nu_2}(X, A)] &= 0, \\
\mathbb{E}[\phi_2(X, A, \Delta, Y) S_{0;\nu_2}(X, A)] &= 0, \\
\mathbb{E}[\phi_1(X) S_{0;\nu_3}(X, A, \Delta)] &= 0, \\
\mathbb{E}[\phi_2(X, A, \Delta, Y) S_{0;\nu_3}(X, A, \Delta)] &= 0, \\
\mathbb{E}[\phi_1(X) S_{0;\nu_4}(X, A, \Delta, Y)] &= 0.
\end{aligned}$$

Exercise 14.4

Show that

$$\begin{aligned}
&\left.\frac{\partial \theta_{\text{ATE}}(\boldsymbol{\nu})}{\partial \nu_1}\right|_{\boldsymbol{\nu}=0} \\
&= \left.\frac{\partial \mathbb{E}_{\nu_1}\left[\mathbb{E}\left(Y \mid \Delta = 1, A = 1, X\right) - \mathbb{E}(Y \mid \Delta = 1, A = 0, X)\right]}{\partial \nu_1}\right|_{\nu_1=0}.
\end{aligned}$$

Exercise 14.5

Show that

$$\left.\frac{\partial \theta_{\text{ATE}}(\boldsymbol{\nu})}{\partial \nu_4}\right|_{\boldsymbol{\nu}=0} = \partial \left\{\mathbb{E}\frac{(2A-1)\Delta \mathbb{E}_{\nu_4}(Y \mid \Delta = 1, A, X)}{\mathbb{P}(A, \Delta = 1 \mid X)}\right\} \Big/ \partial \nu_4 \Bigg|_{\nu_4=0}.$$

15

Longitudinal Data

15.1 Longitudinal Study

In the pharmaceutical industry, studies involving multiple follow-up visits are more commonly conducted and are referred to as *longitudinal studies*. To distinguish from longitudinal studies, those that generate data in the form $(X_i, A_i, \Delta_i, \Delta_i Y_i)$, $i = 1, \ldots, n$, are referred to as *point-treatment studies*.

Longitudinal studies generally provide greater power for evaluating the impact of intercurrent events on the estimation of treatment effects. According to ICH E9(R1), *Addendum on Estimands and Sensitivity Analysis in Clinical Trials*, intercurrent events are defined as:

> "Events occurring after treatment initiation that affect either the interpretation or the existence of the measurements associated with the clinical question of interest."

Assume that one longitudinal study is conducted starting from baseline time $t = 0$, with follow-up visits at $t = 1, \ldots, T$. The primary outcome variable, Y, which may be continuous or binary, is measured at the final visit T. Time-to-event outcome variables will be discussed in Part IV of the book.

To understand how data are collected from baseline $t = 0$ to the final visit $t = T$, the following steps are considered.

At baseline $t = 0$, let $X(0)$ denote the vector of baseline covariates, and let $A(0)$ denote the initial treatment variable. In two-arm studies, $A(0)$ takes the value 0 or 1. For three-arm studies, $A(0)$ is coded using two dummy variables: $(1, 0)$, $(0, 1)$, and $(0, 0)$, representing the three arms. Similarly, in four-arm studies, the coding $(1, 1)$, $(1, 0)$, $(0, 1)$, and $(0, 0)$ is used to represent the four treatment arms. This coding scheme can be extended for studies with more treatment arms.

Let $\Delta(0)$ denote the indicator for whether data are observed at visit $t = 1$, serving as the non-missing indicator. In this chapter, attention is restricted to monotone missingness, which is interpreted as censoring. This interpretation is further motivated by the fact that the results developed here for binary or continuous outcomes can be extended to time-to-event outcomes to be discussed in Chapter 19.

DOI: 10.1201/9781003635468-18

If $\Delta(0) = 1$, indicating that data at visit $t = 1$ are observed, then $X(1)$ is used to denote the measurements obtained at $t = 1$, including time-dependent covariates and any intermediately measured outcome variables.

Next, let $A(1)$ denote the treatment received at $t = 1$, which may depend on prior observed data including $X(0)$, $A(0)$, and $X(1)$. Possible values for $A(1)$ include:

- 1: treatment 1,
- 0: treatment 0,
- `NULL`: no treatment,
- 1+: treatment 1 with an add-on,
- 0+: treatment 0 with an add-on,
- 2: an alternative treatment distinct from 0 and 1.

By allowing $A(1)$ to differ from $A(0)$, intercurrent events can be captured. For instance, the combination "$A(0) = 0$, $A(1) = 1$" indicates a switch from treatment 0 at $t = 0$ to treatment 1 at $t = 1$; "$A(0) = 1$, $A(1) =$ `NULL`" indicates dropout from treatment 1; and "$A(0) = 0$, $A(1) = 2$" indicates the use of rescue medication at $t = 1$.

This process is continued for $t = 1, \ldots, T-1$, resulting in the collection of the following data:

$$X(0), A(0), \Delta(0), X(1), A(1), \Delta(1), \ldots, X(T-1), A(T-1), \Delta(T-1), Y.$$

If $\Delta(0) = \cdots = \Delta(T-1) = 1$, indicating complete follow-up through $t = T$, then Y is measured as the final outcome at visit T. However, if there exists some $0 \leq t_0 \leq T-1$ such that $\Delta(t) = 1$ for $t < t_0$ and $\Delta(t) = 0$ for $t \geq t_0$, then censoring is said to have occurred at time t_0. In such cases, $X(t_0+1), A(t_0+1), \ldots, X(T-1), A(T-1)$, and Y are considered missing.

In the case of a point-treatment study (i.e., when $T = 1$), the collected data consist of $X(0)$, $A(0)$, $\Delta(0)$, and Y.

In some scenarios, although a study is longitudinal in nature, only two timepoints—the baseline ($t = 0$) and the final visit ($t = T$)—are of primary interest. As a result, the data reduce to $X(0)$, $A(0)$, $\Delta(T-1)$, and Y, making the analysis methodologically equivalent to that of a point-treatment study.

15.1.1 Estimand

The treatment sequence up to time t is denoted by $\overline{A}(t) = (A(0), \ldots, A(t))$, and the covariate history up to time t, including baseline covariates, time-dependent covariates, and intermediate outcomes, is denoted by $\overline{X}(t) = (X(0), \ldots, X(t))$, for $t = 0, \ldots, T-1$. In particular, the full treatment sequence and covariate history are denoted by $\overline{A} = (A(0), \ldots, A(T-1))$ and $\overline{X} = (X(0), \ldots, X(T-1))$, respectively. The primary outcome variable measured at time T is denoted by Y.

Let $\overline{a}(t) = (a(0), a(1), \ldots, a(t))$ denote the treatment sequence under investigation up to time t. Two specific treatment sequences of interest are $\overline{1} = (1, \ldots, 1)$ and $\overline{0} = (0, \ldots, 0)$. The potential outcome associated with the treatment sequence $\overline{a} = (a(0), a(1), \ldots, a(T-1))$ is denoted by $Y^{\overline{a}}$. Based on these potential outcomes, the causal estimand of interest can be defined. For instance, the average treatment effect (ATE) of treatment sequence $\overline{1}$ compared with treatment sequence $\overline{0}$ is expressed as

$$\theta^*_{\text{ATE}} = \mathbb{E}\left(Y^{\overline{a}=\overline{1}}\right) - \mathbb{E}\left(Y^{\overline{a}=\overline{0}}\right).$$

Analogous to the primary outcome Y, the variables $X(t)$, $t = 1, \ldots, T-1$, are post-treatment variables. Consequently, their potential outcomes in the potential world are defined as $X^{a(0)}(1)$ and $X^{\overline{a}(t-1)}(t)$, $t = 2, \ldots, T-1$. The stable unit treatment value assumption (SUTVA) is imposed, along with the consistency assumption:

$$\begin{aligned} X^{a(0)}(1) &= X(1), \quad \text{if } A(0) = a(0); \\ X^{\overline{a}(t-1)}(t) &= X(t), \quad \text{if } \overline{A}(t-1) = \overline{a}(t-1),\ t = 2, \ldots, T-1; \\ Y^{\overline{a}} &= Y, \quad \text{if } \overline{A} = \overline{a}. \end{aligned}$$

For convenience, the collection of potential outcomes, including those of the primary outcome variable and those of the time-dependent covariates, is summarized in the following vector:

$$W^{\overline{a}} = \left\{X^{a(0)}(1),\ X^{\overline{a}(t-1)}(t),\ t = 2, \ldots, T-1,\ Y^{\overline{a}}\right\}.$$

15.1.2 Identifiability Assumptions

In this subsection, identifiability assumptions are presented progressively, beginning with simple scenarios and proceeding to more complex ones.

Scenario One: $T = 2$ and No Missing Data

The simplest scenario is considered first, with $T = 2$ and no missing data. A typical directed acyclic graph (DAG) for a study with $T = 2$ is shown in Figure 15.1. The observed data consist of $\{X(0), A(0), X(1), A(1), Y\}$.

It is convenient to begin with the estimand

$$v^*(\overline{a}) = \mathbb{E}(Y^{\overline{a}}),$$

which is referred to as the *value* associated with the treatment sequence $\overline{a}$. On this basis, the ATE estimand can be written as

$$\theta^*_{\text{ATE}} = v^*(\overline{1}) - v^*(\overline{0}).$$

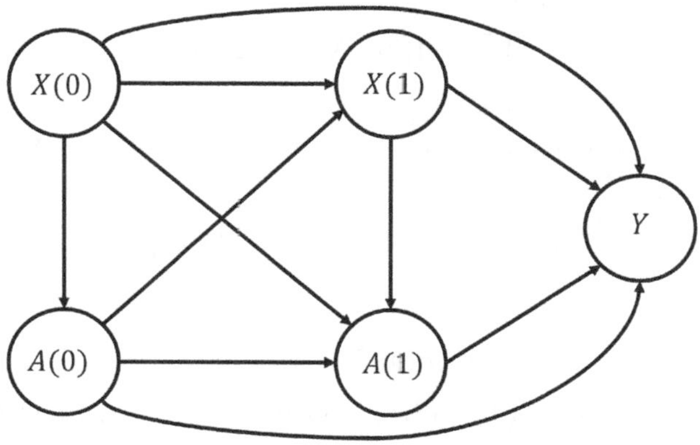

FIGURE 15.1
A DAG for a longitudinal study

The assumptions encoded in the DAG of Figure 15.1 can be made explicit by representing the relationships among $\{X(0), A(0), X(1), A(1), Y\}$ through the following structural causal model (SCM):

$$\begin{aligned} X(0) &= f_{X(0)}(U_{X(0)}), \\ A(0) &= f_{A(0)}(X(0), U_{A(0)}), \\ X(1) &= f_{X(1)}(X(0), A(0), U_{X(1)}), \\ A(1) &= f_{A(1)}(X(0), A(0), X(1), U_{A(1)}), \\ Y &= f_Y(X(0), A(0), X(1), A(1), U_Y), \end{aligned}$$

where $f_{X(0)}, f_{A(0)}, f_{X(1)}, f_{A(1)}$, and f_Y are unknown deterministic functions, and $U_{X(0)}, U_{A(0)}, U_{X(1)}, U_{A(1)}$, and U_Y are mutually independent.

From this SCM, the potential outcomes associated with $\overline{a} = (a(0), a(1))$ are expressed as

$$\begin{aligned} X^{a(0)}(1) &= f_{X(1)}(X(0), a(0), U_{X(1)}), \\ Y^{a(0),a(1)} &= f_Y(X(0), a(0), X^{a(0)}(1), a(1), U_Y). \end{aligned}$$

Conditional on $X(0) = x'(0)$, it follows that

$$A(0) = f_{A(0)}(x'(0), U_{A(0)}),$$

and the potential outcomes take the form

$$\begin{aligned} X^{a(0)}(1) &= f_{X(1)}(x'(0), a(0), U_{X(1)}), \\ Y^{a(0),a(1)} &= f_Y(x'(0), a(0), f_{X(1)}(x'(0), a(0), U_{X(1)}), a(1), U_Y). \end{aligned}$$

Thus, these findings imply that

$$W^{\overline{a}} \perp\!\!\!\perp A(0) \mid X(0) = x'(0),$$

where $W^{\overline{a}} = \{X^{a(0)}(1), Y^{a(0),a(1)}\}$.

Similarly, conditional on $X(0) = x'(0)$, $A(0) = a'(0)$, and $X(1) = x'(1)$, it follows that

$$A(1) = f_{A(1)}(x'(0), a'(0), x'(1), U_{A(1)}),$$

and

$$\begin{aligned} X^{a(0)}(1) &= f_{X(1)}(x'(0), a(0), U_{X(1)}), \\ Y^{a(0),a(1)} &= f_Y(x'(0), a(0), f_{X(1)}(x'(0), a(1), U_{X(1)}), a(1), U_Y). \end{aligned}$$

Hence, these findings imply that

$$W^{\overline{a}} \perp\!\!\!\perp A(1) \mid (X(1), A(0), X(0)) = (x'(1), a'(0), x'(0)).$$

In summary, the following is implied:

$$W^{\overline{a}} \perp\!\!\!\perp A(0) \mid X(0) \quad \text{and} \quad W^{\overline{a}} \perp\!\!\!\perp A(1) \mid (X(1), A(0), X(0)).$$

This property is referred to as the *sequential exchangeability assumption.*

In addition, the positivity assumption is required. The positivity assumption, which is not implied in the DAG, states that

$$\mathbb{P}(\overline{A} = \overline{a} \mid X(0) = x(0)) > 0, \quad \text{for } x(0) \in \text{supp}(X(0)).$$

Scenario Two: $T = 2$ and Missing Data

The case with $T = 2$ and missing data is next considered. A typical DAG for this setting is more complex, as additional missingness indicators $\Delta(0)$ and $\Delta(1)$ must be incorporated into Figure 15.1. When the DAG is too complicated, an SCM provides a more transparent representation of the relationships among $\{X(0), A(0), \Delta(0), \Delta(0)X(1), \Delta(0)A(1), \Delta(1), \Delta(1)Y\}$.

The estimand of interest is again defined as

$$v^*(\overline{a}) = \mathbb{E}(Y^{\overline{a}}).$$

The relationships among the variables are specified by the following SCM:

$$\begin{aligned} X(0) &= f_{X(0)}(U_{X(0)}), \\ A(0) &= f_{A(0)}(X(0), U_{A(0)}), \\ \Delta(0) &= f_{\Delta(0)}(X(0), A(0), U_{\Delta(0)}), \\ X(1) &= f_{X(1)}(X(0), A(0), U_{X(1)}), \\ A(1) &= f_{A(1)}(X(0), A(0), X(1), U_{A(1)}), \\ \Delta(1) &= f_{\Delta(1)}(X(0), A(0), X(1), A(1), U_{\Delta(1)}), \\ Y &= f_Y(X(0), A(0), X(1), A(1), U_Y), \end{aligned}$$

where the functions $f_{\cdot}$'s are unknown deterministic functions and the variables $U_{\cdot}$'s are mutually independent.

From this SCM, the potential outcomes associated with $\overline{a} = (a(0), a(1))$ are given by

$$X^{a(0)}(1) = f_{X(1)}(X(0), a(0), U_{X(1)}),$$
$$Y^{a(0),a(1)} = f_Y(X(0), a(0), X^{a(0)}(1), a(1), U_Y).$$

Conditional on $X(0) = x'(0)$, it follows that

$$A(0) = f_{A(0)}(x'(0), U_{A(0)}),$$
$$\Delta(0) = f_{\Delta(0)}(x'(0), f_{A(0)}(x'(0), U_{A(0)}), U_{\Delta(0)}),$$

and the potential outcomes take the form

$$X^{a(0)}(1) = f_{X(1)}(x'(0), a(0), U_{X(1)}),$$
$$Y^{a(0),a(1)} = f_Y(x'(0), a(0), f_{X(1)}(x'(0), a(1), U_{X(1)}), a(1), U_Y).$$

Thus, the following is implied:

$$W^{\overline{a}} \perp\!\!\!\perp (A(0), \Delta(0)) \mid X(0) = x'(0).$$

Next, conditional on $X(0) = x'(0)$, $A(0) = a'(0)$, and $X(1) = x'(1)$, it follows that

$$A(1) = f_{A(1)}(x'(0), a'(0), x'(1), U_{A(1)}),$$
$$\Delta(1) = f_{\Delta(1)}(x'(0), a'(0), x'(1), f_{A(1)}(x'(0), a'(0), x'(1), U_{A(1)}), U_{\Delta(1)}),$$

and

$$X^{a(0)}(1) = f_{X(1)}(x'(0), a(0), U_{X(1)}),$$
$$Y^{a(0),a(1)} = f_Y(x'(0), a(0), f_{X(1)}(x'(0), a(0), U_{X(1)}), a(1), U_Y).$$

Hence, the following is implied:

$$W^{\overline{a}} \perp\!\!\!\perp (A(1), \Delta(1)) \mid (X(1), A(0), X(0)) = (x'(1), a'(0), x'(0)).$$

In summary, the SCM implies

$$W^{\overline{a}} \perp\!\!\!\perp (A(0), \Delta(0)) \mid X(0) \quad \text{and} \quad W^{\overline{a}} \perp\!\!\!\perp (A(1), \Delta(1)) \mid (X(1), A(0), X(0)).$$

This property is again referred to as the *sequential exchangeability assumption*.

In addition, the positivity assumption is also required:

$$\mathbb{P}(\overline{A} = \overline{a}, (\Delta(0), \Delta(1)) = (1, 1) \mid X(0) = x(0)) > 0, \quad \text{for } x(0) \in \text{supp}(X(0)).$$

Scenario Three: $T \geq 2$ and Missing Data

The general case with $T \geq 2$ and missing data is finally considered. A DAG for this setting is too complex, so an SCM is employed to describe the relationships among the following variables:

$$\{X(0), A(0), \Delta(0), \Delta(0)X(1), \ldots, \Delta(T-2)A(T-1), \Delta(T-1), \Delta(T-1)Y\}.$$

The estimand of interest is defined as

$$v^*(\overline{a}) = \mathbb{E}(Y^{\overline{a}}).$$

Let $\overline{\Delta}(t) = (\Delta(0), \ldots, \Delta(t-1))$ and $\overline{\Delta} = (\Delta(0), \ldots, \Delta(T-1))$. The assumptions from Scenario Two can be generalized to Scenario Three. Specifically, the sequential exchangeability assumption states that

$$W^{\overline{a}} \perp\!\!\!\perp (A(0), \Delta(0)) \mid X(0),$$
$$W^{\overline{a}} \perp\!\!\!\perp (\overline{A}(t), \overline{\Delta}(t)) \mid (\overline{X}(t), \overline{A}(t-1)), \quad t = 1, \ldots, T-1.$$

The positivity assumption states that

$$\mathbb{P}(\overline{A} = \overline{a}, \overline{\Delta} = \overline{1} \mid X(0) = x(0)) > 0, \quad \text{for } x(0) \in \text{supp}(X(0)).$$

More generally, suppose that the estimand is the ATE comparing two treatment sequences $\overline{a}_1$ and $\overline{a}_0$,

$$\theta^*_{\text{ATE}} = \mathbb{E}\left(Y^{\overline{a}_1}\right) - \mathbb{E}\left(Y^{\overline{a}_0}\right).$$

For example, the two treatment sequences under comparison may be $\overline{a}_1 = \overline{1}$ and $\overline{a}_0 = \overline{0}$. For identification, the following assumptions are imposed. The sequential exchangeability assumption requires that

$$\left(W^{\overline{a}_1}, W^{\overline{a}_0}\right) \perp\!\!\!\perp (A(0), \Delta(0)) \mid X(0),$$
$$\left(W^{\overline{a}_1}, W^{\overline{a}_0}\right) \perp\!\!\!\perp (\overline{A}(t), \overline{\Delta}(t)) \mid (\overline{X}(t), \overline{A}(t-1)), \quad t = 1, \ldots, T-1.$$

The positivity assumption requires that

$$\mathbb{P}(\overline{A} = \overline{a}_1, \overline{\Delta} = \overline{1} \mid X(0) = x(0)) > 0, \quad \text{for } x(0) \in \text{supp}(X(0)),$$
$$\mathbb{P}(\overline{A} = \overline{a}_0, \overline{\Delta} = \overline{1} \mid X(0) = x(0)) > 0, \quad \text{for } x(0) \in \text{supp}(X(0)).$$

Gold Coin # 52. *Two types of studies are commonly considered: point-treatment studies and longitudinal studies. A point-treatment study can be regarded as a special case of a longitudinal study with $T = 1$.*

As in point-treatment studies, identification in longitudinal studies relies not only on the SUTVA and the consistency assumption, but also on two additional assumptions: sequential exchangeability and positivity.

15.2 Two Identification Strategies

15.2.1 Standardization Strategy

Scenario One: $T = 2$ and No Missing Data

For the purpose of demonstration, probability mass functions are employed for all variables in $X(t)$ and Y. For $T = 2$, it is obtained that

$$\begin{aligned}
&\mathbb{P}\{Y = y \mid X(0) = x(0), A(0) = a(0), X(1) = x(1), A(1) = a(1)\}\\
\overset{(a)}{=}&\mathbb{P}\{Y^{\overline{a}} = y \mid X(0) = x(0), A(0) = a(0), X(1) = x(1), A(1) = a(1)\}\\
\overset{(b)}{=}&\mathbb{P}\{Y^{\overline{a}} = y \mid X(0) = x(0), A(0) = a(0), X(1) = x(1)\}\\
\overset{(c)}{=}&\mathbb{P}\{Y^{\overline{a}} = y \mid X(0) = x(0), A(0) = a(0), X^{a(0)}(1) = x(1)\}\\
\overset{(d)}{=}&\frac{\mathbb{P}\{Y^{\overline{a}} = y, X^{a(0)}(1) = x(1) \mid X(0) = x(0), A(0) = a(0)\}}{\mathbb{P}\{X^{a(0)}(1) = x(1) \mid X(0) = x(0), A(0) = a(0)\}}\\
\overset{(e)}{=}&\frac{\mathbb{P}\{Y^{\overline{a}} = y, X^{a(0)}(1) = x(1) \mid X(0) = x(0)\}}{\mathbb{P}\{X^{a(0)}(1) = x(1) \mid X(0) = x(0)\}}\\
\overset{(f)}{=}&\mathbb{P}\{Y^{\overline{a}} = y \mid X(0) = x(0), X^{a(0)}(1) = x(1)\},
\end{aligned}$$

where (a) follows from consistency, (b) from sequential exchangeability, (c) again from consistency, (d) from the relationship among joint, marginal, and conditional distributions, (e) from sequential exchangeability applied in both numerator and denominator, and (f) again from the relationship among joint, marginal, and conditional distributions.

It is also obtained that

$$\begin{aligned}
&\mathbb{P}\{X(1) = x(1) \mid X(0) = x(0), A(0) = a(0)\}\\
\overset{(a)}{=}&\mathbb{P}\{X^{a(0)}(1) = x(1) \mid X(0) = x(0), A(0) = a(0)\}\\
\overset{(b)}{=}&\mathbb{P}\{X^{a(0)}(1) = x(1) \mid X(0) = x(0)\},
\end{aligned}$$

where (a) follows from consistency and (b) from sequential exchangeability.

By combining these results, it is shown that

$$\begin{aligned}
&\mathbb{P}\{X(0) = x(0), X^{a(0)}(1) = x(1), Y^{\overline{a}} = y\}\\
=&\mathbb{P}\{X(0) = x(0)\} \times \mathbb{P}\{X^{a(0)}(1) = x(1) \mid X(0) = x(0)\}\\
&\times \mathbb{P}\{Y^{\overline{a}} = y \mid X(0) = x(0), X^{a(0)}(1) = x(1)\}\\
=&\mathbb{P}\{X(0) = x(0)\} \times \mathbb{P}\{X(1) = x(1) \mid X(0) = x(0), A(0) = a(0)\}\\
&\times \mathbb{P}\{Y = y \mid X(0) = x(0), A(0) = a(0), X(1) = x(1), A(1) = a(1)\}.
\end{aligned}$$

Therefore, for $T = 2$, it is established that

$$\mathbb{E}\left(Y^{\overline{a}}\right) = \mathbb{E}_{\overline{X}\sim\mathcal{F}_{\overline{a}}}\left[\mathbb{E}(Y \mid \overline{X}, \overline{A} = \overline{a})\right],$$

where $\mathcal{F}_{\overline{a}}$ is defined as the distribution of $\overline{X}$ given by

$$\begin{aligned}&\mathbb{P}_{\mathcal{F}_{\overline{a}}}(X(0) = x(0), X(1) = x(1))\\ =&\mathbb{P}\{X(0) = x(0)\} \times \mathbb{P}\{X(1) = x(1) \mid X(0) = x(0), A(0) = a(0)\}.\end{aligned}$$

Scenario Two: $T = 2$ and Missing Data

The result can be generalized to the case with missing data. For $T = 2$, it is established that

$$\mathbb{E}\left(Y^{\overline{a}}\right) = \mathbb{E}_{\overline{X}\sim\widetilde{\mathcal{F}}_{\overline{a}}}\left[\mathbb{E}(Y \mid \overline{X}, \overline{A} = \overline{a}, \overline{\Delta} = \overline{1})\right],$$

where $\widetilde{\mathcal{F}}_{\overline{a}}$ is defined as the distribution of $\overline{X}$ given by

$$\begin{aligned}&\mathbb{P}_{\widetilde{\mathcal{F}}_{\overline{a}}}(X(0) = x(0), X(1) = x(1))\\ =&\mathbb{P}\{X(0) = x(0)\} \times \mathbb{P}\{X(1) = x(1) \mid X(0) = x(0), A(0) = a(0), \Delta(0) = 1\}.\end{aligned}$$

Scenario Three: $T \geq 2$ and Missing Data

The result can further be generalized to the case $T \geq 2$:

$$\mathbb{E}\left(Y^{\overline{a}}\right) = \mathbb{E}_{\overline{X}\sim\widetilde{\mathcal{F}}_{\overline{a}}}\left[\mathbb{E}(Y \mid \overline{X}, \overline{A} = \overline{a}, \overline{\Delta} = \overline{1})\right],$$

where $\widetilde{\mathcal{F}}_{\overline{a}}$ is defined as the distribution of $\overline{X}$ given by

$$\begin{aligned}&\mathbb{P}_{\widetilde{\mathcal{F}}_{\overline{a}}}(x(0), \ldots, x(T-1))\\ =&\mathbb{P}\{X(0) = x(0)\} \times \mathbb{P}\{X(1) = x(1) \mid x(0), a(0), \Delta(0) = 1\}\\ &\times \mathbb{P}\{X(2) = x(2) \mid \overline{x}(1), \overline{a}(1), \overline{\Delta}(1) = \overline{1}\} \times \ldots\\ &\times \mathbb{P}\{X(T-1) = x(T-1) \mid \overline{x}(T-2), \overline{a}(T-2), \Delta(T-2) = \overline{1}\}.\end{aligned}$$

Therefore, for θ^*_{ATE}, one formulation of the corresponding statistical estimand is obtained via the standardization strategy,

$$\begin{aligned}\theta_{\text{ATE}} =&\mathbb{E}_{\overline{X}\sim\widetilde{\mathcal{F}}_{\overline{a}_1}}\left[\mathbb{E}(Y \mid \overline{X}, \overline{A} = \overline{a}_1, \overline{\Delta} = \overline{1}\right]\\ &- \mathbb{E}_{\overline{X}\sim\mathcal{F}_{\overline{a}_0}}\left[\mathbb{E}(Y \mid \overline{X}, \overline{A} = \overline{a}_0, \overline{\Delta} = \overline{1}\right],\end{aligned}$$

where $\widetilde{\mathcal{F}}_{\overline{a}_1}$ and $\widetilde{\mathcal{F}}_{\overline{a}_0}$ are defined as in the above.

15.2.2 Weighting Strategy

Scenario One: $T = 2$ and No Missing Data

The case $T = 2$ is considered first. It is obtained that

$$\begin{aligned}
&\mathbb{E}\left[\frac{I(\overline{A}=\overline{a})Y}{\mathbb{P}\{A(0)=a(0)\mid X(0)\}\mathbb{P}\{A(1)=a(1)\mid X(0),A(0),X(1)\}}\right]\\
&\overset{(a)}{=}\mathbb{E}\left[\frac{I(\overline{A}=\overline{a})Y^{\overline{a}}}{\mathbb{P}\{A(0)=a(0)\mid X(0)\}\mathbb{P}\{A(1)=a(1)\mid X(0),A(0),X^{a(0)}(1)\}}\right]\\
&\overset{(b)}{=}\mathbb{E}\left[\frac{\mathbb{E}\{I(\overline{A}=\overline{a})Y^{\overline{a}}\mid X(0),W^{\overline{a}}\}}{\mathbb{P}\{A(0)=a(0)\mid X(0)\}\mathbb{P}\{A(1)=a(1)\mid X(0),A(0)=a(0),X^{a(0)}(1)\}}\right]\\
&\overset{(c)}{=}\mathbb{E}\left[\frac{\mathbb{P}\{\overline{A}=\overline{a}\mid X(0),W^{\overline{a}}\}Y^{\overline{a}}}{\mathbb{P}\{A(0)=a(0)\mid X(0)\}\mathbb{P}\{A(1)=a(1)\mid X(0),A(0)=a(0),X^{a(0)}(1)\}}\right]\\
&\overset{(d)}{=}\mathbb{E}\left[\frac{\mathbb{P}\{A(0)=a(0)\mid X(0),W^{\overline{a}}\}\mathbb{P}\{A(1)=a(1)\mid X(0),A(0)=a(0),W^{\overline{a}}\}Y^{\overline{a}}}{\mathbb{P}\{A(0)=a(0)\mid X(0)\}\mathbb{P}\{A(1)=a(1)\mid X(0),A(0)=a(0),X^{a(0)}(1)\}}\right]\\
&\overset{(e)}{=}\mathbb{E}(Y^{\overline{a}}),
\end{aligned}$$

where the first term is well-defined under the positivity assumption, equality (a) follows from consistency, (b) from the law of iterated expectations, (c) because the expectation of an indicator equals its probability and $Y^{\overline{a}}$ is one component of $W^{\overline{a}}$, (d) from the relationship among joint, marginal, and conditional distributions, and (e) from sequential exchangeability.

In short, it is shown that

$$\mathbb{E}(Y^{\overline{a}})=\mathbb{E}\left[\frac{I(\overline{A}=\overline{a})Y}{\mathbb{P}\{A(0)=a(0)\mid X(0)\}\mathbb{P}\{A(1)=a(1)\mid A(0),\overline{X}(1)\}}\right].$$

Scenario Two: $T = 2$ and Missing Data

The result can be generalized to the case with missing data. For $T = 2$, it is established that

$$\mathbb{E}(Y^{\overline{a}})=\mathbb{E}\left[\frac{I(\overline{A}=\overline{a})I(\overline{\Delta}=\overline{1})Y}{\widetilde{\mathbb{P}}\{A(0)=a(0)\mid X(0)\}\widetilde{\mathbb{P}}\{A(1)=a(1)\mid A(0),\overline{X}(1)\}}\right],$$

where

$$\begin{aligned}
\widetilde{\mathbb{P}}\{A(0)=a(0)\mid X(0)\}&=\mathbb{P}\{A(0)=a(0)\mid X(0)\}\\
&\quad\times\mathbb{P}\{\Delta(0)=1\mid X(0),A(0)\},\\
\widetilde{\mathbb{P}}\{A(1)=a(1)\mid A(0),\overline{X}(1)\}&=\mathbb{P}\{A(1)=a(1)\mid A(0),\overline{X}(1)\}\\
&\quad\times\mathbb{P}\{\Delta(1)=1\mid\overline{X}(1),\overline{A}(1)\}.
\end{aligned}$$

Scenario Three: $T \geq 2$ and Missing Data

The result can further be generalized to the case $T \geq 2$:

$$\mathbb{E}(Y^{\overline{a}}) = \mathbb{E}\left[\frac{I(\overline{A} = \overline{a})I(\overline{\Delta} = \overline{1})Y}{\widetilde{\mathbb{P}}\{A(0) = a(0) \mid X(0)\} \prod_{t=1}^{T-1} \widetilde{\mathbb{P}}\{A(t) = a(t) \mid \overline{A}(t-1), \overline{X}(t)\}}\right],$$

where

$$\begin{aligned}
\widetilde{\mathbb{P}}\{A(0) = a(0) \mid X(0)\} =& \mathbb{P}\{A(0) = a(0) \mid X(0)\} \\
& \times \mathbb{P}\{\Delta(0) = 1 \mid X(0), A(0)\}, \\
\widetilde{\mathbb{P}}\{A(t) = a(t) \mid \overline{A}(t-1), \overline{X}(t)\} =& \mathbb{P}\{A(t) = a(t) \mid \overline{A}(t-1), \overline{X}(t)\} \\
& \times \mathbb{P}\{\Delta(t) = 1 \mid \overline{X}(t), \overline{A}(t)\}.
\end{aligned}$$

Therefore, for θ^*_{ATE}, the other formulation of the corresponding statistical estimand is obtained via the weighting strategy,

$$\begin{aligned}
\theta_{\text{ATE}} = & \mathbb{E}\left[\frac{I(\overline{A} = \overline{a}_1)I(\overline{\Delta} = \overline{1})Y}{\widetilde{\mathbb{P}}\{A(0) = a_1(0) \mid X(0)\} \prod_{t=1}^{T-1} \widetilde{\mathbb{P}}\{A(t) = a_1(t) \mid \overline{A}(t-1), \overline{X}(t)\}}\right] \\
& - \mathbb{E}\left[\frac{I(\overline{A} = \overline{a}_0)I(\overline{\Delta} = \overline{1})Y}{\widetilde{\mathbb{P}}\{A(0) = a_0(0) \mid X(0)\} \prod_{t=1}^{T-1} \widetilde{\mathbb{P}}\{A(t) = a_0(t) \mid \overline{A}(t-1), \overline{X}(t)\}}\right].
\end{aligned}$$

Gold Coin # 53. *For the causal estimand that compares two treatment sequences (a.k.a. static treatment regimes), two strategies are employed for identification: the standardization strategy and the weighting strategy. Through these strategies, two equivalent formulations of the corresponding statistical estimand are obtained.*

15.3 Two Series of Models

15.3.1 Propensity Score Models

To simplify the expression of the estimand, g functions are introduced. First, it is assumed that $\overline{X}(0) = X(0)$, $\overline{A}(0) = A(0)$, and $\overline{X}(-1) = \overline{A}(-1) = \emptyset$.

Second, the propensity score function at time t is defined as

$$g_t\left(a(t) \mid \overline{X}(t), \overline{A}(t-1)\right) = \mathbb{P}\left\{A(t) = a(t) \mid \overline{X}(t), \overline{A}(t-1)\right\},$$

where $t = 0, \ldots, T-1$.

Third, the non-missing probability function at time t is defined as

$$h_t(\overline{X}(t), \overline{A}(t)) = \mathbb{P}\{\Delta(t) = 1 \mid \overline{X}(t), \overline{A}(t)\}.$$

Fourth, the product of g_t and h_t is defined as

$$\begin{aligned}&\widetilde{g}_t(a(t) \mid \overline{X}(t), \overline{A}(t-1)) \\ &= g_t\left(a(t) \mid \overline{X}(t), \overline{A}(t-1)\right) \times h_t\left(\overline{X}(t), (\overline{A}(t-1), a(t))\right).\end{aligned}$$

Hence, the estimand is simplified as

$$\mathbb{E}(Y^{\overline{a}}) = \mathbb{E}\left[\frac{I(\overline{A} = \overline{a})I(\overline{\Delta} = \overline{1})Y}{\prod_{t=0}^{T-1} \widetilde{g}_t(A(t) \mid \overline{X}(t), \overline{A}(t-1))}\right].$$

15.3.2 Outcome Regression Models

The outcome regression function at the final visit is defined as

$$Q(\overline{X}, \overline{A}) = \mathbb{E}(Y \mid \overline{X}, \overline{A} = \overline{a}, \overline{\Delta} = \overline{1}),$$

so that the estimand can be rewritten as

$$\mathbb{E}\left(Y^{\overline{a}}\right) = \mathbb{E}_{\overline{X} \sim \widetilde{\mathcal{F}}_{\overline{a}}}\left[Q(\overline{X}, \overline{A})\right].$$

Since the joint distribution of $\overline{X}$ becomes intractable for moderate or high dimensions, a backward recursion algorithm is applied for identification via the standardization strategy.

Backward Recursion for Standardization

The final visit is considered first:

$$\mathbb{E}(Y^{\overline{a}}) = \mathbb{E}\left\{\mathbb{E}\left[\frac{I(\overline{A} = \overline{a})I(\overline{\Delta} = \overline{1})Y}{\prod_{t=0}^{T-1} \widetilde{g}_t(A(t) \mid \overline{X}(t), \overline{A}(t-1))} \,\middle|\, \overline{X}(T-1), \overline{A}(T-2)\right]\right\},$$

which, by the law of iterated expectations, is expressed as

$$\mathbb{E}\left\{\mathbb{E}\left[\frac{I(\overline{A} = \overline{a})I(\overline{\Delta} = \overline{1})Y}{\prod_{t=0}^{T-1} \widetilde{g}_t\left(A(t) \mid \overline{X}(t), \overline{A}(t-1)\right)} \,\middle|\, \overline{X}(T-1), \overline{A}(T-2)\right]\right\}.$$

Noting that

$$\begin{aligned}&\frac{I(\overline{A} = \overline{a})I(\overline{\Delta} = \overline{1})}{\prod_{t=0}^{T-1} \widetilde{g}_t\left(A(t) \mid \overline{X}(t), \overline{A}(t-1)\right)} \\ &= \frac{I\{\overline{A}(T-2) = \overline{a}(T-2)\}I\{\overline{\Delta}(T-2) = \overline{1}\}}{\prod_{t=0}^{T-2} \widetilde{g}_t\left(A(t) \mid \overline{X}(t), \overline{A}(t-1)\right)} \\ &\quad \times \frac{I(A(T-1) = a(T-1))I(\Delta(T-1) = 1)}{\widetilde{g}_t\left(A(T-1) \mid \overline{X}(T-1), \overline{A}(T-2)\right)},\end{aligned}$$

it follows that

$$\mathbb{E}\left\{\mathbb{E}\left[\frac{I(\overline{A}=\overline{a})I(\overline{\Delta}=\overline{1})Y}{\prod_{t=0}^{T-1}\widetilde{g}_t\left(A(t)\mid\overline{X}(t),\overline{A}(t-1)\right)}\,\middle|\,\overline{X}(T-1),\overline{A}(T-2)\right]\right\}$$
$$=\mathbb{E}\left\{\frac{I\{\overline{A}(T-2)=\overline{a}(T-2)\}I\{\overline{\Delta}(T-2)=\overline{1}\}}{\prod_{t=0}^{T-2}\widetilde{g}_t\left(A(t)\mid\overline{X}(t),\overline{A}(t-1)\right)}\right.$$
$$\left.\times\,\mathbb{E}\left[\frac{I(A(T-1)=a(T-1))I(\Delta(T-1)=1)}{\widetilde{g}_t\left(A(T-1)\mid\overline{X}(T-1),\overline{A}(T-2)\right)}Y\,\middle|\,\overline{X}(T-1),\overline{A}(T-2)\right]\right\}.$$

Using standard results for inverse-probability weighting, the inner conditional expectation is simplified to

$$\mathbb{E}\left[\frac{I(A(T-1)=a(T-1))I(\Delta(T-1)=1)}{\widetilde{g}_t\left(A(T-1)\mid\overline{X}(T-1),\overline{A}(T-2)\right)}Y\,\middle|\,\overline{X}(T-1),\overline{A}(T-2)\right]$$
$$=\mathbb{E}\left[Y\,\middle|\,\overline{X}(T-1),\overline{A}(T-2),A(T-1)=a(T-1),\Delta(T-1)=1\right].$$

The following regression function is then defined:

$$Q_{T-1}(\overline{X}(T-1),\overline{A}(T-1))$$
$$=\mathbb{E}\left[Y\,\middle|\,\overline{X}(T-1),\overline{A}(T-1),\Delta(T-1)=1\right].$$

Thus, the estimand is expressed as

$$\mathbb{E}(Y^{\overline{a}})=\mathbb{E}\left\{\frac{I\{\overline{A}(T-2)=\overline{a}(T-2)\}I\{\overline{\Delta}(T-2)=\overline{1}\}}{\prod_{t=0}^{T-2}\widetilde{g}_t\left(A(t)\mid\overline{X}(t),\overline{A}(t-1)\right)}\right.$$
$$\left.\times\,Q_{T-1}(\overline{X}(T-1),\overline{A}(T-2),a(T-1))\right\}.$$

Next, one step backward is considered. Define the tentative outcome

$$\widetilde{Y}_{T-1}=Q_{T-1}\left(\overline{X}(T-1),\overline{A}(T-2),a(T-1)\right),$$

and repeat the same argument:

$$\mathbb{E}(Y^{\overline{a}})=\mathbb{E}\left\{\frac{I\{\overline{A}(T-2)=\overline{a}(T-2)\}I\{\overline{\Delta}(T-2)=\overline{1}\}\widetilde{Y}_{T-1}}{\prod_{t=0}^{T-2}\widetilde{g}_t\left(A(t)\mid\overline{X}(t),\overline{A}(t-1)\right)}\right\}$$
$$=\mathbb{E}\left\{\frac{I\{\overline{A}(T-3)=\overline{a}(T-3)\}I\{\overline{\Delta}(T-3)=\overline{1}\}}{\prod_{t=0}^{T-3}\widetilde{g}_t\left(A(t)\mid\overline{X}(t),\overline{A}(t-2)\right)}\right.$$
$$\left.\times\,Q_{T-2}(\overline{X}(T-2),\overline{A}(T-3),a(T-2))\right\},$$

where

$$Q_{T-2}\left(\overline{X}(T-2), \overline{A}(T-2)\right) = \mathbb{E}\left[\widetilde{Y}_{T-1} \mid \overline{X}(T-2), \overline{A}(T-2), \Delta(T-2) = 1\right].$$

Repeating $T-2$ steps backward yields

$$\mathbb{E}(Y^{\overline{a}}) = \mathbb{E}\left\{\frac{I\left(A(0) = a(0)\right) I(\Delta(0) = 1)}{\widetilde{g}_0\left(a(0) \mid X(0)\right)} Q_1\left(\overline{X}(1), A(0), a(1)\right)\right\}.$$

Finally, define

$$\widetilde{Y}_1 = Q_1\left(\overline{X}(1), A(0), a(1)\right),$$

so that the estimand becomes

$$\mathbb{E}(Y^{\overline{a}}) = \mathbb{E}\left\{\frac{I\left(A(0) = a(0)\right) I(\Delta(0) = 1)}{\widetilde{g}_0\left(a(0) \mid X(0)\right)} \widetilde{Y}_1\right\} = \mathbb{E}\left\{Q_0\left(X(0), a(0)\right)\right\},$$

where

$$Q_0\left(X(0), A(0)\right) = \mathbb{E}\left[\widetilde{Y}_1 \mid X(0), A(0), \Delta(0) = 1\right].$$

Gold Coin # 54. *Two formulations of the statistical estimand are expressed in terms of a series of propensity score functions and a series of outcome regression functions, respectively.*

15.4 EIFs for the Values of Treatment Regimes

15.4.1 Static Treatment Regimes

The convenient approach is employed to derive the EIF for the following estimand, which is referred to as the value of the static treatment regime $\overline{a}$,

$$v^*(\overline{a}) = \mathbb{E}(Y^{\overline{a}}).$$

The convenient approach consists of two steps. In the first step, subtraction is carried out. This step consists of $T+1$ mini-steps, and these mini-steps are now taken sequentially.

At $t = 0$, since

$$v(\overline{a}) = \mathbb{E}\left\{Q_0\left(X(0), a(0)\right)\right\},$$

the following mean-zero function is defined:

$$\phi_0(O) = Q_0\left(X(0), a(0)\right) - v(\overline{a}).$$

At $t = 1, \ldots, T-1$, since

$$v(\overline{a}) = \mathbb{E}\left\{ \frac{I\{\overline{A}(t-1) = \overline{a}(t-1)\}I\{\overline{\Delta}(t-1) = \overline{1}\}}{\prod_{s=0}^{t-1} \widetilde{g}_s\left(A(s) \mid \overline{X}(s), \overline{A}(s-1)\right)} \times Q_t(\overline{X}(t), \overline{A}(t-1), a(t)) \right\},$$

the following centered function is defined:

$$\phi_t(O) = \frac{I\{\overline{A}(t-1) = \overline{a}(t-1)\}I\{\overline{\Delta}(t-1) = \overline{1}\}}{\prod_{s=0}^{t-1} \widetilde{g}_s\left(A(s) \mid \overline{X}(s), \overline{A}(s-1)\right)} \times \left[Q_t(\overline{X}(t), \overline{A}(t-1), a(t)) - Q_{t-1}(\overline{X}(t-1), \overline{A}(t-1))\right].$$

At $t = T$, since

$$v(\overline{a}) = \mathbb{E}\left[\frac{I(\overline{A} = \overline{a})I(\overline{\Delta} = \overline{1})Y}{\prod_{t=0}^{T-1} \widetilde{g}_t(A(t) \mid \overline{X}(t), \overline{A}(t-1))}\right],$$

the following centered function is defined:

$$\phi_T(O) = \frac{I(\overline{A} = \overline{a})I(\overline{\Delta} = \overline{1})}{\prod_{t=0}^{T-1} \widetilde{g}_t(A(t) \mid \overline{X}(t), \overline{A}(t-1))}\left[Y - Q_{T-1}\left(\overline{X}(T-1), \overline{A}(T-1)\right)\right].$$

In the second step of the convenient approach, addition is carried out. By summing the above $T+1$ centered functions, the EIF of $v(\overline{a})$ is obtained:

$$\phi_{\text{EIF-}\overline{a}}(O) = \sum_{t=0}^{T} \phi_t(O),$$

which is equal to

$$\begin{aligned}
&\left[Q_0\left(X(0), a(0)\right) - v(\overline{a})\right] \\
&+ \sum_{t=1}^{T-1} \frac{I\{\overline{A}(t-1) = \overline{a}(t-1)\}I\{\overline{\Delta}(t-1) = \overline{1}\}}{\prod_{s=0}^{t-1} \widetilde{g}_s\left(A(s) \mid \overline{X}(s), \overline{A}(s-1)\right)} \times \\
&\quad \left[Q_t\left(\overline{X}(t), \overline{A}(t-1), a(t)\right) - Q_{t-1}\left(\overline{X}(t-1), \overline{A}(t-1)\right)\right] \\
&+ \frac{I(\overline{A} = \overline{a})I(\overline{\Delta} = \overline{1})}{\prod_{t=0}^{T-1} \widetilde{g}_t(A(t) \mid \overline{X}(t), \overline{A}(t-1))}\left[Y - Q_{T-1}\left(\overline{X}(T-1), \overline{A}(T-1)\right)\right].
\end{aligned}$$

A proof establishing the validity of the convenient approach described above for deriving the EIF of $v(\overline{a})$ is provided in Exercises 15.1–15.5.

In general, if the estimand is

$$\theta^*_{\text{ATE}} = \mathbb{E}\left(Y^{\overline{a}_1}\right) - \mathbb{E}\left(Y^{\overline{a}_0}\right),$$

then the EIF is expressed as

$$\phi_{\text{EIF-ATE}}(O) = \phi_{\text{EIF-}\overline{a}_1}(O) - \phi_{\text{EIF-}\overline{a}_0}(O).$$

15.4.2 Dynamic Treatment Regimes

The focus so far has been on static treatment regimes (STRs). The discussion is now extended to dynamic treatment regimes (DTRs). A full treatment of DTRs is beyond the scope of this book, and the reader is referred to relevant textbooks.[61] In this subsection, it is merely demonstrated that the convenient approach can also be applied to derive the EIFs for the values of DTRs.

All notations used for STRs are retained. In addition, H_0 is defined as $X(0)$, and $H_t = (X(0), A(0), \ldots, X(t-1), A(t-2), X(t-1))$ is defined as the history before decision t is made, for $t = 1, \ldots, T-1$. A feasible DTR with $T-1$ decisions is expressed as

$$\overline{d} = \{d_0(H_0), \ldots, d_{T-1}(H_{T-1})\},$$

where $d_t(H_t)$ denotes a decision rule at decision t, $t = 0, \ldots, T-1$, mapping the subject's history to a treatment option. The DTR can be re-expressed as

$$\begin{aligned}\overline{d} &= \overline{d}_{T-1}\left(\overline{X}(T-1)\right)\\ &= \left\{d_0\left(X_0\right), d_1\left(\overline{X}(1), d_0(X(0))\right), \ldots, d_{T-1}\left(\overline{X}(T-1), \overline{d}_{T-2}(\overline{X}(T-2))\right)\right\}.\end{aligned}$$

Let $Y^{\overline{d}}$ denote the potential outcome that would have been observed had the patient been treated according to the DTR $\overline{d}$. The causal estimand is defined as the value of $\overline{d}$,

$$v^*(\overline{d}) = \mathbb{E}(Y^{\overline{d}}).$$

Under the corresponding sequential exchangeability and positivity assumptions, the causal estimand $v^*(\overline{d})$ is translated into a statistical estimand with T equivalent forms.

By the weighting strategy, $v^*(\overline{d})$ is expressed as

$$v(\overline{d}) = \mathbb{E}\left\{\frac{I[\overline{A}(T-1) = \overline{d}(\overline{X}(T-1))]Y}{\prod_{t=0}^{T-1} \pi_{d,t}(\overline{X}(t))}\right\},$$

where $\pi_{d,t}(\overline{X}(t))$ is defined as

$$\mathbb{P}\Big\{A(t) = d_t(\overline{X}(t), \overline{d}_{t-1}(\overline{X}(t-1))) \mid \overline{X}(t), \overline{A}(t-1) = \overline{d}_{t-1}(\overline{X}(t-1))\Big\}.$$

A sequence of $T-1$ standardization strategies is then applied. First,

$$v(\overline{d}) = \mathbb{E}\left\{\mathbb{E}\left[\frac{I[\overline{A}(T-1) = \overline{d}_{T-1}(\overline{X}(T-1))]Y}{\prod_{t=1}^{T-1} \pi_{d,t}(\overline{X}(t))} \;\middle|\; \overline{X}(T-1), \overline{A}(T-2)\right]\right\}$$

$$= \mathbb{E}\left\{\frac{I[\overline{A}(T-2) = \overline{d}_{T-2}(\overline{X}(T-2))]Q_{T-1}\left(\overline{X}(T-1), \overline{A}(T-2), d_{T-1}\right)}{\prod_{t=0}^{T-2} \pi_{d,t}(\overline{X}(t))}\right\},$$

where

$$Q_{T-1}(\overline{X}(T-1), \overline{A}(T-1)) = \mathbb{E}[Y \mid \overline{X}(T-1), \overline{A}(T-1)].$$

By defining

$$\widetilde{Y}_{T-1} = Q_{T-1}\left(\overline{X}(T-1), \overline{A}(T-2), d_{T-1}\right),$$

the expression becomes

$$v(\overline{d}) = \mathbb{E}\left\{\frac{I[\overline{A}(T-2) = \overline{d}_{T-2}(\overline{X}(T-2))]\widetilde{Y}_{T-1}}{\prod_{t=0}^{T-2} \pi_{d,t}(\overline{X}(t))}\right\}$$

$$= \mathbb{E}\left\{\frac{I[\overline{A}(T-3) = \overline{d}_{T-3}(\overline{X}(T-3))]Q_{T-2}\left(\overline{X}(T-2), \overline{A}(T-3), d_{T-2}\right)}{\prod_{t=0}^{T-3} \pi_{d,t}(\overline{X}(t))}\right\},$$

where

$$Q_{T-2}(\overline{X}(T-2), \overline{A}(T-2)) = \mathbb{E}[\widetilde{Y}_{T-1} \mid \overline{X}(T-2), \overline{A}(T-2)].$$

Repeating this procedure for $t = T-2, \ldots, 1$, the following is obtained:

$$v(\overline{d}) = \mathbb{E}\left\{\frac{I[\overline{A}(t-1) = \overline{d}_{t-1}(\overline{X}(t-1))]\widetilde{Y}_t}{\prod_{s=0}^{t-1} \pi_{d,s}(\overline{X}(s))}\right\},$$

where

$$\widetilde{Y}_t = Q_t\left(\overline{X}(t), \overline{A}(t-1), d_t\right),$$
$$Q_t(\overline{X}(t), \overline{A}(t)) = \mathbb{E}[\widetilde{Y}_{t+1} \mid \overline{X}(t), \overline{A}(t)].$$

In the final step, it follows that

$$v = \mathbb{E}\left\{\frac{I\left(A(0) = d_0(X(0))\right)\widetilde{Y}_1}{\pi_{d,0}(X(0))}\right\} = \mathbb{E}\left\{Q_0\left(X(0), d_0\right)\right\},$$

where $Q_0(X(0), A(0)) = \mathbb{E}[\widetilde{Y}_1 \mid X(0), A(0)]$.

After these $T+1$ expressions of the statistical estimand $v(\overline{d})$ are obtained, the convenient approach is applied to derive the EIF. The approach again consists of two steps. In the subtraction step, the following functions are defined:

$$\phi_0(O) = Q_0\left(X(0), d_0\right) - v(\overline{d}),$$

$$\begin{aligned}\phi_t(O) =& \frac{I[\overline{A}(t-1) = \overline{d}_{t-1}(\overline{X}(t-1))]}{\prod_{s=1}^{t-1} \pi_{d,s}(\overline{X}(s))} \\ &\times \left[Q_t\left(\overline{X}(t), \overline{A}(t-1), d_t\right) - Q_{t-1}\left(\overline{X}(t-1), \overline{A}(t-1)\right)\right],\end{aligned}$$

for $t = 1, \ldots, T-1$, and

$$\phi_T(O) = \frac{I[\overline{A}(T-1) = \overline{d}(\overline{X}(T-1))][Y - Q_{T-1}(\overline{X}(T-1), \overline{A}(T-1))]}{\prod_{t=0}^{T-1} \pi_{d,t}(\overline{X}(t))}.$$

In the addition step, the EIF is obtained as

$$\phi_{\text{EIF-}v}(O) = \sum_{t=0}^{T} \phi_t(O).$$

Gold Coin # 55. *The convenient approach is applied to derive the corresponding EIFs for the values of both static treatment regimes (STRs) and dynamic treatment regimes (DTRs).*

Thus far, we have completed Part III, which introduced the two primary approaches for deriving efficient influence functions in Chapters 15 and 16. Although the material in Chapters 17–20 may initially seem tedious or unnecessary, it plays a crucial role: solid understanding of the parametric submodels developed in these chapters is essential for grasping the targeted learning methods presented later in Part IV.

15.5 Exercises

Exercise 15.1

Write down the likelihood for the following observation of a participant in a longitudinal study:

$$O = \{X(0), A(0), X(1), A(1), \ldots, X(T-1), A(T-1), Y\}.$$

Exercise 15.2

Continue Exercise 15.1. Specify a family of parametric submodels that pass the truth models for $p^{(t)}, g^t(1)$, $t = 0, \ldots, T-1$, and f.

Exercise 15.3

Continue Exercise 15.2. Obtain the score functions under the above parametric submodels evaluated at $\boldsymbol{\nu} = \mathbf{0}$, where

$$\boldsymbol{\nu} = (\nu^p(0), \ldots, \nu^p(T-1), \nu^g(0), \ldots, \nu^g(T-1), \nu^f)^\top.$$

Exercise 15.4

Continue Exercise 15.3. For a given treatment regime $\overline{a}$, the estimand under the specified parametric submodels is

$$\theta(\boldsymbol{\nu}) = \mathbb{E}_{\boldsymbol{\nu}}(Y^{\overline{a}}),$$

where the expectation is taken under the models parametrized by $\boldsymbol{\nu}$. Obtain the following derivatives:

$$\left.\frac{\partial\theta(\boldsymbol{\nu})}{\partial\nu^p(t)}\right|_{\boldsymbol{\nu}=0}, t = 0, \ldots, T-1,$$
$$\left.\frac{\partial\theta(\boldsymbol{\nu})}{\partial\nu^g(t)}\right|_{\boldsymbol{\nu}=0}, t = 0, \ldots, T-1,$$
$$\left.\frac{\partial\theta(\boldsymbol{\nu})}{\partial\nu^f}\right|_{\boldsymbol{\nu}=0}.$$

Exercise 15.5

Continue Exercise 15.4. Verify the following conditions:

$$\left.\frac{\partial\theta(\boldsymbol{\nu})}{\partial\nu^p(t)}\right|_{\boldsymbol{\nu}=0} = \mathbb{E}\left[\sum_{s=0}^{T}\phi_s(O) \times \left.\frac{\partial\ell}{\partial\nu^p(t)}\right|_{\boldsymbol{\nu}=\mathbf{0}}\right], \quad t = 0, \ldots, T-1,$$

$$\left.\frac{\partial\theta(\boldsymbol{\nu})}{\partial\nu^g(t)}\right|_{\boldsymbol{\nu}=0} = \mathbb{E}\left[\sum_{s=0}^{T}\phi_s(O) \times \left.\frac{\partial\ell}{\partial\nu^g(t)}\right|_{\boldsymbol{\nu}=\mathbf{0}}\right], \quad t = 0, \ldots, T-1,$$

and

$$\left.\frac{\partial\theta(\boldsymbol{\nu})}{\partial\nu^f}\right|_{\boldsymbol{\nu}=0} = \mathbb{E}\left[\sum_{s=0}^{T}\phi_s(O) \times \left.\frac{\partial\ell}{\partial\nu^f}\right|_{\boldsymbol{\nu}=\mathbf{0}}\right].$$

Part IV

An Introduction to Targeted Learning

16

Super Learning

16.1 The Mysterious Number 2

It is sometimes surprising to observe the emergence of the number 2 in a theoretical result. In Chapter 1, the reason for its appearance in the exponent of the Gaussian distribution was discussed. In this section, the appearance of the number 2 in the Akaike Information Criterion (AIC) will be examined. Through this examination, a deeper understanding of the trade-off between goodness of fit and model complexity will be developed.

16.1.1 Model Selection and Variable Selection

The regression function of Y given X is defined as

$$Q_0(X) = \mathbb{E}(Y \mid X).$$

A loss function $L(Y, Q(X))$—such as the $\mathcal{L}_2$ loss, the binary cross-entropy (BCE) loss, or others—is assumed to satisfy the condition that

$$Q_0(\cdot) = \arg\min_{Q(\cdot)} \mathbb{E}[L(Y, Q(X))].$$

A tentative parametric regression function $Q(X; \beta)$ is considered, where β is a finite-dimensional parameter vector. The parameter β_* is defined as the value that minimizes the expected loss:

$$\beta_* = \arg\min_{\beta} \mathbb{E}[L(Y, Q(X; \beta))],$$

which can be estimated by its empirical counterpart,

$$\widehat{\beta}_n = \arg\min_{\beta} \sum_{i=1}^{n} L(Y_i, Q(X_i; \beta)).$$

Consequently, $Q_0(\cdot)$ can then be estimated by

$$\widehat{Q}_n(\cdot) = Q(\cdot; \widehat{\beta}_n).$$

DOI: 10.1201/9781003635468-20

Model Selection

Selecting a model to fit $Q(X)$ is challenging. A variety of models have been proposed by in the field of statistical learning,[56] including parametric models such as linear and logistic regression models, as well as algorithmic models such as classification and regression trees (CART)[62] and random forests.[63]

If $Q_0(X)$ is assumed to take the form $Q(X;\beta)$, then the true parameter value, denoted by β_0, is defined as the value of β that satisfies

$$Q(\cdot;\beta_0) = Q_0(\cdot),$$

implying that $\beta_0 = \beta_*$.

In the absence of access to such an oracle, any model of the form $Q(X;\beta)$ is treated as a tentative model. Due to the abundance of such tentative models, the task of selecting an "optimal" model among the many available choices must be addressed.

In this chapter, two approaches to model selection will be examined: the information-criterion method and the cross-validation procedure.[64]

Given a set of candidate models, comparisons are carried out using either an information criterion or cross-validation. Based on the result of these comparisons, an optimal model is selected.

Variable Selection

Variable selection and tuning are considered subtopics within the broader task of model selection. Even after a model has been selected to approximate $Q_0(X)$, additional refinement is often required.

For example, if the linear model is selected, such as $Q(X;\beta) = \beta^\top X$, decisions must be made regarding which independent variables, and how many, should be included in the vector $X = (1, X^{(1)}, \ldots, X^{(d)})^\top$.

As another example, if the CART model is selected, the tree must be appropriately tuned to prevent overfitting or underfitting. Typically, a large tree is first grown, which is then pruned to an appropriate size.

In the case where the random forest model is selected, several hyperparameters are required to be specified, including the number of trees, the number of features randomly selected at each split, and the maximum number of leaf nodes allowed per tree.

16.1.2 AIC

Let $O_1, \ldots, O_n$ be independent and identically distributed (i.i.d.) observations drawn from a distribution $O \sim f(o;\theta_0)$, where θ_0 denotes an unknown parameter. The term "unknown" is used to indicate that neither the value nor the dimension of θ_0 is known. A collection of candidate models $f(o;\theta(k))$ is considered, where the parameter $\theta(k) \in \Theta(k)$ has dimension k and $\Theta(k)$ represents the associated parameter space. The maximum likelihood estimator (MLE) of

$\theta(k)$ under the model $f(o;\theta(k))$ is denoted by $\widehat{\theta}_n(k)$, and is defined as

$$\widehat{\theta}_n(k) = \arg \max_{\theta(k)\in\Theta(k)} \sum_{i=1}^{n} \log f(O_i;\theta(k)).$$

The question is then posed: What is the distance between the trained candidate model $f(\cdot;\widehat{\theta}_n(k))$ and the true model $f(\cdot;\theta_0)$? To address this, the Kullback–Leibler (KL) divergence between the working model and the true model is invoked. Based on this measure, the candidate model closest to the true model is selected.

In the derivation of the AIC, the distance between $f(\cdot;\widehat{\theta}_n(k))$ and the true model $f(\cdot;\theta_0)$ is defined as

$$D(\widehat{\theta}_n(k)) = \mathbb{E}\left\{-\sum_{i=1}^{n} 2\log f(O_i;\theta(k))\right\}\Bigg|_{\theta(k)=\widehat{\theta}_n(k)}.$$

This distance is closely related to the KL divergence, which is given by

$$\begin{aligned}
D_{\mathrm{KL}}(\widehat{\theta}_n(k),\theta_0) &= \mathbb{E}\left\{\log\left[\frac{f(O;\theta_0)}{f(O;\theta(k))}\right]\right\}\Bigg|_{\theta(k)=\widehat{\theta}_n(k)} \\
&= \mathbb{E}\left\{\log f(O;\theta_0)\right\} - \mathbb{E}\left\{\log f(O;\theta(k))\right\}\big|_{\theta(k)=\widehat{\theta}_n(k)} \\
&= C - \mathbb{E}\left\{\log f(O;\theta(k))\right\}\big|_{\theta(k)=\widehat{\theta}_n(k)},
\end{aligned}$$

where the first term C is constant with respect to k. Therefore, the following two minimization procedures are equivalent:

$$\arg\min_k \mathbb{E}\left[D(\widehat{\theta}_n(k))\right] = \arg\min_k \mathbb{E}\left[D_{\mathrm{KL}}(\widehat{\theta}_n(k),\theta_0)\right].$$

To estimate $\mathbb{E}\left[D(\widehat{\theta}_n(k))\right]$, a naive estimator is given by

$$-\sum_{i=1}^{n} 2\log f(O_i;\widehat{\theta}_n(k)).$$

However, it has been shown that this naive estimator is negatively biased. In certain situations the bias can be approximated by $2k$, where k denotes the dimension of $\theta(k)$. As a result, the AIC estimator is proposed as:[65]

$$\mathrm{AIC} = -\sum_{i=1}^{n} 2\log f(O_i;\widehat{\theta}_n(k)) + 2k.$$

The surprising result—that the bias is approximately twice the number of parameters—can be attributed to the application of two Taylor expansions during the derivation (see Exercise 16.1).

In the subsequent sections, additional instances of the number 2 will be encountered. As these occurrences accumulate, it will become evident that the number 2 serves to regulate the trade-off between *goodness of fit* (a concept introduced by Karl Pearson) and *model complexity* (associated with degrees of freedom, as introduced by R.A. Fisher). This trade-off plays a fundamental role in model selection, variable selection, and model tuning.

Gold Coin # 56. *Many popular model selection or variable selection procedures take the following form:*

$$\textit{goodness of fit} \; + \; 2 \times \textit{model complexity},$$

where the constant "2" governs the trade-off between the two terms.

As a side note, like the famous Fisher–Neyman debate, the disagreement between Pearson (1857–1936) and Fisher (1890–1962) also holds a prominent place in the history of statistics. Yet the trade-off that connects their contributions highlights the magic of the number 2.

Linear Model Example

Consider the following Gaussian linear model:

$$Y = X^\top \beta + \varepsilon,$$

where $X^\top = (1, X^{(1)}, \ldots, X^{(d)})$, and $\varepsilon \sim \mathcal{N}(0, \sigma^2)$.

The maximum likelihood estimator (MLE) of β coincides with the least squares estimator:

$$\widehat{\beta}_n = (\mathbf{X}^\top \mathbf{X})^{-1} \mathbf{X}^\top \mathbf{Y},$$

where

$$\mathbf{X} = \begin{pmatrix} X_1^\top \\ X_2^\top \\ \vdots \\ X_n^\top \end{pmatrix} \quad \text{and} \quad \mathbf{Y} = \begin{pmatrix} Y_1 \\ Y_2 \\ \vdots \\ Y_n \end{pmatrix}.$$

The MLE of σ^2 is given by

$$\widehat{\sigma}_n^2 = \frac{\text{RSS}}{n},$$

where the residual sum of squares (RSS) is defined as

$$\text{RSS} = \sum_{i=1}^{n} \left(Y_i - X_i^\top \widehat{\beta}_n \right)^2.$$

Thus, the AIC can be expressed as (see Exercise 16.2):

$$\text{AIC} = n \log\left(\frac{\text{RSS}}{n}\right) + 2k,$$

where $k = (d+1) + 1 = d + 2$ is the number of parameters.

16.2 Cross-Validation

16.2.1 Training Error and Test Error

The training data, denoted by (X_i, Y_i) for $i = 1, \ldots, n$, are assumed to be drawn from an unknown distribution. Based on these data, the model $Q(X; \beta)$ is fitted by minimizing the empirical loss, through which the estimator $\widehat{\beta}_n$ is obtained. For notational simplicity, the loss function is defined as

$$L(X, Y; \beta) = L\big(Y, Q(X; \beta)\big).$$

The *training error* is then defined as

$$\text{err} = \sum_{i=1}^{n} L(X_i, Y_i; \widehat{\beta}_n).$$

In practice, greater interest is typically placed on how well the trained model performs on unseen data. To assess this, an independent test dataset (X_i^0, Y_i^0) for $i = 1, \ldots, n$ is considered, having been drawn from the same distribution as the training data. The *test error* is defined as

$$\text{Err} = \sum_{i=1}^{n} L(X_i^0, Y_i^0; \widehat{\beta}_n).$$

It is known that the training error underestimates the expected test error. To quantify this underestimation, a Taylor expansion around β_* is considered:

$$\begin{aligned}\sum_{i=1}^{n} L(X_i^0, Y_i^0; \widehat{\beta}_n) &\overset{(1)}{=} \sum_{i=1}^{n} L(X_i^0, Y_i^0; \beta_*) + \sum_{i=1}^{n} (\widehat{\beta}_n - \beta_*)^\top \frac{\partial L(X_i^0, Y_i^0; \beta_*)}{\partial \beta} \\ &\quad + \frac{1}{2} \sum_{i=1}^{n} (\widehat{\beta}_n - \beta_*)^\top \frac{\partial^2 L(X_i^0, Y_i^0; \beta_*)}{\partial \beta \partial \beta^\top} (\widehat{\beta}_n - \beta_*).\end{aligned}$$

Since

$$\mathbb{E}\left[\frac{\partial L(X_i^0, Y_i^0; \beta_*)}{\partial \beta}\right] = 0,$$

it follows that

$$\mathbb{E}\left[\sum_{i=1}^{n} L(X_i^0, Y_i^0; \widehat{\beta}_n)\right] \doteq \mathbb{E}\left[\sum_{i=1}^{n} L(X_i^0, Y_i^0; \beta_*)\right] + \mathbb{E}\left[\frac{1}{2}\sum_{i=1}^{n}(\widehat{\beta}_n - \beta_*)^\top \frac{\partial^2 L(X_i^0, Y_i^0; \beta_*)}{\partial\beta\beta^\top}(\widehat{\beta}_n - \beta_*)\right].$$

Likewise, by expanding around $\widehat{\beta}_n$,

$$\sum_{i=1}^{n} L(X_i, Y_i; \beta_*) \overset{(2)}{=} \sum_{i=1}^{n} L(X_i, Y_i; \widehat{\beta}_n) + \sum_{i=1}^{n}(\beta_* - \widehat{\beta}_n)^\top \frac{\partial L(X_i, Y_i; \widehat{\beta}_n)}{\partial\beta} + \frac{1}{2}\sum_{i=1}^{n}(\beta_* - \widehat{\beta}_n)^\top \frac{\partial^2 L(X_i, Y_i; \widehat{\beta}_n)}{\partial\beta\beta^\top}(\beta_* - \widehat{\beta}_n).$$

Since

$$\sum_{i=1}^{n}\left[\frac{\partial L(X_i, Y_i; \widehat{\beta}_n)}{\partial\beta}\right] = 0$$

and

$$\mathbb{E}\left[\frac{\partial^2 L(X_i, Y_i; \widehat{\beta}_n)}{\partial\beta\beta^\top}\right] \doteq \mathbb{E}\left[\frac{\partial^2 L(X_i, Y_i; \beta_*)}{\partial\beta\beta^\top}\right],$$

it follows that

$$\mathbb{E}\left[\sum_{i=1}^{n} L(X_i, Y_i; \beta_*)\right] \doteq \mathbb{E}\left[\sum_{i=1}^{n} L(X_i, Y_i; \widehat{\beta}_n)\right] + \mathbb{E}\left[\frac{1}{2}\sum_{i=1}^{n}(\widehat{\beta}_n - \beta_*)^\top \frac{\partial^2 L(X_i, Y_i; \beta_*)}{\partial\beta\beta^\top}(\widehat{\beta}_n - \beta_*)\right].$$

Also note that

$$\mathbb{E}\left[\sum_{i=1}^{n} L(X_i^0, Y_i^0; \beta_*)\right] = \mathbb{E}\left[\sum_{i=1}^{n} L(X_i, Y_i; \beta_*)\right].$$

By combining the above two Taylor expansions, (1) and (2), the following approximation is obtained:

$$\mathbb{E}\left[\sum_{i=1}^{n} L(X_i^0, Y_i^0; \widehat{\beta}_n)\right] \doteq \mathbb{E}\left[\sum_{i=1}^{n} L(X_i, Y_i; \widehat{\beta}_n)\right] + \text{gap},$$

where the gap is given by (with the mysterious 2 highlighted in boldface):

$$\text{gap} = \mathbf{2} \times \mathbb{E}\left\{\frac{1}{2}\sqrt{n}(\widehat{\beta}_n - \beta_*)^\top \mathbb{E}\left[\frac{\partial^2 L(X, Y; \beta_*)}{\partial\beta\beta^\top}\right]\sqrt{n}(\widehat{\beta}_n - \beta_*)\right\} > 0,$$

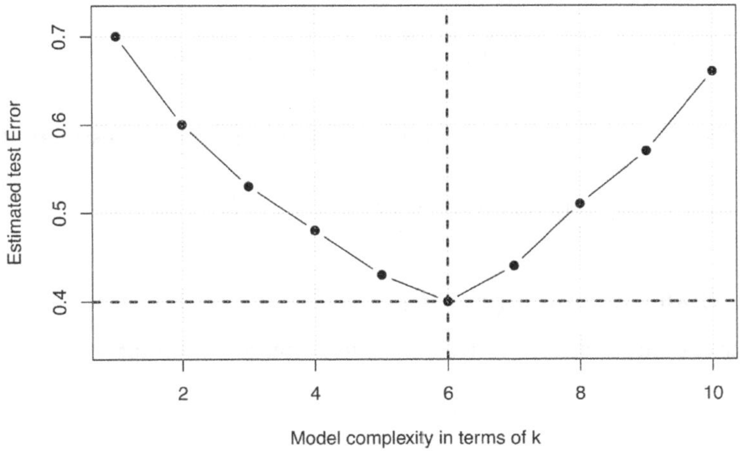

FIGURE 16.1
A hypothetical example of model selection using estimated test error.

This gap can be interpreted as a measure of model complexity. The appearance of the mysterious number 2 in the AIC is revealed here through the expression for the gap, which arises from performing two Taylor expansions in the derivation.

If an estimator for the gap is available, the test error can be estimated by

$$\widehat{\mathbb{E}}(\text{Err}) = \text{err} + \widehat{\text{gap}}.$$

Figure 16.1 displays a hypothetical scenario in which the estimated test error (typically U-shaped) is used to select the optimal model, choosing $k = 6$ based on the minimal estimated test error.

It is worth noting that the cross-validation procedure[64] to be described in the next subsection offers a non-asymptotic alternative for estimating test error—more precisely, the validation error.

16.2.2 Leave-One-Out Cross-Validation

In leave-one-out (LOO) cross-validation, one subject is left out from the dataset at a time to serve as the validation data. The model is then trained on the remaining $n - 1$ subjects, and the performance is evaluated on the held-out subject.

Let $Q(X; \beta)$ denote the model under evaluation. Let $\widehat{\beta}_n^{-i}$ denote the parameter estimator obtained by minimizing the loss over the training data excluding subject i:

$$\widehat{\beta}_n^{-i} = \arg\min_{\beta} \sum_{i' \neq i} L(Y_{i'}, Q(X_{i'}; \beta)).$$

Using this procedure, the LOO cross-validation error can be estimated as

$$\text{LOO-CV} = \sum_{i=1}^{n} L(Y_i, Q(X_i; \widehat{\beta}_n^{-i})).$$

Linear Model Example (cont'd)

Consider the linear model under the Gauss–Markov assumptions:

$$Y = X^\top \beta + \varepsilon,$$

where $X^\top = (1, X^{(1)}, \ldots, X^{(d)})$, $\mathbb{E}(\varepsilon) = 0$, and $\mathbb{V}(\varepsilon) = \sigma^2$.

When the model is trained using the full dataset, the least squares estimator is given by

$$\widehat{\beta}_n = (\mathbf{X}^\top \mathbf{X})^{-1} \mathbf{X}^\top \mathbf{Y}.$$

When subject i is excluded from the dataset, the estimator becomes

$$\widehat{\beta}_n^{-i} = (\mathbf{X}_{-i}^\top \mathbf{X}_{-i})^{-1} \mathbf{X}_{-i}^\top \mathbf{Y}_{-i},$$

where $\mathbf{X}_{-i}$ and $\mathbf{Y}_{-i}$ denote the matrices $\mathbf{X}$ and $\mathbf{Y}$ with the ith row removed, respectively.

Thus, the LOO-CV error for the linear model can be expressed as

$$\text{LOO-CV} = \sum_{i=1}^{n} (Y_i - X_i^\top \widehat{\beta}_n^{-i})^2.$$

This method can be applied to select the "best" subset of independent variables—i.e., the subset that minimizes LOO-CV.

To reduce computational cost, the following identity—often referred to as the *LOO lemma*[66] (see Exercise 16.3)—can be utilized:

$$Y_i - X_i^\top \widehat{\beta}_n^{-i} = \frac{Y_i - X_i^\top \widehat{\beta}_n}{1 - h_{ii}},$$

where h_{ii} is the ith diagonal element of the hat matrix H,

$$H = \mathbf{X}(\mathbf{X}^\top \mathbf{X})^{-1} \mathbf{X}^\top.$$

Historically, the LOO lemma has been considered powerful because it avoids the need to retrain the model n times. With its application, only a single model training is required. While computational efficiency is less critical with modern computing power, the lemma remains useful for understanding the trade-off between goodness of fit and model complexity, as well as the connection between cross-validation and AIC.[67]

Using the LOO lemma, the cross-validation error can be re-expressed as:

$$\text{LOO-CV} = \sum_{i=1}^{n} \frac{(Y_i - X_i^\top \widehat{\beta}_n)^2}{(1 - h_{ii})^2}.$$

Moreover, note that

$$\begin{aligned}\sum_{i=1}^{n} h_{ii} &= \text{trace}(H) = \text{trace}\left\{\mathbf{X}(\mathbf{X}^\top\mathbf{X})^{-1}\mathbf{X}^\top\right\} \\ &= \text{trace}\left\{(\mathbf{X}^\top\mathbf{X})^{-1}\mathbf{X}^\top\mathbf{X}\right\} = \text{trace}(I_{d+1}) = d+1,\end{aligned}$$

where I_{d+1} is the $(d+1) \times (d+1)$ identity matrix.

If h_{ii} is approximated by its average value $(d+1)/n$, the leave-one-out estimate becomes the generalized cross-validation (GCV) estimate:[68]

$$\text{GCV} = \sum_{i=1}^{n} \frac{(Y_i - X_i^\top \widehat{\beta}_n)^2}{[1 - (d+1)/n]^2}.$$

Using the approximation

$$\frac{1}{(1-x)^2} \doteq \frac{1}{1-2x} \doteq 1 + 2x \quad \text{for small } x,$$

we obtain the following:

$$\begin{aligned}\text{GCV} &\doteq \sum_{i=1}^{n} (Y_i - X_i^\top \widehat{\beta}_n)^2 [1 + 2(d+1)/n] \\ &= \sum_{i=1}^{n} (Y_i - X_i^\top \widehat{\beta}_n)^2 + 2(d+1) \cdot \left[\frac{1}{n}\sum_{i=1}^{n} (Y_i - X_i^\top \widehat{\beta}_n)^2\right].\end{aligned}$$

Let the residual variance be estimated by

$$\widehat{\sigma}^2 = \frac{1}{n}\sum_{i=1}^{n} (Y_i - X_i^\top \widehat{\beta}_n)^2.$$

Using this estimate, the GCV can be approximated as

$$\text{GCV} \doteq \text{err} + 2 \times (d+1)\widehat{\sigma}^2,$$

where the training error $\text{err} = \sum_{i=1}^{n} (Y_i - X_i^\top \widehat{\beta}_n)^2$ is the same as the total residual sum of squares in this example.

Hence, at least in the context of linear models, it can be observed that various model selection methods—such as AIC, LOO-CV, and GCV—share a common structure, with the factor "2" reflecting the trade-off between goodness of fit and model complexity.

16.2.3 K-Fold Cross-Validation

The K-fold cross-validation procedure is a generalization of LOO-CV, which corresponds to the special case where $K = n$. In K-fold cross-validation, the dataset is partitioned into K approximately equal-sized subsets, referred to as *folds*. For each iteration, one fold is reserved as validation data, and the remaining $K - 1$ folds are used to train the model. In practice, $K = 5$ or $K = 10$ are commonly used values.

Let the dataset $\mathcal{D}$ be partitioned into K folds as follows:

$$\mathcal{D} = \{\mathcal{D}_1, \ldots, \mathcal{D}_K\}.$$

Let $Q(X;\beta)$ denote the model under evaluation. For each k, let $\widehat{\beta}_n^{-k}$ be the parameter estimate obtained by minimizing the loss over the training data excluding the kth fold:

$$\widehat{\beta}_n^{-k} = \arg\min_{\beta} \sum_{i \in \mathcal{D} \setminus \mathcal{D}_k} L(Y_i, Q(X_i;\beta)).$$

Using this procedure, the validation error is estimated as

$$\mathrm{CV}_K = \sum_{k=1}^{K} \sum_{i \in \mathcal{D}_k} L(Y_i, Q(X_i; \widehat{\beta}_n^{-k})).$$

16.2.4 Tuning

Most predictive models, whether classical or modern, involve one or more *tuning parameters* that control the trade-off between goodness of fit and model complexity.

In the case of generalized linear models (e.g., linear regression and logistic regression), two primary approaches are commonly used for tuning. One approach is *best-subset selection*, in which a subset of independent variables is selected to minimize the validation or test error, as estimated by one of the model selection procedures discussed previously. The second approach is *regularization*, where penalty terms are added to the loss function.

Two widely used regularization techniques are the Lasso penalty[69] and the Ridge penalty.[70] Additionally, the elastic-net penalty[71] combines both the Lasso (i.e., $\mathcal{L}_1$) and Ridge (i.e., $\mathcal{L}_2$) penalties. Although numerous other penalty forms exist, they fall outside the scope of this text.

The Lasso-regularized loss function is defined as

$$L_\lambda(Y, Q(X;\beta)) = L(Y, Q(X;\beta)) + \lambda \|\beta\|_1,$$

where $\|\beta\|_1 = \sum_{j=1}^{d} |\beta^{(j)}|$ for $\beta = (\beta^{(0)}, \beta^{(1)}, \ldots, \beta^{(d)})^\top$.

The Ridge-regularized loss function is defined as

$$L_\lambda(Y, Q(X;\beta)) = L(Y, Q(X;\beta)) + \lambda \|\beta\|_2^2,$$

where $\|\beta\|_2^2 = \sum_{j=1}^{d} {\beta^{(j)}}^2$ for the same β.

For each value of the regularization parameter λ, the validation error can be estimated using K-fold cross-validation:

$$\mathrm{CV}_K(\lambda) = \sum_{k=1}^{K} \sum_{i \in \mathcal{D}_k} L_\lambda(Y_i, Q(X_i; \widehat{\beta}_n^{-k}(\lambda))).$$

The optimal tuning parameter λ is then selected as the value that minimizes the estimated validation error:

$$\lambda_{\mathrm{opt}} = \arg\min_{\lambda} \mathrm{CV}_K(\lambda).$$

For other predictive models, tuning parameters may not be expressed directly through regularization. In such cases, a deeper understanding of the model structure is required to identify the relevant parameters. For instance, in a CART model, tuning parameters include the maximum depth of the tree and the number of variables considered at each split. In a random forest model, tuning parameters typically include the number of trees, the maximum depth of each tree, and others.

Fortunately, most modern statistical and machine learning software (e.g., R packages) are equipped with automated tuning procedures—often relying on cross-validation—as part of their implementation.

Gold Coin # 57. *To estimate the validation error, the dataset is conceptually divided into two parts: a training set and a validation set. The cross-validation is a procedure to estimate the validation error.*

16.3 Super Learner

A variety of methods have been developed for obtaining parametric, nonparametric, or semiparametric estimators of $Q_0(X) = \mathbb{E}(Y \mid X)$. Among these, the *super learner* has been recognized as a promising approach.[72] In contrast to parametric modeling—which depends on a pre-specified parametric form—the super learner is constructed by employing a *library* of predictive models, which may consist of statistical learning or machine learning algorithms.[56] Through this framework, an attempt is made to select the algorithm that achieves optimal performance with respect to a specified loss function.

For instance, the $\mathcal{L}_2$ loss function can be used for a continuous outcome:

$$L(Y, Q(X)) = [Y - Q(X)]^2,$$

whereas the BCE loss can be used for a binary outcome:

$$L(Y, Q(X)) = -\log\left\{Q(X)^Y[1 - Q(X)]^{1-Y}\right\}.$$

In addition, the BCE loss can be used for bounded continuous outcomes.

16.3.1 Discrete Super Learner

In the discrete super learner, cross-validation is employed to identify the "best" algorithm among a library of algorithms. Without cross-validation, a model that overfits the data may be selected. For example, in the case of a continuous outcome, an algorithm that perfectly interpolates the data, i.e., $\widehat{Q}(X_i) = Y_i$ for all $i = 1, \ldots, n$, might be chosen; however, such an algorithm would be useless. By employing cross-validation, a balance between goodness-of-fit and model complexity is achieved.

By K-fold cross-validation, the dataset is partitioned into K approximately equal-sized subsets. Each subset $\mathcal{D}_k$ serves as a validation set once, while the remaining $K-1$ subsets are used as training data. For each fold $k = 1, \ldots, K$ and each algorithm $j = 1, \ldots, J$ in the library, a model $\widehat{Q}^{j,k}$ is fitted based on the training data. The cross-validated risk associated with algorithm j is then estimated as

$$CV^{(j)} = \frac{1}{K}\sum_{k=1}^{K}\frac{1}{\#(\mathcal{D}_k)}\sum_{(X,Y)\in\mathcal{D}_k} L(Y, \widehat{Q}^{j,k}(X)).$$

The procedure is summarized in Table 16.1. Initially, the data are randomly permuted and then divided into K subsets. For each fold k, the kth subset is designated as the validation set, while the remaining subsets are used to

TABLE 16.1
K-fold CV behind a super learner

ID*	k	X	Y	$\widehat{Q}^{1,k}(X)$	...	$\widehat{Q}^{J,k}(X)$
1	1	X_1	Y_1	$\widehat{Q}^{1,1}(X_1)$	...	$\widehat{Q}^{J,1}(X_1)$
...						
n_1	1	X_{n_1}	Y_{n_1}	$\widehat{Q}^{1,1}(X_{n_1})$	...	$\widehat{Q}^{J,1}(X_{n_1})$
n_1+1	2	X_{n_1+1}	Y_{n_1+1}	$\widehat{Q}^{1,2}(X_{n_1+1})$	...	$\widehat{Q}^{J,2}(X_{n_1+1})$
...						
n_1+n_2	2	$X_{n_1+n_2}$	$Y_{n_1+n_2}$	$\widehat{Q}^{1,2}(X_{n_1+n_2})$	...	$\widehat{Q}^{J,2}(X_{n_1+n_2})$
...						
$\overline{n}_{K-1}+1$	K	...	...	...	...	...
...						
n	K	X_n	Y_n	$\widehat{Q}^{1,K}(X_n)$	...	$\widehat{Q}^{J,K}(X_n)$

* Subject IDs are randomly permuted

train each of the J algorithms, resulting in models $\widehat{Q}^{j,k}$, $j = 1, \ldots, J$. These models are applied to the validation set to produce predictions $\widehat{Q}^{j,k}(X_i)$ for each subject $i = \overline{n}_{k-1} + 1, \ldots, \overline{n}_k$, where $\overline{n}_k = \sum_{l=1}^{k} n_l$. These predictions form the last J columns in the table.

When the $\mathcal{L}_2$ loss is used, the cross-validated risk becomes:

$$CV^{(j)} = \frac{1}{K}\sum_{k=1}^{K}\frac{1}{n_k}\sum_{i=\overline{n}_{k-1}+1}^{\overline{n}_k}\left[Y_i - \widehat{Q}^{j,k}(X_i)\right]^2.$$

Using the BCE loss, the corresponding formula is:

$$CV^{(j)} = -\frac{1}{K}\sum_{k=1}^{K}\frac{1}{n_k}\sum_{i=\overline{n}_{k-1}+1}^{\overline{n}_k}\log\left\{\left[\widehat{Q}^{j,k}(X_i)\right]^{Y_i}\left[1 - \widehat{Q}^{j,k}(X_i)\right]^{1-Y_i}\right\}.$$

Based on this cross-validation procedure, the algorithm with the smallest estimated cross-validated risk is selected:[72]

$$j^* = \arg\min_{j=1,\ldots,J} CV^{(j)}.$$

16.3.2 Super Learner

Whereas the discrete super learner selects the "best" individual algorithm among the J candidates, the continuous super learner—also known as the *ensemble super learner* and commonly referred to simply as the *super learner*—constructs an optimally weighted combination of the J algorithms.[72] This is accomplished by minimizing some weighted cross-validated risk:

$$CV(w_1, \ldots, w_J),$$

subject to the constraints $w_j \geq 0$ and $\sum_{j=1}^{J} w_j = 1$.

Under the $\mathcal{L}_2$ loss, the weighted cross-validated risk is given by

$$CV(w_1, \ldots, w_J) = \frac{1}{K}\sum_{k=1}^{K}\frac{1}{\#(\mathcal{D}_k)}\sum_{(X,Y)\in\mathcal{D}_k}\left(Y - \sum_{j=1}^{J} w_j\widehat{Q}^{j,k}(X)\right)^2.$$

Although the ensemble super learner may appear to be more computationally intensive than the discrete version, the optimal weights $w_1, \ldots, w_J$ can, in fact, be efficiently computed using the cross-validated predictions summarized in Table 16.1. For this aim, for the ith subject, let $Z_i^{(j)} = \widehat{Q}^{j,k}(X_i)$, which denotes the cross-validated prediction from the jth candidate algorithm. Define $Z_i = (Z_i^{(1)}, \ldots, Z_i^{(J)})^\top$ as the corresponding covariate vector. Thus, the optimal weights correspond to the coefficients in a constrained linear regression model of the form (See Exercise 16.4):

$$\mathbb{E}(Y \mid Z) = w_1 Z^{(1)} + \cdots + w_J Z^{(J)},$$

subject to the constraints $w_j \geq 0$ and $\sum_{j=1}^{J} w_j = 1$.

Under the BCE loss, the weighted cross-validated risk is given by

$$CV(w_1, \ldots, w_J)$$
$$= \frac{1}{K}\sum_{k=1}^{K} \frac{1}{\#(\mathcal{D}_k)} \sum_{(X,Y)\in\mathcal{D}_k} L\left(Y, \text{expit}\left\{\sum_{j=1}^{J} w_j \,\text{logit}\left[\widehat{Q}^{j,k}(X)\right]\right\}\right)^2,$$

where $\text{logit}(a) = \log\left(\frac{a}{1-a}\right)$ and $\text{expit}(b) = \frac{\exp(b)}{1+\exp(b)}$, for any $a \in (0,1)$ and b.

For the ith subject, let $Z_i^{(j)} = \text{logit}[\widehat{Q}^{j,k}(X_i)]$, which denotes the cross-validated prediction from the jth candidate algorithm. Define $Z_i = (Z_i^{(1)}, \ldots, Z_i^{(J)})^\top$ as the corresponding covariate vector. Thus, the optimal weights are obtained as coefficients in a constrained logistic regression model (See Exercise 16.5):

$$\text{logit}\left\{\mathbb{P}(Y = 1 \mid Z)\right\} = w_1 Z^{(1)} + \cdots + w_J Z^{(J)},$$

again subject to the constraints $w_j \geq 0$ and $\sum_{j=1}^{J} w_j = 1$.

Once the weights have been estimated as $\widehat{w}_1, \ldots, \widehat{w}_J$, the final super learner estimate of $Q(X)$ is given by:

$$\widehat{Q}_{\text{SL}}(X) = \sum_{j=1}^{J} \widehat{w}_j \widehat{Q}^{j,k}(X).$$

Gold Coin # 58. *Two versions of the super learning exist: the discrete super learner and the ensemble super learner.*

Exercises

Exercise 16.1

Heuristically prove that

$$D(\widehat{\theta}_n(k)) \doteq \mathbb{E}\left[-\sum_{i=1}^{n} 2\log f(O_i; \widehat{\theta}_n(k))\right] + 2k.$$

Exercise 16.2

For the Gaussian linear model, show that

$$\text{AIC} = n \log \left(\frac{\text{RSS}}{n} \right) + 2(d+2).$$

Exercise 16.3

Prove the LOO lemma. That is, prove that

$$Y_i - X_i^\top \widehat{\beta}_n^{-i} = \frac{Y_i - X_i^\top \widehat{\beta}_n}{1 - h_{ii}},$$

where h_{ii} is the (i,i) element of the hat matrix H.

Exercise 16.4

Sketch a plan to solve the following optimization problem:

$$\arg\min_{w \in \Omega} CV(w_1, \ldots, w_J),$$

where $\Omega = \{(w_1, \ldots, w_J)^\top \mid w_1 \geq 0, \ldots, w_{j-1} \geq 0, \sum_{j=1}^J w_j = 1\}$ and

$$CV(w_1, \ldots, w_J) = \frac{1}{K} \sum_{k=1}^K \frac{1}{\#(\mathcal{D}_k)} \sum_{(X,Y) \in \mathcal{D}_k} \left(Y - \sum_{j=1}^J w_j \widehat{Q}^{j,k}(X) \right)^2.$$

Exercise 16.5

Sketch a plan to solve the following optimization problem:

$$\arg\min_{w \in \Omega} CV(w_1, \ldots, w_J),$$

where $\Omega = \{(w_1, \ldots, w_J)^\top \mid w_1 \geq 0, \ldots, w_{j-1} \geq 0, \sum_{j=1}^J w_j = 1\}$ and

$$\begin{aligned} &CV(w_1, \ldots, w_J) \\ &= \frac{1}{K} \sum_{k=1}^K \frac{1}{\#(\mathcal{D}_k)} \sum_{(X,Y) \in \mathcal{D}_k} L\left(Y, \text{ expit} \left\{ \sum_{j=1}^J w_j \operatorname{logit} \left[\widehat{Q}^{j,k}(X) \right] \right\} \right)^2. \end{aligned}$$

where L is the BCE loss.

17

Targeted Learning

As discussed in Chapters 8–10, the estimand framework is comprised of three key components: estimand, estimator, and sensitivity analysis. In this chapter, the development of efficient estimators corresponding to a given estimand is discussed. Specifically, targeted learning (TL) approaches will be introduced.[58] As the name implies, TL approaches constitute machine learning methods that are tailored to the target estimands. The application of the TL framework to cohort studies will be explored. As is customary, we begin our discussion with point-treatment cohort studies, treat RCTs as special cases, and then extend the methodology to more general settings involving missing data and longitudinal data.

Of course, readers are referred to the following two books by Mark van der Laan, the pioneer of targeted learning, for a comprehensive introduction:

1. van der Laan and Rose (2011), *Targeted Learning: Causal Inference for Observational and Experimental Data.*
2. van der Laan and Rose (2018), *Targeted Learning in Data Science: Causal Inference for Complex Longitudinal Studies.*

17.1 From Estimand to Estimator

17.1.1 Estimand

A cohort study is considered, generating data $O_i = (X_i, A_i, Y_i)$ for $i = 1, \ldots, n$, assumed to be independent and identically distributed (i.i.d.) with $O = (X, A, Y)$. Here X is a vector of covariates, A is a binary treatment variable, and Y is a binary outcome.

Assume that the following causal estimand—the average treatment effect (ATE)—is of interest:

$$\theta^* = \mathbb{E}(Y^{a=1}) - \mathbb{E}(Y^{a=0}).$$

Under standard identifiability assumptions, two commonly used strategies can be employed to identify this estimand: the standardization strategy and the weighting strategy.

DOI: 10.1201/9781003635468-21

By the standardization strategy, θ^* is identified as

$$\theta_0 = \mathbb{E}[Q_0(X,1) - Q_0(X,0)],$$

where the outcome regression function is defined as

$$Q_0(X,A) = \mathbb{E}(Y \mid X, A).$$

Alternatively, by the weighting strategy, θ^* is identified as:

$$\theta_0 = \mathbb{E}\left[\frac{(2A-1)Y}{g_0(A \mid X)}\right],$$

where the propensity score is defined as

$$g_0(a \mid X) = \mathbb{P}(A = a \mid X), \quad a = 0, 1.$$

17.1.2 Efficient Influence Function

As discussed in Chapters 11–13, the efficient influence function (EIF) $\phi(X, A, Y)$ for the estimation of θ_0 can be expressed in two equivalent forms:

$$\phi(X,A,Y) = Q_0(X,1) - Q_0(X,0) + \frac{(2A-1)}{g_0(A \mid X)}[Y - Q_0(X,A)] - \theta_0,$$

$$\phi(X,A,Y) = \frac{(2A-1)Y}{g_0(A \mid X)} - \left[\frac{Q_0(X,1)}{g_0(1 \mid X)} + \frac{Q_0(X,0)}{g_0(0 \mid X)}\right][A - g_0(1 \mid X)] - \theta_0.$$

17.1.3 Two Paths to Double Robustness

Given that the EIF $\phi(X, A, Y)$ has two equivalent forms, two paths can be pursued for the construction of doubly robust estimators.

From IPW to AIPW

The first path is based on the following representation of the estimand:

$$\theta = \mathbb{E}\left[\frac{(2A-1)Y}{g_0(A \mid X)}\right].$$

An initial estimator $\widehat{g}_n(A \mid X)$ of the propensity score $g_0(A \mid X)$ is assumed to be available. Based on this, the following estimator—known as the inverse probability of treatment weighted (IPW) estimator[30]—is constructed:

$$\widehat{\theta}_{\text{IPW}} = \frac{1}{n}\sum_{i=1}^{n}\left[\frac{(2A_i-1)Y_i}{\widehat{g}_n(A_i \mid X_i)}\right].$$

To augment the IPW estimator and attain double robustness, the following form of the EIF is considered:

$$\phi(X,A,Y) = \frac{(2A-1)Y}{g_0(A \mid X)} - \left[\frac{Q_0(X,1)}{g_0(1 \mid X)} + \frac{Q_0(X,0)}{g_0(0 \mid X)}\right][A - g_0(1 \mid X)] - \theta_0.$$

If an initial estimator $\widehat{Q}_n(X, A)$ for $Q_0(X, A)$ is also available, the augmented inverse probability of treatment weighted (AIPW) estimator[30] is obtained:

$$\widehat{\theta}_{\text{AIPW}} = \frac{1}{n}\sum_{i=1}^{n}\left\{\frac{(2A_i - 1)Y_i}{\widehat{g}_n(A_i \mid X_i)} - \left[\frac{\widehat{Q}_n(X_i, 1)}{\widehat{g}_n(1 \mid X_i)} + \frac{\widehat{Q}_n(X_i, 0)}{\widehat{g}_n(0 \mid X_i)}\right][A_i - \widehat{g}_n(1 \mid X_i)]\right\}.$$

From MLE to TMLE

The second path is based on the alternative representation of the estimand:

$$\theta = \mathbb{E}[Q_0(X, 1) - Q_0(X, 0)].$$

Suppose that an initial estimator $\widehat{Q}_n(X, A)$ for $Q_0(X, A)$ has been obtained. Based on this, the following plug-in estimator—commonly referred to as the maximum likelihood estimator (MLE), the minimum loss estimator (MLE),[58] or, as in Chapter 9, the g-computation estimator[30]—is defined:

$$\widehat{\theta}_{\text{MLE}} = \frac{1}{n}\sum_{i=1}^{n}\left[\widehat{Q}_n(X_i, 1) - \widehat{Q}_n(X_i, 0)\right].$$

To augment the MLE and achieve double robustness, the following form of the EIF is considered:

$$\phi(X, A, Y) = Q_0(X, 1) - Q_0(X, 0) - \theta_0 + \frac{(2A - 1)}{g_0(A \mid X)}[Y - Q_0(X, A)].$$

Using the initial estimator $\widehat{Q}_n$ and the EIF expression above, an updated estimator $\widehat{Q}_n^*$ is obtained through a targeting step. This leads to the construction of the targeted maximum likelihood estimator (TMLE) or the targeted minium likelihood estimator (TMLE),[58] defined as:

$$\widehat{\theta}_{\text{TMLE}} = \frac{1}{n}\sum_{i=1}^{n}\left[\widehat{Q}_n^*(X_i, 1) - \widehat{Q}_n^*(X_i, 0)\right].$$

The next section details how the initial estimator is updated to obtain the targeted estimator.

Gold Coin # 59. *Since the efficient influence function (EIF) of an estimand typically admits two equivalent representations, two distinct paths can be followed to achieve double robustness: one that augments IPW to AIPW, and another that updates MLE to TMLE.*

17.2 Understanding TMLE

The TMLE, denoted by $\widehat{\theta}_{\text{TMLE}}$, is obtained as an updated version of the MLE, $\widehat{\theta}_{\text{MLE}}$. In this section, the theoretical foundation for this update is presented, and it is demonstrated why the updated estimator provides improvements over the initial.

17.2.1 Why an Update Is Needed

Continue the aforementioned cohort study with a binary outcome variable Y.

In the initial step, an initial estimator $\widehat{Q}_n(X, A)$ of $Q_0(X, A)$ is constructed by fitting either a parametric model (e.g., a logistic regression model) or a super learner based on a specified library of models (refer to Chapter 16).

If a logistic regression model $Q(X, A; \beta)$ is employed, the estimation of $Q_0(X, A)$ is carried out via the MLE procedure: the MLE $\widehat{\beta}_n$ of β is first obtained, and $\widehat{Q}_n(X, A) = Q(X, A; \widehat{\beta}_n)$ is then used to estimate $Q_0(X, A)$.

Alternatively, if a super learner is fitted to the regression of $Y \sim X + A$ using the binary cross-entropy (BCE) loss, an MLE $\widehat{Q}_n(X, A)$ of $Q_0(X, A)$ is also produced.

To summarize, regardless of whether logistic regression or super learner is used, an MLE of $Q_0(X, A)$ is obtained. To emphasize the use of MLE, the initial estimator is denoted as $\widehat{Q}_{\text{MLE}}(X, A)$.

The estimand θ_0 is known to be a functional of both the marginal distribution of X, denoted $P_{X,0}$, and the outcome regression function $Q_0(X, A)$. Specifically, it holds that:

$$\theta_0 = \int [Q_0(X, 1) - Q_0(X, 0)]\, d P_{X,0} = \theta(P_{X,0}, Q_0).$$

The nonparametric MLE (NPMLE) of $P_{X,0}$ is the empirical distribution $\widehat{P}_{X,n}$, while the MLE of Q_0 is $\widehat{Q}_{\text{MLE}}$. Accordingly, the MLE of θ_0 is given by:

$$\begin{aligned}
\widehat{\theta}_{\text{MLE}} &= \theta(\widehat{P}_{X,n}, \widehat{Q}_{\text{MLE}}) \\
&= \int \left[\widehat{Q}_{\text{MLE}}(X, 1) - \widehat{Q}_{\text{MLE}}(X, 0)\right] d\widehat{P}_{X,n} \\
&= \frac{1}{n} \sum_{i=1}^{n} \left[\widehat{Q}_{\text{MLE}}(X_i, 1) - \widehat{Q}_{\text{MLE}}(X_i, 0)\right].
\end{aligned}$$

This is the rationale behind referring to the initial estimator as the MLE of θ_0. Moreover, if $\widehat{Q}_{\text{MLE}}$ is consistent for Q_0, then by the continuous mapping theorem, $\widehat{\theta}_{\text{MLE}}$ is also consistent for θ_0.

However, there are at least two reasons why $\widehat{\theta}_{\text{MLE}}$ needs to be updated to obtain $\widehat{\theta}_{\text{TMLE}}$. First, $\widehat{\theta}_{\text{MLE}}$ is only singly robust—its consistency requires a consistent estimator of $Q_0(X, A)$, whereas double robustness is desired—namely, consistency should hold if either the estimator of $Q_0(X, A)$ or the estimator of $g_0(A \mid X)$ is consistent. Second, although $\widehat{\theta}_{\text{MLE}}$ may possess desirable asymptotic properties, it often suffers from small-sample bias because it is not directly targeting toward the estimand θ_0. TMLE is designed to reduce this bias. To achieve these goals, the initial estimator is updated using a least favorable parametric submodel targeted toward the estimand.

17.2.2 How the Update Is Performed

For the purpose of describing the updating process, the initial estimators of $P_{X,0}$, $g_0(A \mid X)$, and $Q_0(X, A)$ are denoted by $\widehat{P}^0_{X,n}$, $\widehat{g}^0_n$, and $\widehat{Q}^0_n$, respectively. For each notation, the subscript 0 denotes the true value of the parameter, while the superscript 0 indicates an initial estimator.

The estimand remains:

$$\theta_0 = \theta(P_{X,0}, Q_0).$$

Since the initial estimators are not targeted toward the estimand, but rather toward general model fit, it is expected that in small samples, $\theta(\widehat{P}^0_{X,n}, \widehat{Q}^0_n)$ will exhibit greater bias than a targeted update $\theta(\widehat{P}^*_{X,n}, \widehat{Q}^*_n)$, provided that the latter is obtained through a process designed to target the estimand θ_0.

Least Favorable Parametric Submodel

The derivation of $\widehat{P}^*_{X,n}$ and $\widehat{Q}^*_n$ such that $\theta(\widehat{P}^*_{X,n}, \widehat{Q}^*_n)$ possesses less bias than $\theta(\widehat{P}^0_{X,n}, \widehat{Q}^0_n)$ involves the use of a least favorable parametric submodel.

The joint distribution of (X, A, Y) is given by:

$$dP_{X,0}(X) \cdot [g_0(1 \mid X)]^A [1 - g_0(1 \mid X)]^{1-A} \cdot [Q_0(X, A)]^Y [1 - Q_0(X, A)]^{1-Y},$$

where $dP_{X,0}$ denotes the density function of $P_{X,0}$ with respect to a base measure given by the product of the counting measure on the discrete covariates and the Lebesgue measure on the continuous covariates.

A parametric submodel $\{dP_X(X; \nu_1), g(1 \mid X; \nu_2), Q(X, A; \nu_3)\}$ is considered, satisfying $dP_X(X; 0) = dP_{X,0}(X)$, $g(1 \mid X; 0) = g_0(1 \mid X)$, and $Q(X, A; 0) = Q_0(X, A)$. Letting

$$\theta(\boldsymbol{\nu}) = \theta(\nu_1, \nu_2, \nu_3) = \theta(P_X(X; \nu_1), Q(X, A; \nu_3)),$$

the Fundamental Theorem of Regularity (FTR) yields:

$$0 = \begin{aligned} \left.\frac{\partial\theta(\boldsymbol{\nu})}{\partial\nu_1}\right|_{\boldsymbol{\nu}=0} &= \mathbb{E}\{\phi(X,A,Y)S_{0;\nu_1}(X)\}, \\ \left.\frac{\partial\theta(\boldsymbol{\nu})}{\partial\nu_2}\right|_{\boldsymbol{\nu}=0} &= \mathbb{E}\{\phi(X,A,Y)S_{0;\nu_2}(X,A)\}, \\ \left.\frac{\partial\theta(\boldsymbol{\nu})}{\partial\nu_3}\right|_{\boldsymbol{\nu}=0} &= \mathbb{E}\{\phi(X,A,Y)S_{0;\nu_3}(X,A,Y)\}, \end{aligned}$$

where $S_{0;\nu_j}$ is the corresponding score function evaluated at $\nu_j = 0,\ j = 1, 2, 3$.
Moreover, $\phi(X, A, Y)$ is decomposed as:

$$\phi(X,A,Y) = D_X + D_{Y|X,A},$$

where

$$\begin{aligned} D_X &= Q_0(X,1) - Q_0(X,0) - \theta_0, \\ D_{Y|X,A} &= \frac{(2A-1)}{g_0(A \mid X)}[Y - Q_0(X,A)]. \end{aligned}$$

Substituting into the FTR conditions and applying the law of iterated expectations (LIE), the derivatives simplify to (See Exercise 17.1):

$$\begin{aligned} \left.\frac{\partial\theta(\boldsymbol{\nu})}{\partial\nu_1}\right|_{\boldsymbol{\nu}=0} &= \mathbb{E}\{D_X \cdot S_{0;\nu_1}(X)\}, \\ \left.\frac{\partial\theta(\boldsymbol{\nu})}{\partial\nu_3}\right|_{\boldsymbol{\nu}=0} &= \mathbb{E}\{D_{Y|X,A} \cdot S_{0;\nu_3}(X,A,Y)\}. \end{aligned}$$

Thus, it is revealed that a least favorable parametric submodel corresponds to the one with the following scores (See Exercise 17.2):

$$\begin{aligned} S_{0;\nu_1}(X) &= D_X, \\ S_{0;\nu_2}(X,A) &= 0, \\ S_{0;\nu_3}(X,A,Y) &= D_{Y|X,A}. \end{aligned}$$

Suitable parametric submodels satisfying this requirement can be constructed as (See Exercise 17.3):

$$\begin{aligned} dP_X(X;\nu_1) &= (1+\nu_1 D_X)\, dP_{X,0}(X), \\ g(A \mid X;\nu_2) &= (1+\nu_2\cdot 0)\, g_0(A \mid X), \\ Q(X,A;\nu_3) &= \frac{\exp\left[q_0(X,A)+\nu_3 H(X,A)\right]}{1+\exp\left[q_0(X,A)+\nu_3 H(X,A)\right]} \\ &\triangleq \operatorname{expit}\left[q_0(X,A)+\nu_3 H(X,A)\right], \end{aligned}$$

where

$$q_0(X,A) = \log\left[\frac{Q_0(X,A)}{1-Q_0(X,A)}\right] \triangleq \text{logit}[Q_0(X,A)],$$
$$H(X,A) = \frac{2A-1}{g_0(A \mid X)}.$$

First Round of Update

The initial estimators $\widehat{P}^0_{X,n}$, $\widehat{g}^0_n$, and $\widehat{Q}^0_n$ are now ready to be updated through the use of a least favorable parametric submodel. The following estimated least favorable parametric submodel is considered:

$$dP_X(X;\nu_1) = (1+\nu_1 D_X)\, d\widehat{P}^0_{X,n}(X),$$
$$g(1 \mid X;\nu_2) = (1+\nu_2 \cdot 0)\,\widehat{g}^0_n(1 \mid X),$$
$$Q(X,A;\nu_3) = \text{expit}\left[\widehat{q}^0_n(X,A) + \nu_3 \widehat{H}^0_n(X,A)\right],$$

where

$$\widehat{q}^0_n(X,A) = \text{logit}\left[\widehat{Q}^0_n(X,A)\right],$$
$$\widehat{H}^0_n(X,A) = \frac{2A-1}{\widehat{g}^0_n(A \mid X)}.$$

Based on this parametric submodel, the MLEs of (ν_1,ν_3) are sought by maximizing the following likelihood:

$$L(\nu_1,\nu_3) = \prod_{i=1}^n dP_X(X_i;\nu_1) \cdot \prod_{i=1}^n [\widehat{g}^0_n(1 \mid X_i)]^{A_i}[1-\widehat{g}^0_n(1 \mid X_i)]^{1-A_i}$$
$$\cdot \prod_{i=1}^n [Q(X_i,A_i;\nu_3)]^{Y_i}[1-Q(X_i,A_i;\nu_3)]^{1-Y_i}.$$

Note that the MLE of ν_1 is equal to 0, since the initial estimator $\widehat{P}^0_{X,n}$ is already the NPMLE. The MLE of ν_3 is given by

$$\widehat{\nu}^1_3 = \arg\max \prod_{i=1}^n [Q(X_i,A_i;\nu_3)]^{Y_i}[1-Q(X_i,A_i;\nu_3)]^{1-Y_i},$$

which is equivalent to the solution of the following estimating equation (see Exercise 17.4):

$$\sum_{i=1}^n \widehat{H}^0_n(X_i,A_i)\left\{Y_i - \text{expit}\left[\widehat{q}^0_n(X_i,A_i) + \widehat{\nu}^1_3 \widehat{H}^0_n(X_i,A_i)\right]\right\} = 0.$$

Thus, $\widehat{\nu}^1_3$ can be estimated by performing a logistic regression of Y on $\widehat{H}^0_n(X,A)$ with an offset term equal to $\widehat{q}^0_n(X,A)$.

In the literature, $\widehat{H}_n^0(X, A)$ is referred to as the *clever covariate*.[54] For convenience we will refer to this logistic regression as the *clever logistic regression*, although this terminology does not appear in the literature.

After the first round of update,

$$\widehat{q}_n^1(X, A) = \widehat{q}_n^0(X, A) + \widehat{\nu}_3^1 \widehat{H}_n^0(X, A),$$
$$\widehat{Q}_n^1(X, A) = \text{expit}[\widehat{q}_1^0(X, A)].$$

Remark:

It is worthwhile to pause here and appreciate the striking simplicity of the updating process, which amounts to a simple yet clever logistic regression,

$$\text{logit}\{\mathbb{P}(Y_i = 1 \mid X_i, A_i)\} = 1 \times \widehat{q}_n^0(X_i, A_i) + v_3 \times H(A_i, X_i),$$

and can be implemented straightforwardly in standard statistical software such as R or SAS.

Note that the clever covariate can be written as

$$H(X, A) = \frac{I(A = 1)}{g(A \mid X)} - \frac{I(A = 0)}{g(A \mid X)}.$$

Thus, we may alternatively define

$$H_0(X, A) = \frac{I(A = 0)}{g(A \mid X)} \quad \text{and} \quad H_1(X, A) = \frac{I(A = 1)}{g(A \mid X)},$$

and consider the logistic regression model

$$\text{logit}\{\mathbb{P}(Y_i = 1 \mid X_i, A_i)\} = \widehat{q}_n^0(X_i, A_i) + v_{30}\, \widehat{H}_{0n}(X_i, A_i) + v_{31}\, \widehat{H}_{1n}(X_i, A_i).$$

Second Round of Update

In this scenario, only ν_3 enters the second round of update, given that $\nu_1 = \nu_2 = 0$ in the first round. To update ν_3, the following estimated least favorable parametric submodel is considered:

$$Q(X, A; \nu_3) = \text{expit}\left[\widehat{q}_n^1(X, A) + \nu_3 \widehat{H}_n^0(X, A)\right].$$

Under this parametric submodel, the MLE of ν_3 is given by

$$\widehat{\nu}_3^2 = \arg\max \prod_{i=1}^{n} [Q(X_i, A_i; \nu_3)]^{Y_i} [1 - Q(X_i, A_i; \nu_3)]^{1-Y_i},$$

which is equivalent to the solution of the following estimating equation:

$$\sum_{i=1}^{n} \widehat{H}_n^0(X_i, A_i) \left\{Y_i - \text{expit}\left[\widehat{q}_n^1(X_i, A_i) + \widehat{\nu}_3^2 \widehat{H}_n^0(X_i, A_i)\right]\right\} = 0.$$

Note that, from the first round,

$$\sum_{i=1}^{n} \widehat{H}_n^0(X_i, A_i) \left\{Y_i - \text{expit}\left[\widehat{q}_n^1(X_i, A_i)\right]\right\} = 0.$$

It follows that $\widehat{\nu}_3^2 = 0$.

Final Results

Therefore, the final updated estimator is defined as

$$\widehat{Q}_n^*(X, A) = \widehat{Q}_n^1(X, A),$$

and the TMLE of θ_0 is given by

$$\widehat{\theta}_{\text{TMLE}} = \frac{1}{n}\sum_{i=1}^{n}\left[\widehat{Q}_n^*(X_i, 1) - \widehat{Q}_n^*(X_i, 0)\right],$$

where

$$\widehat{Q}_n^*(X_i, 1) = \text{expit}\left[\widehat{q}_n^0(X_i, 1) + \widehat{\nu}_3^1 \widehat{H}_n^0(X_i, 1)\right], \quad i = 1, \ldots, n;$$
$$\widehat{Q}_n^*(X_i, 0) = \text{expit}\left[\widehat{q}_n^0(X_i, 0) + \widehat{\nu}_3^1 \widehat{H}_n^0(X_i, 0)\right], \quad i = 1, \ldots, n.$$

To estimate this variance, the estimated EIF is defined as

$$\phi_n(X, A, Y) = \widehat{Q}_n^*(X, 1) - \widehat{Q}_n^*(X, 0) - \widehat{\theta}_{\text{TMLE}} - \frac{2A - 1}{\widehat{g}_n^0(A \mid X)}\left[Y - \widehat{Q}_n^*(X, A)\right].$$

An estimator of $\mathbb{V}(\phi(X, A, Y))$ is then given by

$$\widehat{\sigma}_n^2 = \frac{1}{n}\sum_{i=1}^{n} \phi_n^2(X_i, A_i, Y_i).$$

And an estimator of the standard error of $\widehat{\theta}_{\text{TMLE}}$ is given by $\widehat{\sigma}_n/\sqrt{n}$.

Gold Coin # 60. *The TMLE procedure is composed of two steps: (1) the construction of an initial estimator, and (2) the derivation of a targeted estimator. Beginning with the initial estimator, an update is performed in the direction of the score function of a least favorable parametric submodel. The resulting estimator is endowed with desirable asymptotic properties, including double robustness and efficiency, and is often found to be less biased in small samples.*

17.2.3 Brief Roadmap for TMLE

Defining the Estimand

The target estimand is defined as $\theta_0 = \theta(P_0)$, where $P_0(O)$ denotes the true joint distribution of the observation O.

Deriving the EIF

The EIF is derived. Then, this EIF is examined to identify a least favorable parametric submodel $P(O;\nu)$ such that $P(O;0) = P_0(O)$ when $\nu = 0$.

Obtaining an Initial Estimator

A loss function is defined, and an initial estimator of P_0 is constructed using the super learner approach, yielding $\widehat{P}_n^0$. An initial estimator of θ_0 is then obtained as $\widehat{\theta}_n^0 = \theta(\widehat{P}_n^0)$.

Updating the Initial Estimator

Based on $\widehat{P}_n^0$, a least favorable parametric submodel $P^0(O;\nu)$ is constructed so that $P^0(O;0) = \widehat{P}_n^0$ when $\nu = 0$. Using this submodel, the MLE of ν, denoted $\widehat{\nu}^1$, is obtained. This results in the first updated estimator of P_0, $\widehat{P}_n^1 = P^0(O;\widehat{\nu}^1)$. Accordingly, the first targeted update of the estimator of θ is given by $\widehat{\theta}_n^1 = \theta(\widehat{P}_n^1)$.

Repeating the Updating Process

Given the current estimator $\widehat{P}_n^1$, a new least favorable parametric submodel $P^1(O;\nu)$ is constructed such that $P^1(O;0) = \widehat{P}_n^1$. The MLE of ν based on this submodel, denoted $\widehat{\nu}^2$, is obtained, leading to the second update of the estimator of P_0, $\widehat{P}_n^2 = P^1(O;\widehat{\nu}^2)$. The second targeted update of the estimator of θ is then computed as $\widehat{\theta}_n^2 = \theta(\widehat{P}_n^2)$. This updating process is repeated until the estimate satisfies $\widehat{\nu}^k = 0$ in the kth iteration.

Completing the TMLE

Upon completion of the final update, the final estimator of P_0 is given by $\widehat{P}_n^* = \widehat{P}_n^k = P^{k-1}(O;\widehat{\nu}^k)$. The final targeted estimator of θ is then obtained as $\widehat{\theta}_{\text{TMLE}} = \theta(\widehat{P}_n^*)$.

17.3 Using TMLE

17.3.1 ATE for a Binary Outcome

Although the content in this subsection overlaps with that of Subsections 17.2.1 and 17.2.2, it is revisited here by following the above TMLE roadmap in Subsection 17.2.3. This walkthrough is intended to reinforce understanding of the TMLE steps in the context of estimating the ATE estimand for a binary outcome. Subsequently, the discussion will be extended to the estimation of the ATE for a bounded continuous outcome, as well as to the estimation of the average treatment effect on the treated (ATT) and average treatment effect on the controls (ATC).

Defining the Estimand

Let the observed data be denoted as $O_i = (X_i, A_i, Y_i)$ for $i = 1, \ldots, n$, where observations are assumed to be i.i.d. with $O = (X, A, Y)$, and where Y is a binary outcome coded as 0/1.

The target estimand is defined as

$$\theta_0 = \mathbb{E}[Q_0(X,1) - Q_0(X,0)] = \theta(P_{X,0}, Q_0),$$

where $P_{X,0}$ is the true distribution of X, and $Q_0(X, A) = \mathbb{E}(Y \mid X, A)$ is the true outcome regression function.

Deriving the EIF

The EIF for estimating θ_0 is given by

$$\begin{aligned}\phi(X,A,Y) &= Q_0(X,1) - Q_0(X,0) - \theta_0 + \frac{(2A-1)}{g_0(A \mid X)}[Y - Q_0(X,A)] \\ &= D_X + D_{Y|X,A},\end{aligned}$$

where the components are defined as

$$\begin{aligned}D_X &= Q_0(X,1) - Q_0(X,0) - \theta_0, \\ D_{Y|X,A} &= \frac{(2A-1)}{g_0(A \mid X)}[Y - Q_0(X,A)].\end{aligned}$$

It can be verified that $\mathbb{E}[D_X] = 0$ and $\mathbb{E}[D_{Y|X,A} \mid X, A] = 0$. Furthermore, $D_{Y|X,A}$ can be expressed as

$$D_{Y|X,A} = H(X, A; g_0)[Y - Q_0(X,A)],$$

where

$$H(X, A; g_0) = \frac{(2A-1)}{g_0(A \mid X)}$$

is referred to as the *clever covariate*.

The EIF is then examined to construct the following least favorable parametric submodel:

$$dP_X(X;\nu_1) = (1+\nu_1 D_X)\,dP_{X,0}(X),$$
$$Q(X,A;\nu_3) = \text{expit}\left\{\text{logit}[Q_0(X,A)] + \nu_3 H(X,A;g_0)\right\}.$$

Obtaining an Initial Estimator

The empirical distribution $\widehat{P}^0_{X,n}$ is used as the initial estimator of $P_{X,0}$. The BCE loss is employed in the super learner procedure to estimate Q_0 and g_0, resulting in initial estimators $\widehat{Q}^0_n$ and $\widehat{g}^0_n$, respectively. Based on these, the initial estimator of θ_0 is given by

$$\widehat{\theta}^0_n = \theta(\widehat{P}^0_{X,n}, \widehat{Q}^0_n) = \frac{1}{n}\sum_{i=1}^n \left[\widehat{Q}^0_n(X_i,1) - \widehat{Q}^0_n(X_i,0)\right].$$

Updating the Initial Estimator

Given the current estimators $\widehat{P}^0_{X,n}$, $\widehat{Q}^0_n$, and $\widehat{g}^0_n$, the following least favorable parametric submodel is constructed:

$$dP_X(X;\nu_1) = \left\{1 + \nu_1[\widehat{Q}^0_n(X,1) - \widehat{Q}^0_n(X,0) - \widehat{\theta}^0_n]\right\} dP_{X,0}(X),$$
$$Q(X,A;\nu_3) = \text{expit}\left\{\text{logit}[\widehat{Q}^0_n(X,A)] + \nu_3 H(X,A;\widehat{g}^0_n)\right\}.$$

The MLEs of (ν_1,ν_3) are obtained by maximizing the following likelihood:

$$\begin{aligned}(\widehat{\nu}^1_1,\widehat{\nu}^1_3) = \arg\max \Bigg\{ &\prod_{i=1}^n dP_X(X_i;\nu_1) \\ &\times \prod_{i=1}^n [\widehat{g}^0_n(1\mid X_i)]^{A_i}[1-\widehat{g}^0_n(1\mid X_i)]^{1-A_i} \\ &\times \prod_{i=1}^n [Q(X_i,A_i;\nu_3)]^{Y_i}[1-Q(X_i,A_i;\nu_3)]^{1-Y_i}\Bigg\}.\end{aligned}$$

It follows that $\widehat{\nu}^1_1 = 0$, and $\widehat{\nu}^1_3$ solves the estimating equation:

$$\sum_{i=1}^n H(X_i,A_i;\widehat{g}^0_n)\left[Y_i - \text{expit}\left\{\text{logit}[\widehat{Q}^0_n(X_i,A_i)] + \widehat{\nu}^1_3 H(X_i,A_i;\widehat{g}^0_n)\right\}\right] = 0.$$

The updated estimators are given by:

$$\widehat{P}^1_{X,n}(X) = \widehat{P}^0_{X,n}(X),$$
$$\widehat{g}^1_n(1\mid X) = \widehat{g}^0_n(1\mid X),$$
$$\widehat{Q}^1_n(X,A) = \text{expit}\left\{\text{logit}[\widehat{Q}^0_n(X,A)] + \widehat{\nu}^1_3 H(X,A;\widehat{g}^0_n)\right\}.$$

Repeating the Updating Process

The updating step is repeated. The new MLEs, $(\widehat{\nu}_1^2, \widehat{\nu}_3^2)$, are found with $\widehat{\nu}_1^2 = 0$, and $\widehat{\nu}_3^2$ satisfies:

$$\sum_{i=1}^{n} H(X_i, A_i; \widehat{g}_n^0) \left[Y_i - \text{expit}\left\{\text{logit}[\widehat{Q}_n^1(X_i, A_i)] + \widehat{\nu}_3^2 H(X_i, A_i; \widehat{g}_n^0)\right\}\right] = 0.$$

Given that

$$\sum_{i=1}^{n} H(X_i, A_i; \widehat{g}_n^0) \left[Y_i - \text{expit}\left\{\text{logit}[\widehat{Q}_n^1(X_i, A_i)]\right\}\right] = 0,$$

it follows that $\widehat{\nu}_3^2 = 0$, indicating that no further update is required after the first iteration.

Completing the TMLE

Let $\widehat{Q}_n^* = \widehat{Q}_n^1$ be the final estimator of Q_0. Since no update was applied to $P_{X,0}$, the final estimator remains $\widehat{P}_{X,n}^0$. Therefore, the TMLE of θ_0 is

$$\widehat{\theta}_{\text{TMLE}} = \theta(\widehat{P}_{X,n}, \widehat{Q}_n^*) = \frac{1}{n}\sum_{i=1}^{n} \left[\widehat{Q}_n^*(X_i, 1) - \widehat{Q}_n^*(X_i, 0)\right].$$

A Comparison with AIPW Estimation:

As a bonus, note that because

$$\sum_{i=1}^{n} \frac{(2A_i - 1)}{\widehat{g}_n^0(A_i \mid X_i)}[Y_i - \widehat{Q}_n^*(X_i, A_i)] = 0,$$

$\widehat{Q}_n^*$ solves the following estimating equation:

$$\sum_{i=1}^{n} \left\{Q(X_i, 1) - Q(X_i, 0) - \theta(\widehat{P}_{X,n}, Q) + \frac{(2A_i - 1)}{\widehat{g}_n^0(A_i \mid X_i)}[Y_i - Q(X_i, A_i)]\right\} = 0.$$

This highlights a key distinction between TMLE and AIPW. In TMLE, the solution $\widehat{Q}_n^*$ is first obtained by solving the EIF-based equation, after which the substitution estimator $\theta(\widehat{P}_{X,n}, \widehat{Q}_n^*)$ is used. In contrast, the AIPW estimator is obtained by solving directly for θ in the following estimating equation:

$$\sum_{i=1}^{n} \left\{\widehat{Q}_n^0(X_i, 1) - \widehat{Q}_n^0(X_i, 0) - \theta + \frac{(2A_i - 1)}{\widehat{g}_n^0(A_i \mid X_i)}[Y_i - \widehat{Q}_n^0(X_i, A_i)]\right\} = 0.$$

17.3.2 ATE for a Bounded Continuous Outcome

Consider a continuous outcome Y and suppose that it is known that $Y \in [a, b]$ for some $a < b$.

Defining the Estimand

Let the observed data be denoted as $O_i = (X_i, A_i, Y_i)$ for $i = 1, \dots, n$, where observations are assumed to be i.i.d. with $O = (X, A, Y)$, and where Y is a bounded continuous variable with $Y \in [a, b]$ for some $a < b$.

The target estimand is defined as

$$\theta_0 = \mathbb{E}[Q_0(X, 1) - Q_0(X, 0)] = \theta(P_{X,0}, Q_0),$$

where $P_{X,0}$ is the true distribution of X, and $Q_0(X, A) = \mathbb{E}(Y \mid X, A)$ is the true outcome regression function.

Let

$$\widetilde{Y} = \frac{Y - a}{b - a}.$$

Then $\widetilde{Y} \in [0, 1]$. Accordingly, let

$$\widetilde{\theta}_0 = \mathbb{E}[\mathbb{E}(\widetilde{Y} \mid X, A = 1) - \mathbb{E}(\widetilde{Y} \mid X, A = 0)].$$

Then, $\theta_0 = (b - a)\widetilde{\theta}_0$. Thus, if the TMLE estimator $\widehat{\theta}_n$ is obtained for $\widetilde{\theta}_0$, then $(b - a)\widehat{\theta}_n$ is the TMLE estimator for θ_0.

Therefore, without loss of generality, hereafter assume that $Y \in [0, 1]$. This allows us to extend the TMLE method, originally developed for binary outcomes, to accommodate bounded continuous outcomes. For this purpose, the BCE loss function can be considered instead of the $\mathcal{L}_2$ loss function. That is, consider the following loss function for the estimating $Q_0(X, A)$,

$$L(O, Q) = -Y \log[Q(X, A)] - (1 - Y) \log[1 - Q(X, A)].$$

A Remark:

There is a subtle point regarding the acronym "TMLE." For a binary outcome, the BCE loss coincides with the negative log-likelihood, so the acronym "TMLE" can be interpreted as both *targeted maximum likelihood estimator* and *targeted minimum loss estimator*. However, for a bounded continuous variable, the BCE loss no longer corresponds exactly to the likelihood. In such cases, "TMLE" should be understood as referring to the *targeted minimum loss estimator*.

Deriving the EIF

The EIF for estimating θ_0 is given by

$$\begin{aligned}\phi(X, A, Y) &= Q_0(X, 1) - Q_0(X, 0) - \theta_0 + \frac{(2A - 1)}{g_0(A \mid X)}[Y - Q_0(X, A)] \\ &= D_X + D_{Y|X,A},\end{aligned}$$

where the components are defined as

$$D_X = Q_0(X,1) - Q_0(X,0) - \theta_0,$$
$$D_{Y|X,A} = H(X,A;g_0)[Y - Q_0(X,A)],$$

where

$$H(X,A;g_0) = \frac{(2A-1)}{g_0(A \mid X)}.$$

Then the following least favorable parametric submodel is constructed:

$$dP_X(X;\nu_1) = (1 + \nu_1 D_X)\, dP_{X,0}(X),$$
$$Q(X,A;\nu_3) = \text{expit}\left\{\text{logit}[Q_0(X,A)] + \nu_3 H(X,A;g_0)\right\}.$$

Obtaining an Initial Estimator

The empirical distribution $\widehat{P}^0_{X,n}$ is used as the initial estimator of $P_{X,0}$. The BCE loss is employed in the super learner procedure to estimate Q_0 and g_0, resulting in initial estimators $\widehat{Q}^0_n$ and $\widehat{g}^0_n$, respectively. Based on these, the initial estimator of θ_0 is given by

$$\widehat{\theta}^0_n = \theta(\widehat{P}^0_{X,n}, \widehat{Q}^0_n) = \frac{1}{n}\sum_{i=1}^{n}\left[\widehat{Q}^0_n(X_i,1) - \widehat{Q}^0_n(X_i,0)\right].$$

Updating the Initial Estimator

Given the current estimators $\widehat{P}^0_{X,n}$, $\widehat{Q}^0_n$, and $\widehat{g}^0_n$, the following least favorable parametric submodel is constructed:

$$dP_X(X;\nu_1) = \left(1 + \nu_1[\widehat{Q}^0_n(X,1) - \widehat{Q}^0_n(X,0) - \widehat{\theta}^0_n]\right) dP_{X,0}(X),$$
$$Q(X,A;\nu_3) = \text{expit}\left\{\text{logit}[\widehat{Q}^0_n(X,A)] + \nu_3 H(X,A;\widehat{g}^0_n)\right\}.$$

The MLEs of (ν_1, ν_3) are obtained by the following minimization:

$$(\widehat{\nu}^1_1, \widehat{\nu}^1_3) = -\arg\min \log\left\{\prod_{i=1}^{n} dP_X(X_i;\nu_1) \times \prod_{i=1}^{n}[\widehat{g}^0_n(1 \mid X_i)]^{A_i}[1-\widehat{g}^0_n(1 \mid X_i)]^{1-A_i} \times \prod_{i=1}^{n}[Q(X_i,A_i;\nu_3)]^{Y_i}[1-Q(X_i,A_i;\nu_3)]^{1-Y_i}\right\}.$$

It follows that $\widehat{\nu}^1_1 = 0$, and $\widehat{\nu}^1_3$ solves the estimating equation (Note: recall the clever logistic regression!):

$$\sum_{i=1}^{n} H(X_i,A_i;\widehat{g}^0_n)\left[Y_i - \text{expit}\left\{\text{logit}[\widehat{Q}^0_n(X_i,A_i)] + \widehat{\nu}^1_3 H(X_i,A_i;\widehat{g}^0_n)\right\}\right] = 0.$$

The updated estimators are given by:

$$\begin{aligned}
\widehat{P}^1_{X,n}(X) &= \widehat{P}^0_{X,n}(X),\\
\widehat{g}^1_n(1 \mid X) &= \widehat{g}^0_n(1 \mid X),\\
\widehat{Q}^1_n(X,A) &= \text{expit}\left\{\text{logit}[\widehat{Q}^0_n(X,A)] + \widehat{\nu}^1_3 H(X,A;\widehat{g}^0_n)\right\}.
\end{aligned}$$

Repeating the Updating Process

The updating step is repeated. The new MLEs, $(\widehat{\nu}^2_1, \widehat{\nu}^2_3)$, are found with $\widehat{\nu}^2_1 = 0$, and $\widehat{\nu}^2_3$ satisfies:

$$\sum_{i=1}^{n} H(X_i, A_i; \widehat{g}^0_n)\left[Y_i - \text{expit}\left\{\text{logit}[\widehat{Q}^1_n(X_i,A_i)] + \widehat{\nu}^2_3 H(X_i,A_i;\widehat{g}^0_n)\right\}\right] = 0.$$

Given that

$$\sum_{i=1}^{n} H(X_i, A_i; \widehat{g}^0_n)\left[Y_i - \text{expit}\left\{\text{logit}[\widehat{Q}^1_n(X_i,A_i)]\right\}\right] = 0,$$

it follows that $\widehat{\nu}^2_3 = 0$, indicating that no further update is required after the first iteration.

Completing the TMLE

Let $\widehat{Q}^*_n = \widehat{Q}^1_n$ be the final estimator of Q_0. Since no update was applied to $P_{X,0}$, the final estimator remains $\widehat{P}^0_{X,n}$. Therefore, the TMLE of θ_0 is

$$\widehat{\theta}_{\text{TMLE}} = \theta(\widehat{P}_{X,n}, \widehat{Q}^*_n) = \frac{1}{n}\sum_{i=1}^{n}\left[\widehat{Q}^*_n(X_i,1) - \widehat{Q}^*_n(X_i,0)\right].$$

17.3.3 ATT and ATC

Let Y be a binary variable encoded as 0/1, or a continuous variable bounded within $[0,1]$. In this subsection, attention is restricted to the ATT estimand. The derivation for the ATC estimand follows identically, with $A = 1$ replaced by $A = 0$ throughout.

Defining the Estimand

The ATT estimand is defined as

$$\theta_0 = \mathbb{E}_{X|A=1}[Q_0(X,1) - Q_0(X,0)] = \theta(P_{X|A=1,0}, Q_0),$$

where $P_{X|A=1,0}$ is the true distribution of X conditional on $A = 1$, and $Q_0(X,A) = \mathbb{E}(Y \mid X, A)$ is the true regression function of Y given (X, A).

Deriving the EIF

The corresponding EIF for the ATT estimand is given by

$$\begin{aligned}\phi(X,A,Y) &= \frac{I(A=1)}{\mathbb{P}(A=1)}\left[Q_0(X,1)-Q_0(X,0)-\theta_0\right] \\ &\quad + \left[\frac{I(A=1)}{\mathbb{P}(A=1)} - \frac{I(A=0)g_0(1\mid X)}{\mathbb{P}(A=1)g_0(0\mid X)}\right]\left[Y-Q_0(X,A)\right].\end{aligned}$$

This EIF can equivalently be expressed as

$$\begin{aligned}\phi(X,A,Y) &= \frac{g_0(1\mid X)}{\mathbb{P}(A=1)}\left[Q_0(X,1)-Q_0(X,0)-\theta_{\text{ATT},0}\right] \\ &\quad + \frac{I(A=1)-g_0(1\mid X)}{\mathbb{P}(A=1)}\left[Q_0(X,1)-Q_0(X,0)-\theta_0\right] \\ &\quad + \left[\frac{I(A=1)}{\mathbb{P}(A=1)} - \frac{I(A=0)g_0(1\mid X)}{\mathbb{P}(A=1)g_0(0\mid X)}\right]\left[Y-Q_0(X,A)\right] \\ &= D_X + D_{A\mid X} + D_{Y\mid X,A},\end{aligned}$$

where the three components are defined as

$$\begin{aligned}D_X &= \frac{g_0(1\mid X)}{\mathbb{P}(A=1)}\left[Q_0(X,1)-Q_0(X,0)-\theta_0\right], \\ D_{A\mid X} &= \frac{I(A=1)-g_0(1\mid X)}{\mathbb{P}(A=1)}\left[Q_0(X,1)-Q_0(X,0)-\theta_0\right], \\ D_{Y\mid X,A} &= \left[\frac{I(A=1)}{\mathbb{P}(A=1)} - \frac{I(A=0)g_0(1\mid X)}{\mathbb{P}(A=1)g_0(0\mid X)}\right]\left[Y-Q_0(X,A)\right].\end{aligned}$$

Thus, construct the following least favorable parametric submodel:

$$\begin{aligned}dP_X(X;\nu_1) &= (1+\nu_1 D_X)dP_{X,0}(X), \\ g(1\mid X;\nu_2) &= \text{expit}\left\{\text{logit}[g_0(1\mid X)] + \nu_2 H_2(X;Q_0)\right\}, \\ Q(X,A;\nu_3) &= \text{expit}\left\{\text{logit}[Q_0(X,A)] + \nu_3 H_3(X,A;g_0)\right\},\end{aligned}$$

where the clever covariates are defined by

$$\begin{aligned}H_2(X;Q_0) &= Q_0(X,1)-Q_0(X,0)-\theta(P_{X\mid A=1,0},Q_0), \\ H_3(X,A;g_0) &= I(A=1) - \frac{I(A=0)g_0(1\mid X)}{g_0(0\mid X)}.\end{aligned}$$

Obtaining an Initial Estimator

The empirical distribution $\widehat{P}^0_{X,n}$ is used as the initial estimator of $P_{X,0}$. Likewise, the conditional empirical distributions $\widehat{P}^0_{X\mid A=a,n}$ are employed as initial estimators of $P_{X\mid A=a,0}$ for $a=0,1$. Super learning with the BCE loss is applied to obtain the initial estimators $\widehat{Q}^0_n$ and $\widehat{g}^0_n$ of Q_0 and g_0, respectively.

The initial estimator of the ATT estimand is then given by

$$\widehat{\theta}_n^0 = \frac{1}{\sum_{i=1}^n I(A_i = 1)} \sum_{i=1}^n I(A_i = 1) \left[\widehat{Q}_n^0(X_i, 1) - \widehat{Q}_n^0(X_i, 0)\right].$$

Updating the Initial Estimator

Using the initial estimators, construct the following least favorable submodel:

$$\begin{aligned} dP_X(X; \nu_1) &= \left\{1 + \nu_1[\widehat{Q}_n^0(X,1) - \widehat{Q}_n^0(X,0) - \widehat{\theta}_n^0]\right\} d\widehat{P}_{X,n}^0, \\ g(1 \mid X; \nu_2) &= \text{expit}\left\{\text{logit}[\widehat{g}_n^0(1 \mid X)] + \nu_2 H_2(X; \widehat{Q}_n^0)\right\}, \\ Q(X, A; \nu_3) &= \text{expit}\left\{\text{logit}[\widehat{Q}_n^0(X, A)] + \nu_3 H_3(X, A; \widehat{g}_n^0)\right\}, \end{aligned}$$

with

$$\begin{aligned} H_2(X; \widehat{Q}_n^0) &= \widehat{Q}_n^0(X,1) - \widehat{Q}_n^0(X,0) - \widehat{\theta}_n^0, \\ H_3(X, A; \widehat{g}_n^0) &= I(A=1) - \frac{I(A=0)\widehat{g}_n^0(1 \mid X)}{\widehat{g}_n^0(0 \mid X)}. \end{aligned}$$

The MLEs of ν_1, ν_2, and ν_3 are obtained by solving

$$\begin{aligned} (\widehat{\nu}_1^1, \widehat{\nu}_2^1, \widehat{\nu}_3^1) = \arg\max \Bigg\{ & \prod_{i=1}^n dP_X(X_i, \nu_1) \\ & \times \prod_{i=1}^n [\widehat{g}_n^0(1 \mid X_i; \nu_2)]^{A_i} [1 - \widehat{g}_n^0(1 \mid X_i; \nu_2)]^{1-A_i} \\ & \times \prod_{i=1}^n [Q(X_i, A_i; \nu_3)]^{Y_i} [1 - Q(X_i, A_i; \nu_3)]^{1-Y_i} \Bigg\}. \end{aligned}$$

Note that $\widehat{\nu}_1^1 = 0$ given $\widehat{P}_{X,n}^0$ is the NPMLE. In addition, the parameters $\widehat{\nu}_2^1$ and $\widehat{\nu}_3^1$ solve the following estimating equations (Note: recall the clever logistic regression!):

$$\sum_{i=1}^n H_2(X_i; \widehat{Q}_n^0) \left[A_i - \text{expit}\left\{\text{logit}[\widehat{g}_n^0(1 \mid X_i)] + \widehat{\nu}_2^1 H_2(X_i; \widehat{Q}_n^0)\right\}\right] = 0,$$

$$\sum_{i=1}^n H_3(X_i, A_i; \widehat{g}_n^0) \left[Y_i - \text{expit}\left\{\text{logit}[\widehat{Q}_n^0(X_i, A_i)] + \widehat{\nu}_3^1 H_3(X_i, A_i; \widehat{g}_n^0)\right\}\right] = 0.$$

The updated estimators are then given by

$$\begin{aligned} \widehat{P}_{X,n}^1(X) &= \widehat{P}_{X,n}^0(X), \\ \widehat{g}_n^1(1 \mid X) &= \text{expit}\left\{\text{logit}[\widehat{g}_n^0(1 \mid X)] + \widehat{\nu}_2^1 H_2(X; \widehat{Q}_n^0)\right\}, \\ \widehat{Q}_n^1(X, A) &= \text{expit}\left\{\text{logit}[\widehat{Q}_n^0(X, A)] + \widehat{\nu}_3^1 H_3(X, A; \widehat{g}_n^0)\right\}, \\ \widehat{\theta}_n^1 &= \theta(\widehat{P}_{X|A=1,n}, \widehat{Q}_n^1). \end{aligned}$$

Repeating the Updating Process

The above procedure is repeated iteratively, resulting in

$$\begin{aligned}
\widehat{P}^k_{X,n}(X) &= \widehat{P}^0_{X,n}(X),\\
\widehat{g}^k_n(1 \mid X) &= \text{expit}\left\{\text{logit}[\widehat{g}^{k-1}_n(1 \mid X)] + \widehat{\nu}^k_2 H_2(X; \widehat{Q}^{k-1}_n)\right\},\\
\widehat{Q}^k_n(X, A) &= \text{expit}\left\{\text{logit}[\widehat{Q}^{k-1}_n(X, A)] + \widehat{\nu}^k_3 H_3(X, A; \widehat{g}^{k-1}_n)\right\},\\
\widehat{\theta}^k_n &= \theta(\widehat{P}_{X|A=1,n}, \widehat{Q}^k_n),
\end{aligned}$$

for $k = 2, 3, \ldots$.

Completing the TMLE

Upon convergence of the iterative procedure, define the final estimators as $\widehat{P}^*_{X,n}(X) = \widehat{P}^0_{X,n}(X)$, $\widehat{g}^*_n(1 \mid X) = \widehat{g}^k_n(1 \mid X)$, and $\widehat{Q}^*_n(X, A) = \widehat{Q}^k_n(X, A)$, satisfying

$$\sum_{i=1}^n H_2(X_i; \widehat{Q}^*_n)\,[A_i - \widehat{g}^*_n(1 \mid X_i)] = 0,$$

$$\sum_{i=1}^n H_3(X_i, A_i; \widehat{g}^*_n)\left[Y_i - \widehat{Q}^*_n(X_i, A_i)\right] = 0.$$

The final TMLE of the ATT estimand is therefore given by

$$\widehat{\theta}_{\text{TMLE}} = \theta(\widehat{P}_{X|A=1,n}, \widehat{Q}^*_n) = \frac{\sum_{i=1}^n I(A_i = 1)\left[\widehat{Q}^*_n(X_i, 1) - \widehat{Q}^*_n(X_i, 0)\right]}{\sum_{i=1}^n I(A_i = 1)}.$$

A Comparison with AIPW Estimation:

Because

$$\sum_{i=1}^n \left\{\frac{I(A_i = 1)}{\widehat{\mathbb{P}}(A = 1)}\left[\widehat{Q}^*_n(X_i, 1) - \widehat{Q}^*_n(X_i, 0) - \theta(\widehat{P}_{X|A=1,n}, \widehat{Q}^*_n)\right]\right\} = 0$$

and

$$\begin{aligned}
0 &= \sum_{i=1}^n H_3(X_i, A_i; \widehat{g}^*_n)\left[Y_i - \widehat{Q}^*_n(X_i, A_i)\right]\\
&= \sum_{i=1}^n \left[\frac{I(A_i = 1)}{\widehat{\mathbb{P}}(A = 1)} - \frac{I(A_i = 0)\widehat{g}^*_n(1|X_i)}{\widehat{\mathbb{P}}(A = 1)\widehat{g}^*_n(0|X_i)}\right]\left[Y_i - \widehat{Q}^*_n(X_i, A_i)\right],
\end{aligned}$$

it implies that $\widehat{g}_n^*$ and $\widehat{Q}_n^*$ are the solutions of the following equation,

$$\sum_{i=1}^{n}\left\{\frac{I(A_i=1)}{\widehat{\mathbb{P}}(A=1)}\left[Q(X_i,1)-Q(X_i,0)-\theta(\widehat{P}_{X|A=1,n},Q)\right]\right.$$
$$\left.+\left[\frac{I(A_i=1)}{\widehat{\mathbb{P}}(A=1)}-\frac{I(A_i=0)g(1|X_i)}{\widehat{\mathbb{P}}(A=1)g(0|X_i)}\right][Y_i-Q(X_i,A_i)]\right\}=0,$$

where $\widehat{\mathbb{P}}(A=1)=\sum_{i=1}^{n}I(A_i=1)/n$.

This demonstrates the essential difference between the TMLE estimation and the AIPW estimation. The AIPW uses the solution of θ to the following equation as the estimator of θ,

$$\sum_{i=1}^{n}\left\{\frac{I(A_i=1)}{\widehat{\mathbb{P}}(A=1)}\left[\widehat{Q}_n^0(X_i,1)-\widehat{Q}_n^0(X_i,0)-\theta\right]\right.$$
$$\left.+\left[\frac{I(A_i=1)}{\widehat{\mathbb{P}}(A=1)}-\frac{I(A_i=0)\widehat{g}_n^0(1|X_i)}{\widehat{\mathbb{P}}(A=1)\widehat{g}_n^0(0|X_i)}\right][Y_i-\widehat{Q}_n^0(X_i,A_i)]\right\}=0.$$

Thus,

$$\widehat{\theta}_{\text{AIPW}}=\sum_{i=1}^{n}\left\{\frac{I(A_i=1)}{\widehat{\mathbb{P}}(A=1)}\left[\widehat{Q}_n^0(X_i,1)-\widehat{Q}_n^0(X_i,0)\right]\right.$$
$$\left.+\left[\frac{I(A_i=1)}{\widehat{\mathbb{P}}(A=1)}-\frac{I(A_i=0)\widehat{g}_n^0(1|X_i)}{\widehat{\mathbb{P}}(A=1)\widehat{g}_n^0(0|X_i)}\right][Y_i-\widehat{Q}_n^0(X_i,A_i)]\right\}.$$

An Alternative Path for Updating:

An alternative least favorable parametric submodel can be constructed using an expanded set of covariates:

$$dP_X(X;\nu_1)=(1+D_X)dP_{X,0}(X),$$
$$g(1\mid X;\nu_2)=\text{expit}\left\{\text{logit}[g_0(1\mid X)]+\nu_{21}H_{21}(X;Q_0)+\nu_{22}H_{22}(X;Q_0)\right\},$$
$$Q(X,A;\nu_3)=\text{expit}\left\{\text{logit}[Q_0(X,A)]+\nu_{31}H_{31}(X,A;g_0)+\nu_{32}H_{32}(X,A;g_0)\right\},$$

with

$$\begin{aligned}H_{21}(X;Q_0)&=Q_0(X,1)-Q_0(X,0),\\H_{22}(X;Q_0)&=1,\\H_{31}(X,A;g_0)&=I(A=1),\\H_{32}(X,A;g_0)&=\frac{I(A=0)g_0(1\mid X)}{g_0(0\mid X)}.\end{aligned}$$

Using this parametric submodel to update the initial estimator $\widehat{\theta}_n^0$, an alternative version of the TMLE estimation can be derived—which is more straightforward to understand:

$$\widehat{\theta}_{\text{TMLE}} = \frac{\sum_{i=1}^{n} I(A_i = 1)\left[Y_i - \widehat{Q}_n^*(X_i, 0)\right]}{\sum_{i=1}^{n} I(A_i = 1)}.$$

In fact, if the above parametric submodel is used for updating, eventually it leads to the following equations:

$$\begin{aligned}
\sum_{i=1}^{n} H_{21}(X_i; \widehat{Q}_n^*)\,[A_i - \widehat{g}_n^*(1 \mid X_i)] &= 0,\\
\sum_{i=1}^{n} H_{22}(X_i; \widehat{Q}_n^*)\,[A_i - \widehat{g}_n^*(1 \mid X_i)] &= 0,\\
\sum_{i=1}^{n} H_{31}(X_i, A_i; \widehat{g}_n^*)\left[Y_i - \widehat{Q}_n^*(X_i, A_i)\right] &= 0,\\
\sum_{i=1}^{n} H_{32}(X_i, A_i; \widehat{g}_n^*)\left[Y_i - \widehat{Q}_n^*(X_i, A_i)\right] &= 0.
\end{aligned}$$

Plugging in the expressions of clever covariates, the above equations become:

$$\begin{aligned}
\sum_{i=1}^{n} [\widehat{Q}_n^*(X_i, 1) - \widehat{Q}_n^*(X_i, 0)]\,[A_i - \widehat{g}_n^*(1 \mid X_i)] &= 0,\\
\sum_{i=1}^{n} [A_i - \widehat{g}_n^*(1 \mid X_i)] &= 0,\\
\sum_{i=1}^{n} I(A_i = 1)\left[Y_i - \widehat{Q}_n^*(X_i, A_i)\right] &= 0,\\
\sum_{i=1}^{n} \frac{I(A_i = 0)\widehat{g}_n^*(1 \mid X_i)}{\widehat{g}_n^*(0 \mid X_i} \left[Y_i - \widehat{Q}_n^*(X_i, A_i)\right] &= 0.
\end{aligned}$$

In particular, the second equation implies that

$$\widehat{\mathbb{P}}(A = 1) = \frac{1}{n}\sum_{i=1}^{n} I(A_i = 1) = \frac{1}{n}\sum_{i=1}^{n} \widehat{g}_n^*(1 \mid X_i),$$

while the third equation implies that

$$\sum_{i=1}^{n} I(A_i = 1)Y_i = \sum_{i=1}^{n} I(A_i = 1)\widehat{Q}_n^*(X_i, A_i).$$

TABLE 17.1
An alternative viewpoint of the TMLE for the ATT/ATC

ID	X	A	if treated	if untreated	treatment effect
1	X_1	1	Y_1	$\widehat{Q}_n^*(X_1, 0)$	$Y_1 - \widehat{Q}_n^*(X_1, 0)$
2	X_2	1	Y_2	$\widehat{Q}_n^*(X_2, 0)$	$Y_2 - \widehat{Q}_n^*(X_2, 0)$
...					
n_1	X_{n_1}	1	Y_{n_1}	$\widehat{Q}_n^*(X_{n_1}, 0)$	$Y_{n_1} - \widehat{Q}_n^*(X_{n_1}, 0)$
n_1+1	X_{n_1+1}	0	$\widehat{Q}_n^*(X_{n_1+1}, 1)$	Y_{n_1+1}	$\widehat{Q}_n^*(X_{n_1+1}, 1) - Y_{n_1+1}$
n_1+2	X_{n_1+2}	0	$\widehat{Q}_n^*(X_{n_1+2}, 1)$	Y_{n_1+2}	$\widehat{Q}_n^*(X_{n_1+2}, 1) - Y_{n_1+2}$
...					
n	X_n	0	$\widehat{Q}_n^*(X_n, 1)$	Y_n	$\widehat{Q}_n^*(X_n, 1) - Y_n$

Thus,

$$
\begin{aligned}
\widehat{\theta}_{\text{TMLE}} &= \theta(\widehat{P}_{X|A=1,n}, \widehat{Q}_n^*) \\
&= \frac{\sum_{i=1}^{n} I(A_i = 1)\left[\widehat{Q}_n^*(X_i, 1) - \widehat{Q}_n^*(X_i, 0)\right]}{\sum_{i=1}^{n} I(A_i = 1)} \\
&= \frac{\sum_{i=1}^{n} I(A_i = 1)\left[Y_i - \widehat{Q}_n^*(X_i, 0)\right]}{\sum_{i=1}^{n} I(A_i = 1)}. \quad \square
\end{aligned}
$$

This expression provides a clear interpretation of the TMLE for the ATT.[73] As illustrated in Table 17.1, for each subject in the treated arm ($A = 1$), the difference between the observed outcome Y_i and the predicted counterfactual outcome $\widehat{Q}_n^*(X_i, 0)$ represents an estimate of that subject's individual treatment effect. Averaging these individual treatment-effect estimates over all treated subjects yields the TMLE estimate of the ATT estimand.

Similarly, this expression provides an interpretation of the TMLE for the ATC. As illustrated in Table 17.1, for each subject in the control arm ($A = 0$), the difference between the predicted counterfactual outcome $\widehat{Q}_n^*(X_i, 1)$ and the observed outcome Y_i represents an estimate of that subject's individual treatment effect. Averaging these individual treatment-effect estimates over all controlled subjects yields the TMLE estimate of the ATC estimand.

> **Gold Coin # 61.** *The TMLE procedure can be applied to both binary outcomes and bounded continuous outcomes. Furthermore, it is applicable to the ATE estimand, as well as the ATT and ATC estimands.*

17.4 Two Extensions

Two extensions are considered: one for missing data, discussed in Chapter 14, and the other for longitudinal data, discussed in Chapter 15.

17.4.1 Missing Data

Consider data $\{O_i = (X_i, A_i, \Delta_i, \Delta_i Y_i)\}_{i=1}^n$, assumed to be i.i.d. copies of $O = (X, A, \Delta, \Delta Y)$, where Y is either a binary outcome coded as 0/1 or a bounded continuous outcome in $[0, 1]$.

Defining the Estimand

The target estimand is defined as

$$\theta_0 = \mathbb{E}[\mathbb{E}(Y \mid X, A = 1, \Delta = 1) - \mathbb{E}(Y \mid X, A = 0, \Delta = 1)] = \theta(P_{X,0}, \widetilde{Q}_0),$$

where $P_{X,0}$ denotes the distribution of X, and

$$\widetilde{Q}_0(X, A) = \mathbb{E}(Y \mid X, A, \Delta = 1)$$

is the outcome regression conditional on X, A, and $\Delta = 1$. Additionally, define

$$g_0(a \mid X) = \mathbb{P}(A = a \mid X), \quad a = 0, 1,$$
$$h_0(X, A) = \mathbb{P}(\Delta = 1 \mid X, A).$$

Deriving the EIF

From Chapter 14, the EIF for estimating θ_0 is given by

$$\begin{aligned}\phi(O) &= \mathbb{E}(Y \mid X, A = 1, \Delta = 1) - \mathbb{E}(Y \mid X, A = 0, \Delta = 1) - \theta_0 \\ &\quad + \frac{(2A - 1)\Delta}{\mathbb{P}(A \mid X)\mathbb{P}(\Delta = 1 \mid X, A)} [Y - \mathbb{E}(Y \mid X, A, \Delta = 1)] \\ &= D_X + D_{Y|X,A,\Delta},\end{aligned}$$

where

$$\begin{aligned}D_X &= \widetilde{Q}_0(X, 1) - \widetilde{Q}_0(X, 0) - \theta_0, \\ D_{Y|X,A,\Delta} &= \frac{(2A - 1)\Delta}{g_0(A \mid X)h_0(X, A)} [Y - \widetilde{Q}_0(X, A)].\end{aligned}$$

Note that $\mathbb{E}(D_X) = 0$, and $\mathbb{E}(D_{Y|X,A,\Delta} \mid X, A, \Delta = 1) = 0$. Additionally,

$$D_{Y|X,A,\Delta} = H(X, A, \Delta; g_0, h_0)[Y - \widetilde{Q}_0(X, A)],$$

where

$$H(X, A, \Delta; g_0, h_0) = \frac{(2A - 1)\Delta}{g_0(A \mid X)h_0(X, A)}$$

is referred to as the clever covariate.

Based on the above EIF, a least favorable parametric submodel can be constructed as:

$$dP_X(X; \nu_1) = (1 + \nu_1 D_X)\, dP_{X,0}(X),$$
$$\widetilde{Q}(X, A; \nu_4) = \text{expit}\left\{\text{logit}[\widetilde{Q}_0(X, A)] + \nu_4 H(X, A, \Delta; g_0, h_0)\right\}.$$

Obtaining an Initial Estimator

An initial estimator $\widehat{P}^0_{X,n}$ for $P_{X,0}$ is taken to be the empirical distribution of X. A super learner with the BCE loss is used to estimate $\widetilde{Q}_0$, g_0, and h_0, yielding initial estimators $\widehat{Q}^0_n$, $\widehat{g}^0_n$, and $\widehat{h}^0_n$, respectively.

The initial estimator of θ_0 is then computed as

$$\widehat{\theta}^0_n = \theta(\widehat{P}^0_{X,n}, \widehat{Q}^0_n) = \frac{1}{n}\sum_{i=1}^{n}\left[\widehat{Q}^0_n(X_i, 1) - \widehat{Q}^0_n(X_i, 0)\right].$$

Updating the Initial Estimator

Using the current estimators $\widehat{P}^0_{X,n}$, $\widehat{g}^0_n$, $\widehat{h}^0_n$, and $\widehat{Q}^0_n$, the following submodels are constructed:

$$dP_X(X; \nu_1) = \left(1 + \nu_1\left[\widehat{Q}^0_n(X, 1) - \widehat{Q}^0_n(X, 0) - \widehat{\theta}^0_n\right]\right) dP^0_X(X),$$
$$\widetilde{Q}(X, A; \nu_4) = \text{expit}\left\{\text{logit}[\widehat{Q}^0_n(X, A)] + \nu_4 H(X, A, \Delta; \widehat{g}^0_n, \widehat{h}^0_n)\right\}.$$

The MLEs of ν_1 and ν_4 are obtained by solving:

$$\begin{aligned}(\widehat{\nu}^1_1, \widehat{\nu}^1_4) = \arg\max\Bigg\{ \prod_{i=1}^{n} & dP_X(X_i; \nu_1) \cdot [\widehat{g}^0_n(1 \mid X_i)]^{A_i}[1 - \widehat{g}^0_n(1 \mid X_i)]^{1-A_i} \\ & \cdot [\widehat{h}^0_n(X_i, A_i)]^{\Delta_i}[1 - \widehat{h}^0_n(X_i, A_i)]^{1-\Delta_i} \\ & \cdot [\widetilde{Q}(X_i, A_i; \nu_4)]^{\Delta_i Y_i}[1 - \widetilde{Q}(X_i, A_i; \nu_4)]^{\Delta_i(1-Y_i)}\Bigg\}.\end{aligned}$$

It follows that $\widehat{\nu}^1_1 = 0$, and $\widehat{\nu}^1_4$ is the solution to the following estimating equation (Note: recall the clever logistic regression!):

$$\begin{aligned}&\sum_{i=1}^{n} H(X_i, A_i, \Delta_i; \widehat{g}^0_n, \widehat{h}^0_n) \\ &\quad \cdot \left[Y_i - \text{expit}\left\{\text{logit}[\widehat{Q}^0_n(X_i, A_i)] + \widehat{\nu}^1_4 H(X_i, A_i, \Delta_i; \widehat{g}^0_n, \widehat{h}^0_n)\right\}\right] = 0.\end{aligned}$$

The updated estimators are then given by:

$$\begin{aligned}
\widehat{P}^1_{X,n}(X) &= \widehat{P}^0_{X,n}(X),\\
\widehat{g}^1_n(1 \mid X) &= \widehat{g}^0_n(1 \mid X),\\
\widehat{h}^1_n(X, A) &= \widehat{h}^0_n(X, A),\\
\widehat{Q}^1_n(X, A) &= \text{expit}\left\{\text{logit}[\widehat{Q}^0_n(X, A)] + \widehat{\nu}^1_4 H(X, A, \Delta; \widehat{g}^0_n, \widehat{h}^0_n)\right\}.
\end{aligned}$$

Repeating the Updating Process

This updating process is repeated. Note that $\widehat{\nu}^2_1 = 0$, and $\widehat{\nu}^2_4$ solves the following estimating equation:

$$\begin{aligned}
&\sum_{i=1}^{n} H(X_i, A_i, \Delta_i; \widehat{g}^1_n, \widehat{h}^1_n)\\
&\quad \cdot \left[Y_i - \text{expit}\left\{\text{logit}[\widehat{Q}^1_n(X_i, A_i)] + \widehat{\nu}^2_4 H(X_i, A_i, \Delta_i; \widehat{g}^1_n, \widehat{h}^1_n)\right\}\right] = 0.
\end{aligned}$$

Because $\widehat{g}^1_n = \widehat{g}^0_n$ and $\widehat{h}^1_n = \widehat{h}^0_n$, the equation simplifies to:

$$\begin{aligned}
&\sum_{i=1}^{n} H(X_i, A_i, \Delta_i; \widehat{g}^0_n, \widehat{h}^0_n)\\
&\quad \cdot \left[Y_i - \text{expit}\left\{\text{logit}[\widehat{Q}^1_n(X_i, A_i)] + \widehat{\nu}^2_4 H(X_i, A_i, \Delta_i; \widehat{g}^0_n, \widehat{h}^0_n)\right\}\right] = 0.
\end{aligned}$$

Since it has already been established that

$$\sum_{i=1}^{n} H(X_i, A_i, \Delta_i; \widehat{g}^0_n, \widehat{h}^0_n) \left[Y_i - \text{expit}\left\{\text{logit}[\widehat{Q}^1_n(X_i, A_i)]\right\}\right] = 0,$$

it follows that $\widehat{\nu}^2_4 = 0$. Hence, no further updating is required.

Completing the TMLE

Let $\widehat{Q}^*_n = \widehat{Q}^1_n$ denote the final estimator of $\widetilde{Q}_0$. The final estimator of $P_{X,0}$ remains $\widehat{P}^0_n$. The TMLE of θ_0 is given by:

$$\widehat{\theta}_{\text{TMLE}} = \theta(\widehat{P}_{X,n}, \widehat{Q}^*_n) = \frac{1}{n}\sum_{i=1}^{n}\left[\widehat{Q}^*_n(X_i, 1) - \widehat{Q}^*_n(X_i, 0)\right],$$

where

$$\begin{aligned}
\widehat{Q}^*_n(X_i, 1) &= \text{expit}\left\{\text{logit}[\widehat{Q}^0_n(X_i, 1)] + \widehat{\nu}^1_4 H(X_i, 1, 1; \widehat{g}^0_n, \widehat{h}^0_n)\right\},\\
\widehat{Q}^*_n(X_i, 0) &= \text{expit}\left\{\text{logit}[\widehat{Q}^0_n(X_i, 0)] + \widehat{\nu}^1_4 H(X_i, 0, 1; \widehat{g}^0_n, \widehat{h}^0_n)\right\}.
\end{aligned}$$

17.4.2 Longitudinal Data

Consider a longitudinal study with T follow-up visits that generate the following:

$$X(0), A(0), \Delta(0), \ldots, \Delta(T-2)X(T-1), \Delta(T-2)A(T-1), \ \Delta(T-1), \Delta(T-1)Y.$$

Defining the Estimand

The potential outcome under treatment sequence $\overline{a} = (a(0), a(1), \ldots, a(T-1))$ is denoted by $Y^{\overline{a}}$. For a given $\overline{a}$, the estimand of interest is defined as

$$\theta_0 = \mathbb{E}\left(Y^{\overline{a}}\right) = \mathbb{E}_{\overline{X} \sim \widetilde{\mathcal{F}}_{\overline{a}}}\left[\mathbb{E}(Y \mid \overline{X}, \overline{A} = \overline{a}, \overline{\Delta} = \overline{1})\right],$$

where $\widetilde{\mathcal{F}}_{\overline{a}}$ is the distribution of $\overline{X}$ given in Chapter 15.

An alternative expression, following a backward fashion, of the above estimand is more useful. Define

$$\begin{aligned}
\widetilde{Q}_{T-1}(\overline{X}(T-1), \overline{A}(T-1)) &= \mathbb{E}\left[Y \,\middle|\, \overline{X}(T-1), \overline{A}(T-1), \Delta(T-1) = 1\right], \\
\widetilde{Y}_{T-1} &= \widetilde{Q}_{T-1}\left(\overline{X}(T-1), \overline{A}(T-2), a(T-1)\right), \\
\widetilde{Q}_{T-2}\left(\overline{X}(T-2), \overline{A}(T-2)\right) &= \mathbb{E}\left[\widetilde{Y}_{T-1} \mid \overline{X}(T-2), \overline{A}(T-2), \Delta(T-2) = 1\right], \\
\widetilde{Y}_{T-2} &= \widetilde{Q}_{T-2}\left(\overline{X}(T-2), \overline{A}(T-3), a(T-2)\right), \\
&\cdots \\
\widetilde{Y}_1 &= \widetilde{Q}_1\left(\overline{X}(1), A(0), a(1)\right), \\
\widetilde{Q}_0\left(X(0), A(0)\right) &= \mathbb{E}\left[\widetilde{Y}_1 \mid X(0), A(0), \Delta(0) = 1\right].
\end{aligned}$$

The estimand can be expressed as

$$\theta_0 = \mathbb{E}\left[\widetilde{Q}_0\left(X(0), a(0)\right)\right].$$

Deriving the EIF

From Chapter 15, the EIF for estimating θ_0 is given by

$$\begin{aligned}
\phi(O) = &\left[\widetilde{Q}_0\left(X(0), a(0)\right) - \theta_0\right] \\
&+ \sum_{t=1}^{T-1} \frac{I\{\overline{A}(t-1) = \overline{a}(t-1)\} I\{\overline{\Delta}(t-1) = \overline{1}\}}{\prod_{s=0}^{t-1} \widetilde{g}_s\left(A(s) \mid \overline{X}(s), \overline{A}(s-1)\right)} \times \\
&\quad \left[\widetilde{Q}_t\left(\overline{X}(t), \overline{A}(t-1), a(t)\right) - \widetilde{Q}_{t-1}\left(\overline{X}(t-1), \overline{A}(t-1)\right)\right] \\
&+ \frac{I(\overline{A} = \overline{a}) I(\overline{\Delta} = \overline{1})}{\prod_{t=0}^{T-1} \widetilde{g}_t(A(t) \mid \overline{X}(t), \overline{A}(t-1))} \left[Y - \widetilde{Q}_{T-1}\left(\overline{X}(T-1), \overline{A}(T-1)\right)\right],
\end{aligned}$$

where $\widetilde{g}_t(a(t) \mid \overline{X}(t), \overline{A}(t-1))$ is the product of the propensity score function and the non-missing probability function,

$$g_t\left(a(t) \mid \overline{X}(t), \overline{A}(t-1)\right) \times h_t\left(\overline{X}(t), (\overline{A}(t-1), a(t))\right),$$

for $t = 0, \ldots, T-1$, where

$$g_t\left(a(t) \mid \overline{X}(t), \overline{A}(t-1)\right) = \mathbb{P}\left\{A(t) = a(t) \mid \overline{X}(t), \overline{A}(t-1)\right\},$$
$$h_t(\overline{X}(t), \overline{A}(t)) = \mathbb{P}\{\Delta(t) = 1 \mid \overline{X}(t), \overline{A}(t)\}.$$

Obtaining an Initial Estimator

An initial estimator $\widehat{P}_{X(0),n}$ for $P_{X(0),0}$ is taken to be the empirical distribution of $X(0)$. Super learners with the BCE loss are used to estimate g_t and h_t, obtaining $\widehat{g}_{t,n}$ and $\widehat{h}_{t,n}$, $t = 0, \ldots, T-1$. Let $\widetilde{g}_{t,n} = \widehat{g}_{t,n} \times \widehat{h}_{t,n}$.

Super learners with the BCE loss are used to estimate $\widetilde{Q}_{T-1}, \widetilde{Q}_{T-2}, \ldots, \widetilde{Q}_0$, but in a backward fashion:

- $\widehat{Q}^0_{T-1,n}$ is trained using a super learner for $Y \sim \overline{X}(T-1) + \overline{A}(T-1) \mid \Delta(T-1) = 1$, and $\widetilde{Y}^0_{T-1,i}, i = 1, \ldots, n$, are obtained;
- $\widehat{Q}^0_{T-2,n}$ is trained using a super learner for $\widetilde{Y}^0_{T-1} \sim \overline{X}(T-2) + \overline{A}(T-2) \mid \Delta(T-2) = 1$, and $\widetilde{Y}^0_{T-2,i}, i = 1, \ldots, n$, are obtained;
- $\cdots$
- $\widehat{Q}^0_{1,n}$ is trained using a super learner for $\widetilde{Y}^0_2 \sim \overline{X}(1) + \overline{A}(1) \mid \Delta(1) = 1$, and $\widetilde{Y}^0_{1,i}, i = 1, \ldots, n$, are obtained;
- $\widehat{Q}^0_{0,n}$ is trained using a super learner for $\widetilde{Y}^0_1 \sim X(0) + A(0) \mid \Delta(1) = 1$.

Note that in the notation of $\widehat{Q}^0_{0,n}$, the zero in the subscript means that it is an estimator of $\widetilde{Q}_0$ at $t = 0$, while the zero in the superscript means that it is an initial estimator.

The initial estimator of θ_0 is then computed as

$$\widehat{\theta}^0_n = \theta(\widehat{P}_{X,n}, \widehat{Q}^0_{0,n}) = \frac{1}{n}\sum_{i=1}^{n} \widehat{Q}^0_{0,n}(X_i(0), a(0)).$$

Updating the Initial Estimator

Following the backward fashion, once again, $\widehat{Q}^0_{T-1,n}, \widehat{Q}^0_{T-2,n}, \ldots, \widehat{Q}^0_{1,n}, \widehat{Q}^0_{0,n}$ are updated as follows:

- $\widehat{Q}^*_{T-1,n}$ is obtained by performing logistic regression of Y on the clever covariate $\widehat{H}^0_T = \frac{I(\overline{A}=\overline{a})I(\overline{\Delta}=\overline{1})}{\prod_{t=0}^{T-1}\widetilde{g}_{t,n}(A(t)|\overline{X}(t),\overline{A}(t-1))}$ with an offset term equal to $\widehat{Q}^0_{T-1,n}$, and $\widetilde{Y}^*_{T-1,i}, i = 1, \ldots, n$, are obtained based on $\widehat{Q}^*_{T-1,n}$;

- $\widehat{Q}^*_{T-2,n}$ is obtained by performing logistic regression of $\widetilde{Y}^*_{T-1}$ on the clever covariate $\widehat{H}^0_{T-1} = \frac{I(\overline{A}(T-2)=\overline{a}(T-2))I(\overline{\Delta}(T-2)=\overline{1})}{\prod_{t=0}^{T-2}\widetilde{g}_{t,n}(A(t)|\overline{X}(t),\overline{A}(t-1))}$ with an offset term equal to $\widehat{Q}^0_{T-2,n}$, and $\widetilde{Y}^*_{T-2,i}, i = 1, \ldots, n$, are obtained based on $\widehat{Q}^*_{T-2,n}$;
- $\cdots$
- $\widehat{Q}^*_{1,n}$ is obtained by performing logistic regression of $\widetilde{Y}^*_2$ on the clever covariate $\widehat{H}^0_2 = \frac{I(\overline{A}(1)=\overline{a}(1))I(\overline{\Delta}(1)=\overline{1})}{\prod_{t=0}^{1}\widetilde{g}_{t,n}(A(t)|\overline{X}(t),\overline{A}(t-1))}$ with an offset term equal to $\widehat{Q}^0_{1,n}$, and $\widetilde{Y}^*_{1,i}, i = 1, \ldots, n$, are obtained based on $\widehat{Q}^*_{1,n}$;
- $\widehat{Q}^*_{0,n}$ is obtained by performing logistic regression of $\widetilde{Y}^*_1$ on the clever covariate $\widehat{H}^0_1 = \frac{I(A(0)=a(0))I(\Delta(1)=1)}{\widetilde{g}_{0,n}(A(0)|X(0))}$ with an offset term equal to $\widehat{Q}^0_{0,n}$.

Repeating the Updating Process

No further updating is needed for this EIF (See Exercise 17.5).

Completing the TMLE

The TMLE estimator of θ_0 is finally given by:

$$\widehat{\theta}_{\text{TMLE}} = \theta(\widehat{P}_{X,n}, \widehat{Q}^*_{0,n}) = \frac{1}{n}\sum_{i=1}^{n} \widehat{Q}^*_{0,n}(X_i(0), a(0)).$$

Gold Coin # 62. *The TMLE procedure is applicable to both point-treatment studies and longitudinal studies.*

17.5 Exercises

Exercise 17.1

Assume that $S_{0;\nu_j}$ for $j = 1, 2, 3$ are the score functions evaluated at $\nu_j = 0$, and $\phi(X, A, Y) = D_X + D_{Y|X,A}$ where $D_X = Q_0(X, 1) - Q_0(X, 0) - \theta_0$ and $D_{Y|X,A} = \frac{(2A-1)}{g_0(A|X)}[Y - Q_0(X, A)]$. Show that

$$\mathbb{E}\{D_{Y|X,A}\, S_{0;\nu_1}(X)\} = 0 \quad \text{and} \quad \mathbb{E}\{D_X\, S_{0;\nu_3}(X, A, Y)\} = 0.$$

Exercise 17.2

Show that the least favorable parametric submodel corresponds to the one with the following scores:

$$\begin{aligned} S_{0;\nu_1}(X) &= D_X, \\ S_{0;\nu_2}(X, A) &= 0, \\ S_{0;\nu_3}(X, A, Y) &= D_{Y|X,A}. \end{aligned}$$

Exercise 17.3

Find the score functions of the following parametric submodel at $\nu = 0$:

$$\begin{aligned} dP(X; \nu_1) &= (1 + \nu_1 D_X)\, d\,P_0(X), \\ g(A \mid X; \nu_2) &= (1 + \nu_2 \cdot 0)\, g_0(A \mid X), \\ Q(X, A; \nu_3) &= \text{expit}\left[q_0(X, A) + \nu_3 H(X, A)\right]. \end{aligned}$$

Exercise 17.4

Consider the following submodel for Q:

$$Q(X, A; \nu_3) = \text{expit}\left[\widehat{q}_n^0(X, A) + \nu_3 \widehat{H}_n^0(X, A)\right].$$

Show that the MLE of ν_3,

$$\widehat{\nu}_3^1 = \arg\max \prod_{i=1}^{n} [Q(X_i, A_i; \nu_3)]^{Y_i} [1 - Q(X_i, A_i; \nu_3)]^{1-Y_i},$$

is equivalent to the solution of the following estimating equation:

$$\sum_{i=1}^{n} \widehat{H}_n^0(X_i, A_i) \left\{ Y_i - \text{expit}\left[\widehat{q}_n^0(X_i, A_i) + \widehat{\nu}_3^1 \widehat{H}_n^0(X_i, A_i)\right]\right\} = 0.$$

Exercise 17.5

Refer to the TMLE procedure described in Subsection 17.4.2 as the *longitudinal TMLE* (LTMLE). Show that, in the estimation of $\theta_0 = \mathbb{E}(Y^{\overline{a}})$ using LTMLE, no further updates are required after the initial targeting step.

18

Implementation

This marks a turning point in the book. The contents of Chapters 1–17 are mature, while the contents of Chapters 18–20 are still evolving over time. Hence, the material in these final chapters reflects only the current status at the time of writing.

18.1 Two Software Environments

In the pharmaceutical industry, SAS and R are the two most widely used software environments for conducting statistical analyses. Below is a brief comparison of the two.

SAS is commonly preferred for standard statistical analyses and routine reporting. It is widely considered for producing robust and validated outputs. Two popular procedures in SAS 9.4 for implementing the classical causal inference methods discussed in Part II of this book are the `PSMATCH` procedure and the `CAUSALTRT` procedure.[44] However, SAS is a commercial software with high licensing costs, a concern particularly for start-ups. It is not open source, and its flexibility for methodological development is relatively limited.

In contrast, R is often favored for its advanced capabilities in causal inference, particularly in integrating machine learning techniques such as the super learner algorithm. Numerous R packages exist for implementing the causal inference methods covered in Parts II and IV of this book, and users can easily access them, as R is free and open source. In addition, R is powerful for flexible programming and simulation studies, making it well-suited for innovative research and development.

The following two sections will demonstrate how to implement the super learner (SL) and targeted maximum likelihood estimation (TMLE) discussed in Chapters 16 and 17 using SAS and R, respectively.

Gold Coin # 63. *In the pharmaceutical industry, SAS and R are the two most widely used software environments for conducting statistical analyses and producing statistical outputs.*

DOI: 10.1201/9781003635468-22

18.2 SAS Procedures for TMLE

SAS rebranded and modernized its platform under the name SAS Viya, moving away from the SAS 9.X version line. SAS Viya is a cloud-native, scalable platform designed for artificial intelligence (AI), machine learning (ML), data management, and analytics. Viya Workbench is a lightweight, cloud-native development environment built on top of SAS Viya.[74]

In this section, two SAS procedures, the SUPERLEARNER procedure and the CAEFFECT procedure, are briefly introduced for implementing SL and TMLE. For complete descriptions, the reader is referred to their user manual.[74]

18.2.1 Procedure SUPERLEARNER

The SUPERLEARNER procedure implements the super learner algorithm in SAS Viya.[72] This ensemble modeling framework has been extensively discussed in the literature[72,54,75] and is briefly discussed in Chapter 16.

To train a super learner model, you must define a library of predictive models, also known as *base learners.* These base learners can range from simple parametric regression models to complex nonparametric machine learning algorithms. The final output is an ensemble predictive model that maps a vector of predictor variables to an outcome variable, also known as *target variable*, which can be either continuous or binary.

Here is the basic syntax structure of the SUPERLEARNER procedure:

```
PROC SUPERLEARNER data <options>;
    BASELEARNER 'name' model-type <(model-options)>;
    CROSSVALIDATION <options>;
    TARGET variable </option>;
    INPUT variables </option>;
    OUTPUT OUT=libref.data-table <options>;
RUN;
```

The BASELEARNER statement allows you to specify and configure individual base learner models. At least two BASELEARNER statements are required. The available base learner models are:

- BART specifies a Bayesian additive regression trees model,
- BNET specifies a Bayesian network model,
- FACTMAC specifies a factorization machine model,
- FOREST specifies a forest model,
- GAMMOD specifies a generalized additive model based on low-rank regression splines,

- GAMSELECT specifies a generalized additive model with model selection,
- GPCLASS specifies a Gaussian process classification model,
- GPREG specifies a Gaussian process regression model,
- GRADBOOST specifies a gradient boosting model,
- LIGHTGRADBOOST specifies a light gradient boosting machine model,
- LOGSELECT specifies a logistic regression model with model selection,
- REGSELECT specifies an ordinary least squares regression model with model selection,
- SVMACHINE specifies a support vector machine model,
- TREESPLIT specifies a tree-based statistical model.

The `CROSSVALIDATION` statement controls the specifications of the K-fold cross-validation process. If omitted, `PROC SUPERLEARNER` selects the number of folds using an internal algorithm. The `TARGET` statement specifies the target variable (outcome variable, endpoint) for the super learner model, while the `INPUT` statement identifies one or more predictor variables of a common type. The `OUTPUT` statement generates a data table containing individual-level predicted values after model fitting.

18.2.2 Procedure CAEFFECT

The `CAEFFECT` procedure is a model-agnostic tool for estimating causal effects. While it does not directly model the treatment or outcome variables, it accounts for confounding by incorporating precomputed predictions—such as those generated by the `SUPERLEARNER` procedure—provided as input.

The `CAEFFECT` procedure supports the implementation of all four estimators discussed in Chapter 17: inverse probability of treatment weighting (IPW), augmented IPW (AIPW), maximum likelihood estimation (MLE) or g-computation, and targeted MLE (TMLE). These correspond to the following options in the procedure:

- `PROC CAEFFECT data METHOD=IPW;`
- `PROC CAEFFECT data METHOD=AIPW;`
- `PROC CAEFFECT data METHOD=REGADJ;`
- `PROC CAEFFECT data METHOD=TMLE;`

Consider the `TMLE` option as an example. The first step of the TMLE method involves obtaining an initial estimator $\widehat{Q}_n^0(X, A)$ using Super Learner, which yields predictions of the form $\{(\widehat{Q}_n(X_i, 1), \widehat{Q}_n(X_i, 0)) : i = 1, \ldots, n\}$.

Concurrently, an initial estimator of the propensity score, $\widehat{g}_n^0(X)$, is also obtained using Super Learner, producing predictions $\{(\widehat{g}_n(1 \mid X_i), \widehat{g}_n(0 \mid X_i)) : i = 1, \ldots, n\}$.

These predicted values are then appended to the original dataset, resulting in an enriched dataset containing both outcome and treatment model predictions. This enriched dataset is then used as the data to the `CAEFFECT` procedure, which is invoked using syntax like the following:

```
PROC CAEFFECT data METHOD=TMLE;
    TREATVAR treatment-variable;
    OUTCOMEVAR outcome-variable;
    POM TREATLEV='treatment-level'
        <PREDOUT=predicted-outcome-value>
        <TREATPROB=predicted-treatment-probability>;
    DIFFERENCE <REFLEV=(reference-list)>;
RUN;
```

The `TREATVAR` statement specifies the treatment variable, while the `OUTCOMEVAR` statement specifies the outcome variable. The `POM` statement identifies the potential outcome mean to estimate: the `PREDOUT=` option specifies the input variable containing the predicted potential outcome values for the given treatment level, and the `TREATPROB=` option specifies the input variable containing the predicted probability of receiving the specified treatment level. The DIFFERENCE statement specifies causal effects to estimate on the difference scale.

> **Gold Coin # 64.** *The two steps that constitute TMLE—the initial estimation step and the targeting step—can be implemented using two SAS procedures: the **SUPERLEARNER** procedure for the initial step and the **CAEFFECT** procedure for the targeting step.*

18.3 R Packages for TMLE

The SAS procedures described in the previous section have several limitations: (1) They are not freely available; (2) The TMLE procedure must be implemented in two separate steps rather than as a unified process; (3) They do not support handling missing data; (4) They are not equipped to handle longitudinal data.

This reflects the current state of SAS procedure development. It is possible that future versions will be applicable for handling missing data and longitudinal data. Given these limitations, R can be considered instead.

The following two subsections will introduce two well-maintained R packages, the `tmle` package and the `ltmle` package, which are developed for analyzing point-treatment data and longitudinal data respectively. These packages integrate with Super Learner directly, eliminating the need to run Super Learner as a separate step.

Gold Coin # 65. *Two well-maintained R packages, `tmle` and `ltmle`, are available for implementing Targeted Maximum Likelihood Estimation (TMLE) in point-treatment and longitudinal studies, respectively.*

Although the SuperLearner package will be called internally by the `tmle` and `ltmle` packages to be described soon, it is still worthwhile to briefly introduce it first. The SuperLearner package implements both the discrete super learner and the ensemble super learner.[76] Extensive documentation and guides for using Super Learner are available online. For example, see the official vignette: https://cran.r-project.org/web/packages/SuperLearner/vignettes/Guide-to-SuperLearner.html.

At the time of writing, the SuperLearner package includes 40 built-in base learner models, which are listed as follows:

```
install.packages("SuperLearner")
library(SuperLearner)
listWrappers()
## All prediction algorithm wrappers in SuperLearner:
## [1]  "SL.bartMachine"        "SL.bayesglm"
## [3]  "SL.biglasso"           "SL.caret"
## [5]  "SL.caret.rpart"        "SL.cforest"
## [7]  "SL.earth"              "SL.gam"
## [9]  "SL.gbm"                "SL.glm"
## [11] "SL.glm.interaction"    "SL.glmnet"
## [13] "SL.ipredbagg"          "SL.kernelKnn"
## [15] "SL.knn"                "SL.ksvm"
## [17] "SL.lda"                "SL.leekasso"
## [19] "SL.lm"                 "SL.loess"
## [21] "SL.logreg"             "SL.mean"
## [23] "SL.nnet"               "SL.nnls"
## [25] "SL.polymars"           "SL.qda"
## [27] "SL.randomForest"       "SL.ranger"
## [29] "SL.ridge"              "SL.rpart"
## [31] "SL.rpartPrune"         "SL.speedglm"
## [33] "SL.speedlm"            "SL.step"
## [35] "SL.step.forward"       "SL.step.interaction"
## [37] "SL.stepAIC"            "SL.svm"
## [39] "SL.template"           "SL.xgboost"
```

In addition to these built-in learners, custom base learner models can be created. For example, the following code defines a learner based on `SL.ranger` with a specific number of trees:

```
learners = create.Learner("SL.ranger", params = list(num.trees =
    1000))
```

18.3.1 Package "tmle"

The `tmle` package is a well-maintained R package for implementing TMLE.[77] It can automatically estimate point-treatment effects for both continuous and binary outcomes. The package provides estimates of the average treatment effect (ATE), the average treatment effect among the treated (ATT), and the average treatment effect among the controls (ATC). For binary outcomes, it offers effect estimates in terms of risk difference, risk ratio, and odds ratio. For each estimand, the package reports the point estimate, standard error, confidence interval, and p-value.

A complete description of the `tmle` package is available on its CRAN homepage: `https://cran.r-project.org/web/packages/tmle/index.html`.

Below is the main `tmle()` function from the package:

```
tmle(Y, A, W, Z=NULL, Delta = rep(1,length(Y)), Q = NULL, Q.Z1 =
    NULL, Qform = NULL, Qbounds = NULL, Q.SL.library = c("SL.glm",
     "tmle.SL.dbarts2", "SL.glmnet"), cvQinit = TRUE, g1W = NULL,
    gform = NULL, gbound = NULL, pZ1=NULL, g.Zform = NULL, pDelta1
     = NULL, g.Deltaform = NULL, g.SL.library = c("SL.glm", "tmle.
    SL.dbarts.k.5", "SL.gam"), g.Delta.SL.library = c("SL.glm", "
    tmle.SL.dbarts.k.5", "SL.gam"), family = "gaussian",
    fluctuation = "logistic", alpha = 0.9995, id=1:length(Y), V.Q
    = 10, V.g=10, V.Delta=10, V.Z = 10, verbose = FALSE, Q.
    discreteSL=FALSE, g.discreteSL=FALSE, g.Delta.discreteSL=FALSE
    , prescreenW.g=TRUE, min.retain = NULL, target.gwt = FALSE,
    automate=FALSE, obsWeights = NULL, alpha.sig = 0.05, B = 1,
    evalATT = TRUE)
```

Below is a brief explanation of some key arguments of the `tmle()` function:[77]

- `Y`—continuous or binary outcome variable
- `A`—binary treatment indicator, 1 for treatment, 0 for control
- `W`—vector, matrix, or dataframe containing baseline covariates
- `Delta`—indicator of missing outcome or treatment assignment. 1 for observed, 0 for missing
- `Qform`—optional regression formula for estimation of $\mathbb{E}(Y \mid A, W)$, suitable for call to glm
- `Q.SL.library`—optional vector of prediction algorithms to use for `SuperLearner` estimation of initial Q

- `g1W`—optional vector of conditional treatment assingment probabilities, $\mathbb{P}(A = 1 \mid W)$
- `gform`—optional regression formula of the form $A \sim W$, if specified this overrides the call to `SuperLearner`
- `gbound`—value between $(0, 1)$ for truncation of predicted probabilities
- `g.SL.library`—optional vector of prediction algorithms to use for `SuperLearner` estimation of `g1W`
- `pDelta1`—optional $n \times 2$ matrix of conditional probabilities for missingness mechanism
- `g.Deltaform`—optional regression formula of the form $Delta \sim A + W$, if specified this overrides the call to `SuperLearner`
- `g.Delta.SL.library`—optional vector of prediction algorithms to use for `SuperLearner` estimation of `pDelta1`
- `family`—family specification for working regression models, generally '`gaussian`' for continuous outcomes (default), '`binomial`' for binary outcomes

Remark:

In terms of implementing TMLE, randomized controlled trials (RCTs) can be viewed as a special case of cohort studies, noting that in RCTs the propensity score $g(1 \mid W)$ is known by design and can therefore be correctly specified as a function of W. Due to double robustness, since the model for propensity score function can be specified correctly for sure, the TMLE is consistent for sure regardless of whether the outcome regression function is correctly specified or not. This "oracle" propensity score in RCTs can be incorporated using either the `g1W` or `gform` option.

A Generic Example

Consider a point-treatment study that generates data

$$(X_i, A_i, \Delta_i, \Delta_i Y_i), \quad i = 1, \ldots, n.$$

Let W_i be a function of X_i, where $X_i = (X_{i1}, \ldots, X_{id})^\top$ and $W_i = (W_{i1}, \ldots, W_{ip})^\top$. For example (See Exercise 18.1), suppose that X_{i1} and X_{i2} represent height and weight, respectively. Then W_{i1} could be the body mass index (BMI), defined as $W_{i1} = X_{i2}/X_{i1}^2$. As another example, suppose $X_{\cdot 3} = (X_{13}, \ldots, X_{n3})^\top$ has missing values. Then, to handle missing values in $X_{\cdot 3}$, two variables are defined: $W_{i2} = X_{i3}$ with a single imputation for the missing value, and $W_{i3} = I(X_{i3} = \texttt{NA})$, the missingness indicator.

After the above steps of obtaining $W_i = (W_{i1}, \ldots, W_{ip})^\top$, the data are

$$(W_i, A_i, \Delta_i, \Delta_i Y_i), \quad i = 1, \ldots, n.$$

Next, specify a library of base learners for the outcome regression function $Q(W, A) = \mathbb{E}(Y \mid W, A)$, for example, `Q.SL.library = c("SL.glm", "tmle.SL.dbarts2", "SL.glmnet")`, which is the default.

Similarly, specify a library of base learners for the propensity score function $g(1 \mid W) = \mathbb{P}(A = 1 \mid W)$, for example, `g.SL.library = c("SL.glm", "tmle.SL.dbarts.k.5", "SL.gam")`, as by default.

Lastly, specify a library for the non-missingness mechanism (i.e., the probability that an observation is not missing), defined by $h(W, A) = \mathbb{P}(\Delta = 1 \mid W, A)$, for example, `g.Delta.SL.library = c("SL.glm", "SL.glmnet", "SL.gam")`.

After completing all the preparatory steps, the following R function can be implemented to perform the estimation:

```
tmle(Y, A, W, Delta,
     Q.SL.library = c("SL.glm", "tmle.SL.dbarts2", "SL.glmnet"),
     g.SL.library = c("SL.glm", "tmle.SL.dbarts.k.5", "SL.gam"),
     g.Delta.SL.library = c("SL.glm", "SL.glmnet", "SL.gam"),
     family = "gaussian")
```

An Important Note:

Note that the above libraries are selected for demonstration purposes. In practice, users of the `tmle` package may refer to the flowchart recently proposed by Phillips *et al.* (2023)[75] to guide the specification of Super Learner libraries and the specification of the number of folds in the cross-validation.

18.3.2 Package "ltmle"

The `ltmle` package is a well-maintained R package for using TMLE to analyze longitudinal data.[78] It can be used to estimate treatment effects of static or dynamic treatment regimes for continuous outcomes, binary outcomes, and time-to-event outcomes (i.e., survival outcomes). A complete description of the `ltmle` package is available online: `https://cran.r-project.org/web/packages/ltmle/index.html`.

Below is the main `ltmle()` function from the package:

```
ltmle(
  data,
  Anodes,
  Cnodes = NULL,
  Lnodes = NULL,
  Ynodes,
  survivalOutcome = NULL,
  Qform = NULL,
```

```
  gform = NULL,
  abar,
  rule = NULL,
  gbounds = c(0.01, 1),
  Yrange = NULL,
  deterministic.g.function = NULL,
  stratify = FALSE,
  SL.library = "glm",
  SL.cvControl = list(),
  estimate.time = TRUE,
  gcomp = FALSE,
  iptw.only = FALSE,
  deterministic.Q.function = NULL,
  variance.method = "tmle",
  observation.weights = NULL,
  id = NULL
)
```

Below is a brief explanation of some key arguments of the `ltmle()` function:[78]

- `data`—data frame following the time-ordering of the nodes
- `Anodes`—column names or indicies in `data` of treatment nodes
- `Cnodes`—column names or indicies in `data` of censoring nodes
- `Lnodes`—column names or indicies in `data` of time-dependent covariate nodes
- `Ynodes`—column names or indicies in `data` of outcome nodes
- `survivalOutcome`—If TRUE, then `Ynodes` are indicators of an event, and if Y at some timepoint is 1, then all following should be 1
- `Qform`—character vector of regression formulas for Q
- `gform`—character vector of regression formulas for g or a character vector or a matrix/array for A nodes and C nodes
- `abar`—binary vector or matrix of counterfactual treatment or a list of length 2
- `rule`—function to be applied to each row (a named vector) of data that returns a numeric vector of length `numAnodes`
- `gbounds`—lower and upper bounds on estimated cumulative probabilities for g-factors
- `SL.library`—optional character vector of libraries to pass to the `SuperLearner` function. `NULL` indicates the `glm` function should be called instead of `SuperLearner`. `'default'` indicates a standard set of libraries. May be separately specified for Q and g

- `SL.cvControl`—optional list to be passed as `cvControl` to `SuperLearner`
- `gcomp`—if `TRUE`, run the maximum likelihood based g-computation estimate instead of TMLE
- `iptw.only`—if `TRUE`, only IPTW is run, which is faster

A Specific Example

Consider a longitudinal study with three follow-up timepoints ($T = 3$), three baseline covariates, and two time-dependent covariates. Let $A = (A(0), A(1), A(2))^\top$ denote the treatment sequence, $C = (C(0), C(1), C(2))^\top$ the censoring indicators, and Y the outcome variable measured at each follow-up timepoint.

A simulated dataset is generated with the following columns and orders (see Exercise 18.2) :

```
L0.a, L0.b, L0.c # Baseline covariates at t = 0
A0, C0           # Treatment and censoring status after t = 0
L1.a, L1.b, Y1   # Time-dependent covariates and outcome at t = 1
A1, C1           # Treatment and censoring status after t = 1
L2.a, L2.b, Y2   # Time-dependent covariates and outcome at t = 2
A2, C2           # Treatment and censoring status after t = 2
Y3               # Final outcome at t = 3
```

Here, $C = (C(0), C(1), C(2))^\top$ is equivalent to $\Delta = (\Delta(0), \Delta(1), \Delta(2))^\top$ as used in the preceding chapter. Specifically, each $C(t)$ is a categorical variable with two levels: "censored" and "uncensored," which corresponds to a binary variable $\Delta(t)$, where $\Delta(t) = 0$ indicates a missing (censored) observation and $\Delta(t) = 1$ indicates an observed (uncensored) value. Based on this definition of C, the `ltmle` package can be applied to analyze time-to-event outcomes in the next chapter.

Before applying the `ltmle` function, four types of nodes are specified. Although baseline covariates are not included in `Lnodes`, they are automatically treated as baseline covariates by default.

```
Lnodes <- c("L1.a", "L1.b", "L2.a", "L2.b")
Anodes <- c("A0", "A1", "A2")
Cnodes <- c("C0", "C1", "C2")
Ynodes <- c("Y1", "Y2", "Y3")
```

In the `ltmle` function, the `SL.library` argument is used to specify the set of prediction algorithms passed to the `SuperLearner` function. Two examples are shown below.

```
# Example 1
SL.library = list(Q = c("SL.glm", "SL.stepAIC", "SL.glmnet"),
                  g = c("SL.glm", "SL.stepAIC", "SL.glmnet"))

# Example 2
SL.library = list(Q = NULL,
                  g = c("SL.glm", "SL.stepAIC", "SL.glmnet"))
```

In Example 1, a list of Super Learner libraries is provided for both the outcome regression (Q) and treatment/censoring mechanism (g) models. In Example 2, `NULL` is passed for the Q function, indicating that the default `glm` function is used instead of `SuperLearner`.

After completing all preparatory steps, the causal estimand

$$\mathbb{E}(Y^{\overline{a}=\overline{1}}) - \mathbb{E}(Y^{\overline{a}=\overline{0}})$$

which compares the expected outcomes under two static treatment regimes, $\overline{1}$ (always treated) and $\overline{0}$ (never treated), can be estimated using the following `ltmle` function (See Exercise 18.3):

```
ltmle(
  data,
  Anodes = Anodes,
  Cnodes = Cnodes,
  Lnodes = Lnodes,
  Ynodes = Ynodes,
  abar = list(treatment = c(1, 1, 1), control = c(0, 0, 0)),
  SL.library = SL.library
)
```

Dynamic Treatment Regimes

A static treatment regime (STR) is specified using the `abar` argument, whereas a dynamic treatment regime (DTR) is specified using the `rule` argument. The `rule` argument takes a function that operates on a single row of the data and returns a binary vector. Each entry in this vector corresponds to a treatment node, indicating whether the individual would receive treatment at that node under the specified dynamic regime.[78]

Continue the above example dataset with $T = 3$. A possible regime might assign treatment at time $t = 0$ to all individuals, assign treatment at $t = 1$ only to those for whom the variable `L1.a` exceeds a threshold (e.g., 0.8), and assign treatment at $t = 2$ only if `L2.a` exceeds another threshold (e.g., 1.0). The value of the regime can be estimated using the following function (see Exercise 18.4):

```
ltmle(..., rule = function(row)
    c(1, ifelse(row["L1.a"] > 0.8, 1, 0),
         ifelse(row["L2.a"] > 1., 1, 0)),...)
```

More Than Two Arms

Consider a three-arm clinical trial or a cohort study with three distinct cohorts. Let $(0, 0)$ denote treatment arm 1, $(1, 0)$ denote treatment arm 2, and $(0, 1)$ denote the control arm. In this setup, the treatment variable `A0` is expanded into two dummy variables: `A0.a` and `A0.b`. Similarly, `A1` becomes `A1.a` and `A1.b`, and `A2` becomes `A2.a` and `A2.b`.

The corresponding nodes can be defined as:

```
Lnodes <- c("L1.a", "L1.b", "L2.a", "L2.b")
Anodes <- c("A0.a", "A0.b", "A1.a", "A1.b", "A2.a", "A2.b")
Cnodes <- c("C0", "C1", "C2")
Ynodes <- c("Y1", "Y2", "Y3")
```

To estimate the causal effect comparing treatment 1 with the control treatment, the `abar` argument can be specified as:

```
ltmle(..., abar = list(treatment = rep(c(1, 0), 3),
                       control = rep(c(0, 0), 3)), ...)
```

To estimate the causal effect comparing treatment 2 with the control treatment, the `abar` argument can be specified as:

```
ltmle(..., abar = list(treatment = rep(c(0, 1), 3),
                       control = rep(c(0, 0), 3)), ...)
```

This approach can be extended to studies with four or more treatment arms (see Exercise 18.5).

18.4 Exercises

Exercise 18.1

Prepare the dataset below for use with the `tmle` package by creating a BMI variable and a missingness indicator for the `Age` variable. Assume that the sample mean age is 40 years.

ID	Height (m)	Weight (kg)	Age (yrs)	Treatment	Outcome
1	1.70	65	NA	1	18
2	1.60	85	35	0	NA
...					
200	1.80	58	42	1	14

Exercise 18.2

Use the following R code to generate a dataset. Then, write down the corresponding structural causal model (SCM) that represents the data-generating process.

```
set.seed(18)
n.Ex2 <- 200
L0.a <- runif(n.Ex2)
L0.b <- runif(n.Ex2)
L0.c <- runif(n.Ex2)
```

```
A0 <- rbinom(n.Ex2, 1, plogis(-1 + L0.a + L0.b))
C0 <- factor(c("uncensored", "censored")[((runif(n.Ex2) + 0.2 *
    A0) > 0.9) + 1])
L1.a <- L0.a + (1 - L0.b) * runif(n.Ex2) + 0.3 * A0
L1.b <- -1 + L0.b + (0.5 + A0) * runif(n.Ex2)
Y1 <- (L0.c + (L0.a * A0 + runif(n.Ex2))/2)/2
A1 <- rbinom(n.Ex2, 1, plogis(-1.5 + L0.a + L0.b + A0))
C1 <- factor(c("uncensored", "censored")[((runif(n.Ex2) + 0.2 *
    A1) > 0.9) + 1])
L2.a <- L1.a + (1 - L1.b) * runif(n.Ex2) + 0.3 * A1
L2.b <- -1 + L1.b + (0.5 + A1) * runif(n.Ex2)
Y2 <- (Y1 + (L1.a/2.3 * A1 + runif(n.Ex2))/2)/2
A2 <- rbinom(n.Ex2, 1, plogis(-1.5 + L1.a + L1.b + A1))
C2 <- factor(c("uncensored", "censored")[((runif(n.Ex2) + 0.2 *
    A2) > 0.9) + 1])
Y3 <- (Y2 + (L2.a/3.5 * A2 + runif(n.Ex2))/2)/2
data.Ex2 <- data.frame(L0.a, L0.b, L0.c, A0, C0, L1.a, L1.b, Y1,
    A1, C1, L2.a, L2.b, Y2, A2, C2, Y3)
```

Remark. The R code presented above is adapted from the example code provided in the original `ltmle` package paper.[78]

Exercise 18.3

Using the dataset from Exercise 18.2, compute an estimate of the causal effect $\mathbb{E}(Y^{\bar{1}} - Y^{\bar{0}})$, along with a 95% confidence interval.

Exercise 18.4

Using the dataset from Exercise 18.2, compute an estimate of the value of the following dynamic treatment regime (DTR):

```
ltmle(..., rule = function(row)
    c(1, ifelse(row["L1.a"] > 0.8, 1, 0),
         ifelse(row["L2.a"] > 1., 1, 0)),...)
```

Modify the thresholds from `(0.8, 1.0)` to a set of thresholds given by `seq(0.5, 0.8, 0.1)` and `seq(0.7, 1.0, 0.1)`. Then, identify the optimal DTR that corresponds to the highest estimated value.

Exercise 18.5

Suppose A is the treatment variable in a multi-arm study. For a four-arm study, represent A using two dummy variables, $A.a$ and $A.b$. For a five-arm study, use three dummy variables: $A.a$, $A.b$, and $A.c$.

19

Intercurrent Events

19.1 Missing Data and Intercurrent Events

Missing data and intercurrent events present major challenges throughout the planning, design, conduct, analysis, and interpretation of clinical trials and real-world studies.

The ICH E9(R1), *Addendum on Estimands and Sensitivity Analysis in Clinical Trials*, provides the following definitions:

> **Intercurrent Events:**
>
> "Events occurring after treatment initiation that affect either the interpretation or the existence of the measurements associated with the clinical question of interest. It is necessary to address intercurrent events when describing the clinical question of interest in order to precisely define the treatment effect to be estimated."
>
> **Missing Data:**
>
> "Data that would be meaningful for the analysis of a given estimand but were not collected. These should be distinguished from data that do not exist or are not considered meaningful due to an intercurrent event."

From these definitions, intercurrent events and missing data are closely intertwined. Consequently, their handling should be integrated into both the definition of the target estimand and the construction of its estimators.

The ICH E9(R1) outlines five key attributes that define an estimand:

1. "The **treatment** condition of interest and, as appropriate, the alternative treatment condition to which comparison will be made."

2. "The **population** of patients targeted by the clinical question."

3. "The **variable** (or endpoint) to be obtained for each patient that is required to address the clinical question."

4. The proposed strategies for **handling intercurrent events**. (Remark: This attribute will be discussed in detail soon.)

DOI: 10.1201/9781003635468-23

5. "A **population-level summary** for the variable should be specified, providing a basis for comparison between treatment conditions."

Regarding the fourth attribute, the ICH E9(R1) addendum describes several strategies for handling intercurrent events:

- *Treatment policy* strategy
- *Hypothetical* strategy
- *Composite variable* strategy
- *While-on-treatment* strategy
- *Principal stratum* strategy

While the ICH E9(R1) guideline discusses these strategies in the context of binary and continuous outcome variables, it does not provide specific guidance for time-to-event outcomes. The literature on handling intercurrent events for binary and continuous outcomes is well developed,[79] whereas guidance for time-to-event outcomes remains limited. Accordingly, the following two sections address the application of these strategies separately for: (1) binary and continuous outcomes, and (2) time-to-event outcomes.

Gold Coin # 66. *Intercurrent events and missing data are closely intertwined. Consequently, their handling should be integrated into both the definition of the target estimand and the construction of its estimators.*

19.2 Binary and Continuous Outcomes

Continue the setting of a longitudinal study that generates the following data:

$$L(0), A(0), C(0), Y(1), L(1), A(1), C(1), \ldots, A(T-1), C(T-1), Y(T),$$

to be aligned with variable names in the `ltmle` package, $L(0)$ denotes the vector of baseline covariates, $A(t-1)$ the treatment variable, $C(t-1)$ the censoring indicator, $Y(t)$ the outcome variable, $L(t)$ the vector of time-dependent covariates, $t = 1, \ldots T-1$, and $Y(T)$ the final outcome variable. Let $\overline{A} = (A(0), \ldots, A(T-1))$ and $\overline{C} = (C(0), \ldots, C(T-1))$.

Assume that $A(0) \in \{0, 1\}$, where $A(0) = 1$ denotes assignment to treatment 1 and $A(0) = 0$ denotes assignment to treatment 0. Let the target causal estimand be defined as

$$\theta^* = \mathbb{E}\big[Y^{\overline{a}=\overline{1},\overline{c}=\overline{0}}(T)\big] - \mathbb{E}\big[Y^{\overline{a}=\overline{0},\overline{c}=\overline{0}}(T)\big],$$

where $Y^{\overline{a}=\overline{1},\overline{c}=\overline{0}}(T)$ and $Y^{\overline{a}=\overline{0},\overline{c}=\overline{0}}(T)$ represent the potential outcomes at time-point T, if observed, under static treatment regimes $\overline{1}$ and $\overline{0}$, respectively.

In the absence of intercurrent events, estimation of θ^* can be implemented using the `ltmle` function, as discussed in Chapter 18.

To handle intercurrent events, define t_0, which is a random variable, as the earliest timepoint at which treatment deviates from the initial assignment:

$$A(t) = A(0), \quad \text{for } t = 0, \ldots, t_0 - 1; \qquad A(t_0) \neq A(0).$$

Here, the interpretation of some selected scenarios for $A(t_0)$ is as follows:

- $A(t_0) = 1 - A(0)$: a treatment switch occurs at time t_0,
- $A(t_0) =$ `NULL`: a treatment discontinuation occurs at time t_0,
- $A(t_0) = 2$: a medical rescue with an alternative treatment (distinct from treatments 0 and 1) occurs at time t_0.

19.2.1 Treatment Policy Strategy

According to the ICH E9(R1) addendum, under the treatment policy strategy, "the occurrence of the intercurrent event is considered irrelevant in defining the treatment effect of interest: the value for the variable of interest is used regardless of whether or not the intercurrent event occurs."

To apply this strategy, revise the treatment variables as:

$$\widetilde{A}(t) = A(0), \quad t = 0, \ldots, T - 1,$$

regardless of whether the intercurrent event occurs. That is, the treatment sequence $(\widetilde{A}(0), \ldots, \widetilde{A}(T-1))$ is fully determined by the initial treatment $A(0)$, and thus, the two are equivalent.

Estimand

Under the treatment policy strategy, the target causal estimand is:

$$\widetilde{\theta}^* = \mathbb{E}\left[Y^{a(0)=1,\overline{c}=\overline{0}}(T)\right] - \mathbb{E}\left[Y^{a(0)=0,\overline{c}=\overline{0}}(T)\right].$$

Estimator

Based on the L nodes, $\widetilde{A}$ nodes, C nodes, and Y nodes, estimation of $\widetilde{\theta}^*$ can be performed using the `ltmle` function.

19.2.2 Hypothetical Strategy

According to ICH E9(R1), under the hypothetical strategy, "a scenario is envisaged in which the intercurrent event would not occur: the value of the variable to reflect the clinical question of interest is the value which the variable would have taken in the hypothetical scenario defined."

To apply this strategy, revise the censoring variables as:

$$\widetilde{C}(t) = \begin{cases} C(t), & \text{for } t = 0, \ldots, t_0 - 1, \\ \texttt{censored}, & \text{for } t = t_0, \ldots, T - 1, \end{cases}$$

where t_0 is the time at which an intercurrent event occurs. Let $\widetilde{C} = (\widetilde{C}(0), \ldots, \widetilde{C}(T-1))$. By convention, if $\widetilde{C}(t) = \texttt{censored}$, then $Y(t') = \texttt{NA}$ for all $t' = t+1, \ldots, T$.

Estimand

Under the hypothetical strategy, the target causal estimand is:

$$\widetilde{\theta}^* = \mathbb{E}\left[Y^{\overline{a}=\overline{1}, \widetilde{c}=\overline{0}}(T)\right] - \mathbb{E}\left[Y^{\overline{a}=\overline{0}, \widetilde{c}=\overline{0}}(T)\right],$$

where $\widetilde{c} = (\widetilde{c}(0), \ldots, \widetilde{c}(T-1))$.

Estimator

Based on the L nodes, A nodes, $\widetilde{C}$ nodes, and Y nodes, estimation of $\widetilde{\theta}^*$ can be implemented using the `ltmle` function.

19.2.3 Composite Variable Strategy

According to ICH E9(R1), under the composite variable strategy, "an intercurrent event is considered in itself to be informative about the patient's outcome and is therefore incorporated into the definition of the variable."

To apply this strategy, revise the outcome variable as:

$$\widetilde{Y}(t) = \begin{cases} Y(t), & \text{for } t = 0, \ldots, t_0 - 1, \\ \xi(Y(t), I\{A(t_0) \neq A(0)\}), & \text{for } t = t_0, \ldots, T - 1, \end{cases}$$

where t_0 is the time at which an intercurrent event occurs, and ξ is a function that combines the outcome with the occurrence of the intercurrent event.

For example, if $Y(t)$ is binary with 1 indicating failure, define $\widetilde{Y}(t) = 1$ if $Y(t) = 1$ or an intercurrent event occurs at t. If $Y(t)$ is continuous and M denotes the worst possible value, define $\widetilde{Y}(t) = M$ if an intercurrent event occurs at t.

Estimand

The target causal estimand under this strategy is:

$$\widetilde{\theta}^* = \mathbb{E}\left[\widetilde{Y}^{\overline{a}=\overline{1}, \overline{c}=\overline{0}}(T)\right] - \mathbb{E}\left[\widetilde{Y}^{\overline{a}=\overline{0}, \overline{c}=\overline{0}}(T)\right].$$

Estimator

Based on the L nodes, A nodes, C nodes, and $\widetilde{Y}$ nodes, estimation of $\widetilde{\theta}^*$ can be implemented using the `ltmle` function.

19.2.4 While-on-Treatment Strategy

According to ICH E9(R1), under the while-on-treatment strategy, "response to treatment prior to the occurrence of the intercurrent event is of interest. [...] If a variable is measured repeatedly, its values up to the time of the intercurrent event may be considered relevant for the clinical question."

To apply this strategy, revise the outcome variables as:

$$\widetilde{Y} = \begin{cases} Y(t_0), & \text{if an intercurrent event occurs at } t_0, \\ Y(T), & \text{if no intercurrent events occur,} \end{cases}$$

Alternately, we can treat the individually estimated slope as a new outcome variable. Let $Y(0)$ denote the outcome measured at baseline. Define

$$\widetilde{Y} = \begin{cases} \frac{Y(t_0)-Y(0)}{t_0}, & \text{if an intercurrent event occurs at } t_0, \\ \frac{Y(T)-Y(0)}{T}, & \text{if no intercurrent event occurs.} \end{cases}$$

Estimand

The target causal estimand is:

$$\widetilde{\theta}^* = \mathbb{E}\left[\widetilde{Y}^{a(0)=1}\right] - \mathbb{E}\left[\widetilde{Y}^{a(0)=0}\right].$$

Estimator

Based on the data $\{(L_i(0), A_i(0), \widetilde{Y}_i) : i = 1, \ldots, n\}$, the parameter $\widetilde{\theta}^*$ can be estimated using the `tmle` function.

19.2.5 Principal Stratum Strategy

According to ICH E9(R1), under the principal stratum strategy: "The target population might be taken to be the principal stratum in which an intercurrent event would occur. Alternatively, the target population might be taken to be the principal stratum in which an intercurrent event would not occur. The clinical question of interest relates to the treatment effect only within the principal stratum."

Estimand

To apply this strategy, one must specify a principal stratum of interest,[80] denoted by $\widetilde{\mathcal{P}}$. The target causal estimand is defined as:

$$\widetilde{\theta}^* = \mathbb{E}_{\widetilde{\mathcal{P}}}\left[Y^{\overline{a}=\overline{1},\overline{c}=\overline{0}}(T)\right] - \mathbb{E}_{\widetilde{\mathcal{P}}}\left[Y^{\overline{a}=\overline{0},\overline{c}=\overline{0}}(T)\right],$$

where the expectation is taken over the principal stratum $\widetilde{\mathcal{P}}$.

Estimator

A full discussion of estimation under the principal stratum strategy is beyond the scope of this chapter. However, a practical approach involves using multiple imputation to impute both potential outcomes and principal stratum membership.[81] For each imputed dataset, the `ltmle` function can be used to estimate $\widetilde{\theta}^*$. Finally, the Rubin's rule can be applied to combine the resulting estimates and produce a final estimate of $\widetilde{\theta}^*$.

> **Gold Coin # 67.** *Five strategies are proposed for handling intercurrent events. Regardless of which strategy, or combination of strategies, is applied in defining the target estimand, the corresponding estimator should be aligned accordingly.*

19.3 Time-to-Event Outcome

Let Y denote the time from a defined starting point (e.g., randomization or treatment initiation) to the occurrence of a specific event (e.g., death, relapse, or hospitalization). For a time-to-event outcome, a major challenge is *censoring*, which arises when the exact event time is unknown (e.g., when a patient drops out or the study ends before the event occurs).

Similar to binary and continuous outcomes, defining a sound estimand for a time-to-event outcome requires specifying five key attributes. The fifth attribute involves choosing a suitable population-level summary.

For time-to-event outcomes, due to historical reasons,[82] the hazard ratio (HR) has traditionally been the most commonly used population-level summary. However, there has been substantial discussion regarding the limitations of HR.[83] As pointed out by Hernán (2010), "endowing a HR with a causal interpretation is risky for two key reasons: the HR may change over time, and the HR has a built-in selection bias."

Therefore, two alternative population-level summaries have been proposed: the *restricted mean survival time* (RMST) and the *survival rate* (SR). When RMST is used, the time-to-event outcome is treated like a continuous outcome. In contrast, when SR is used, the time-to-event outcome is treated as a binary outcome, defined by whether the event occurs within a fixed time window.

19.3.1 Two Population-level Summaries

Restricted Mean Survival Time

Let $Y^{a(0)=1}$ and $Y^{a(0)=0}$ denote the potential time-to-event outcomes had a subject received treatment 1 or treatment 0, respectively. A target estimand is the difference in RMST:[84,85]

$$\theta^*_{\text{RMST}} = \mathbb{E}(Y^{a(0)=1} \wedge \tau) - \mathbb{E}(Y^{a(0)=0} \wedge \tau),$$

where τ is a pre-specified cutoff time and the symbol $\wedge$ denotes the minimum of two values.

An alternative expression for the target estimand θ^*_{RMST} can be derived using the survival function. Let $S^{a(0)}(t) = \mathbb{P}(Y^{a(0)} > t)$ denote the survival function of the potential outcome $Y^{a(0)}$. Then,

$$\mathbb{E}(Y^{a(0)} \wedge \tau) = \int_0^\tau S^{a(0)}(t)\, dt,$$

which corresponds to the area under the survival curve over the interval $[0, \tau]$. Therefore, the estimand θ^*_{RMST} can be written as:

$$\theta^*_{\text{RMST}} = \int_0^\tau S^{a(0)=1}(t)\, dt - \int_0^\tau S^{a(0)=0}(t)\, dt.$$

To estimate θ^*_{RMST}, a targeted learning framework based on pseudo-observations has been proposed.[86,87] A detailed discussion of this approach is omitted here, as the focus is instead on the second population-level summary: the survival rate.

Survival Rate

If survival rate is considered as the population-level summary in the estimand definition, the estimand is

$$\theta^*_{\text{SR}} = \mathbb{E}\left[I(Y^{a(0)=1} > \tau)\right] - \mathbb{E}\left[I(Y^{a(0)=0} > \tau)\right],$$

where τ is a pre-specified cutoff time.

To estimate the estimand θ^*_{SR}, the discretization is performed. Theoretically, the time-to-event outcome variable Y is a continuous variable that is subject to being censored. Practically, it can be considered as a discrete variable in units of days, weeks, or months, assuming that the clinical study is monitoring the subjects at daily, weekly, or monthly basis. Let $t = 0, 1, 2, \ldots, T$ index the time in a given unit, where $t = 0$ indicates the time zero (e.g., the time when the treatment is initiated) and $t = T$ indicates the final follow-up time (e.g., time τ). The discretization method has been used widely in the literature of survival analysis.[88,89,54]

Let W_i be the vector of baseline characteristics for subject i measured at $t = 0$. Let A_i be the treatment (say, 1 for the investigational treatment and 0

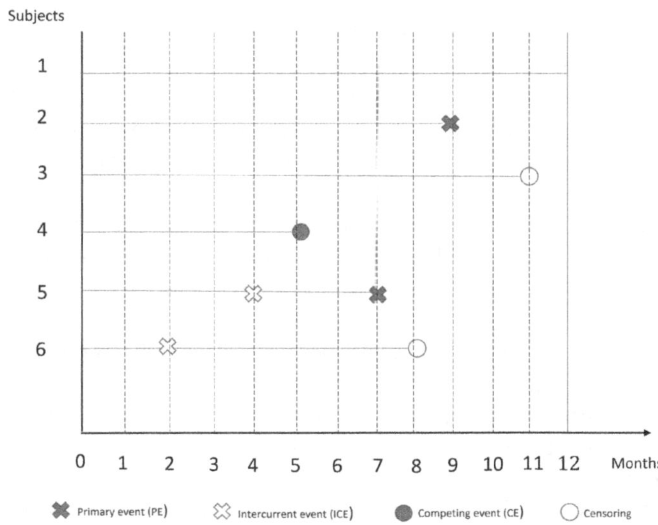

FIGURE 19.1
Discretization of the time-to-event outcome variable

for the control treatment) assigned to subject i at $t = 0$. Let $C_i(t-1)$ be the censoring status at the follow-up t, $t = 1, \ldots, T$; in particular, $C_i(T-1)$ is the censoring status for the final follow-up time T. Let $Y_i(t)$ be the outcome variable at t, where $Y_i(t) = 0$ means the primary event hasn't occurred at t and $Y_i(t) = 1$ means the primary event has occurred at t, $t = 1, \ldots, T$. One convention is that if $Y_i(t) = 1$ then $Y_i(t') = 1$ for $t' \geq t$. The other convention is that if $C_i(t-1) = \texttt{censored}$ then $Y_i(t') = \texttt{NA}$ for $t' \geq t$.

The above data structure is for a time-independent treatment variable. It can be generalized to the data structure for a time-dependent treatment variable, defined as $A_i(t)$, $t = 0, \ldots, T-1$, to handle settings where there are intercurrent events such as treatment switches, treatment discontinuations, rescue medications, and so on. For the time-dependent treatment, let $L_i(0) = W_i$ and $L_i(t)$, $t = 1, \ldots, T-1$ be the time-dependent covariates. Therefore, the time order of these variables is

$$L(0), A(0), C(0), Y(1), L(1), A(1), C(1), \ldots, A(T-1), C(T-1), Y(T).$$

Therefore, the estimand θ^*_{SR} is equal to

$$\begin{aligned}\theta^*_{\text{SR}} &= \mathbb{P}\left[Y^{a(0)=1}(T) = 0\right] - \mathbb{P}\left[Y^{a(0)=0}(T) = 0\right] \\ &= \mathbb{E}\left[Y^{a(0)=0}(T)\right] - \mathbb{E}\left[Y^{a(0)=1}(T)\right].\end{aligned}$$

Figure 19.1 displays six hypothetical subjects, under the assumption that the primary event is death from cancer.

The first subject is alive at the end of the study (assumed to be 12 months). As no censoring occurs, the censoring indicators are set to $C_1(0) = \cdots = C_1(11) =$ `uncensored`. Since the subject remains alive throughout the study period, the event indicators are given by $Y_1(1) = \cdots = Y_1(12) = 0$.

The second subject dies before the study concludes. As no censoring occurs, $C_2(0) = \cdots = C_2(11) =$ `uncensored`. Death is observed at month 9, so $Y_2(1) = \cdots = Y_2(8) = 0$, and $Y_2(9) = \cdots = Y_2(12) = 1$.

The third subject is censored prior to the end of the study (e.g., due to loss to follow-up for lack of efficacy). Censoring occurs at month 11, leading to $C_3(0) = \cdots = C_3(9) =$ `uncensored` and $C_3(10) = C_3(11) =$ `censored`. As the subject is known to be alive prior to month 10, the event indicators are given by $Y_3(1) = \cdots = Y_3(10) = 0$, while post-censoring outcomes are unknown: $Y_3(11) = Y_3(12) =$ `NA`.

The fourth subject experiences a competing event (e.g., death due to other causes), which precludes the occurrence of the primary event. This competing event is treated as a terminal intercurrent event. Strategies for handling such events will be discussed in the next subsection. The recorded values of $C_4(t-1)$ and $Y_4(t)$, for $t = 1, \ldots, 12$, depend on the specific strategy employed.

The fifth subject experiences the primary event before the end of the study, but a non-terminal intercurrent event (e.g., treatment discontinuation) occurs prior to the primary event.

The sixth subject is censored before the study ends, and a non-terminal intercurrent event (e.g., initiation of rescue medication) occurs prior to censoring. The fifth and sixth subjects are analogous to the second and third subjects, respectively, except that a non-terminal intercurrent event occurs.

See Exercise 19.1 for details on defining the $C(t)$ and $Y(t)$ variables when intercurrent events are ignored for the fifth and sixth subjects. Consequently, a dataset comprising all six hypothetical subjects can be constructed (see Exercise 19.2).

19.3.2 Handling Intercurrent Events

This subsection discusses five strategies for handling intercurrent events (ICEs) in studies with time-to-event outcomes.

According to ICH E9(R1), under the *while-on-treatment* strategy, "response to treatment prior to the occurrence of the intercurrent event is of interest." However, in the context of time-to-event outcomes, this strategy poses challenges. Specifically, the outcome of interest cannot be observed at the time of a non-terminable intercurrent event, and a terminable intercurrent event precludes the occurrence of the primary event. As a result, the while-on-treatment strategy is discussed in ICH E9(R1) only in relation to quantitative and categorical outcomes, but not for time-to-event outcomes.

For completeness, this subsection introduces a new strategy—the *competing risk strategy*—either as an additional strategy or as a replacement for the while-on-treatment strategy in the context of time-to-event data.[90]

Treatment Policy Strategy

The treatment policy strategy can be applied to handle non-terminal ICEs. Suppose that the two initial treatments under comparison are coded as 0 and 1, $A(t) =$ `NULL` indicates that the subject has discontinued the initially assigned treatment at time t, and $A(t) = 2$ denotes the initiation of rescue medication at that time.

As an illustrative example, consider the 5th subject in Figure 19.1. The treatment history is given by $A_5(0) = A_5(1) = A_5(2) = A_5(3)$, followed by $A_5(4) = A_5(5) = A_5(6) =$ `NULL` or 2, depending on whether the ICE occurring at month 4 corresponds to treatment discontinuation or rescue medication.

Under the treatment policy strategy, the occurrence of the ICE is considered irrelevant to the definition of the estimand. Accordingly, the treatment nodes (`Anodes`) are updated so that $A_5(0) = \cdots = A_5(6)$, effectively reflecting continuous treatment throughout. Since no censoring occurs over the 12-month follow-up period, $C_5(0) = \cdots = C_5(11) =$ `uncensored`. The outcome variable is defined as $Y_5(1) = \cdots = Y_5(6) = 0$ and $Y_5(7) = \cdots = Y_5(12) = 1$. See Exercise 19.3 for further details.

Estimation under this strategy can be performed using the `ltmle` function in R, with the argument `survivalOutcome = TRUE`, using the appropriately updated `Anodes`.

Hypothetical Strategy

The hypothetical strategy can be applied to handle non-terminal ICEs, such as the administration of rescue medication. Under this strategy, a hypothetical scenario is envisaged in which the ICE under consideration does not occur.

For example, for the 6th subject in Figure 19.1, imagine a hypothetical setting where rescue medication is unavailable, thus preventing the occurrence of ICE. Operationally, this strategy corresponds to "censoring" the subject's data at the time the ICE occurs. For this subject, who is considered "censored" at month 2 by the hypothetical strategy, the censoring indicators are then updated as $C_6(0) =$ `uncensored` and $C_6(1) = \cdots = C_6(11) =$ `censored`. The outcome values are recorded as $Y_6(1) = 0$ and $Y_6(2) = \cdots = Y_6(12) =$ `NA`. See Exercise 19.4 for further details.

Estimation under this strategy can be performed using the `ltmle` function in R, with the argument `survivalOutcome = TRUE`, and by appropriately updating the `Cnodes` to reflect the censoring mechanism.

Composite Variable Strategy

The composite variable strategy can be applied to handle both terminal and non-terminal ICEs. Under this approach, the outcome definition is modified to incorporate a composite event, which may include either the primary event (PE) or the ICE under consideration, whichever occurs first. To implement this strategy, only the outcome nodes (`Ynodes`) need to be updated.

As an illustrative example, consider the 4th subject in Figure 19.1. Under the composite variable strategy, the composite event is observed at month 5. Accordingly, the outcome indicators are defined as $Y_4(1) = \cdots = Y_4(4) = 0, \quad Y_4(5) = \cdots = Y_4(12) = 1$, with no censoring assumed over the follow-up period, i.e., $C_4(0) = \cdots = C_4(11) = \texttt{uncensored}$. See Exercise 19.5 for further details.

Estimation under this strategy can be performed using the `ltmle` function in R, with the argument `survivalOutcome = TRUE`, and by updating the `Ynodes` to reflect the newly defined composite outcome.

Competing Risk Strategy

To handle an ICE that is considered a competing event (CE), the competing risk strategy may be applied. In the competing risks literature,[91] a common population-level summary is the cumulative incidence function (CIF), which quantifies the probability of experiencing a particular type of event in the presence of competing risks.

Before applying this strategy, the time to PE, denoted by Y_{PE}, is discretized as $Y_{\text{PE}}(t)$, for $t = 1, \ldots, T$, while the time to CE, denoted by Y_{CE}, is discretized as $Y_{\text{CE}}(t)$, for $t = 1, \ldots, T$. Under the competing risk strategy, a two-dimensional outcome variable is defined as

$$Y(t) = \big(Y_{\text{PE}}(t),\ Y_{\text{CE}}(t)\big),$$

with the convention that both components cannot simultaneously equal 1. Specifically, if $Y_{\text{CE}}(t) = 1$, then $Y_{\text{CE}}(t') = 1$ and $Y_{\text{PE}}(t') = 0$ for all $t' \geq t$; conversely, if $Y_{\text{PE}}(t) = 1$, then $Y_{\text{PE}}(t') = 1$ and $Y_{\text{CE}}(t') = 0$ for all $t' \geq t$.

Let $Y_{\text{PE}}^{\overline{a}}(T)$ denote the potential outcome at time T under treatment regime $\overline{a}$. The estimand in terms of the difference in CIF is then defined as

$$\theta^* = \mathbb{E}\left[Y_{\text{PE}}^{\overline{a}=\overline{1}}(T)\right] - \mathbb{E}\left[Y_{\text{PE}}^{\overline{a}=\overline{0}}(T)\right],$$

which is the difference in CIFs at time T, assuming all individuals in the population follow treatment regimes $\overline{1}$ and $\overline{0}$, respectively.

Estimation under this strategy can be performed using the `survtmle` package (https://github.com/benkeser/survtmle).

Principal Stratum Strategy

The principal stratum strategy is applicable to settings where the PE is non-terminal (e.g., hospital readmission) and the ICE is terminal (e.g., death), or to settings where the PE is terminal (e.g., death) and the ICE is non-terminal (e.g., infection). In the following, the former case is considered.

Let $Y_{\text{ICE}}(t)$ denote the indicator of the occurrence of the terminal ICE at time t, for $1 \leq t \leq T$. According to ICH E9(R1), "it is important to distinguish 'principal stratification', which is based on potential intercurrent events [...] from subsetting based on actual intercurrent events."

Let $Y^{\overline{1}}_{\text{ICE}}(T)$ and $Y^{\overline{0}}_{\text{ICE}}(T)$ represent the potential outcomes for the occurrence of the terminal ICE at time T under treatment regimes $\overline{1}$ and $\overline{0}$, respectively. Consider the following principal stratum:

$$\widetilde{\mathcal{P}}_{\text{AA}} = \left\{ i : Y^{\overline{1}}_{\text{ICE},i}(T) = 0,\ Y^{\overline{0}}_{\text{ICE},i}(T) = 0 \right\},$$

which includes individuals who would be alive at time T under both treatment regimes. The label "AA" refers to being alive under both potential scenarios. Note that three other principal strata also exist—namely, "AD", "DA", and "DD"—corresponding to other combinations of potential outcomes.

The estimand of in terms of survival rate difference is then defined as

$$\theta^* = \mathbb{E}_{\widetilde{\mathcal{P}}_{\text{AA}}} \left[1 - Y^{\overline{1}}_{\text{PE}}(T) \right] - \mathbb{E}_{\widetilde{\mathcal{P}}_{\text{AA}}} \left[1 - Y^{\overline{0}}_{\text{PE}}(T) \right],$$

where the expectation is taken over the principal stratum $\widetilde{\mathcal{P}}_{\text{AA}}$, rather than over the full population. In the survival analysis literature, this estimand is referred to as the *survivor average causal effect* (SACE).[39,92]

19.4 Exercises

Exercise 19.1

Define the $C(t)$ and $Y(t)$ variables under the strategy that intercurrent events are ignored for the fifth and sixth subjects depicted in Figure 19.1.

Exercise 19.2

Construct a dataset of the $C(t)$ and $Y(t)$ variables for the six subjects shown in Figure 19.1.

Exercise 19.3

Assume that $A_5(0) = 1$ and $A_6(0) = 1$. Also assume that the intercurrent event for the fifth subject is treatment discontinuation (denoted by `NULL`), and for the sixth subject, it is rescue medication (denoted by 2). Define the values of $\{\widetilde{A}(t), C(t), Y(t+1)\}_{t=0}^{T-1}$ under the treatment policy strategy for the fifth and sixth subjects depicted in Figure 19.1.

Exercise 19.4

Continue Exercise 19.3. Define the values of $\{A(t), \widetilde{C}(t), Y(t+1)\}_{t=0}^{T-1}$ under the hypothetical strategy (i.e., imagine a scenario in which the intercurrent events would not occur) for the fifth and sixth subjects depicted in Figure 19.1.

Exercise 19.5

Continue Exercise 19.3. Define the values of $\{A(t), C(t), \widetilde{Y}(t+1)\}_{t=0}^{T-1}$ under the composite variable strategy for the fifth and sixth subjects depicted in Figure 19.1.

20

Fusion of Two Cultures

20.1 Two Cultures of Statistical Modeling

In a famous paper,[93] Leo Breiman (1928–2005)—the inventor of Classification and Regression Trees (CART), Bagging, and Random Forests—states:

> "There are two cultures in the use of statistical modeling to reach conclusions from data. One assumes that the data are generated by a given stochastic data model. The other uses algorithmic models and treats the data mechanism as unknown. The statistical community has been committed to the almost exclusive use of data models. This commitment has led to irrelevant theory, questionable conclusions, and has kept statisticians from working on a large range of interesting current problems. Algorithmic modeling, both in theory and practice, has developed rapidly in fields outside statistics. It can be used both on large complex data sets and as a more accurate and informative alternative to data modeling on smaller data sets. If our goal as a field is to use data to solve problems, then we need to move away from exclusive dependence on data models and adopt a more diverse set of tools." —Leo Breiman (2001).

This statement was made in 2001. Back then, classical yet simple data models (such as linear model, logistic model, generalized linear model, and Cox proportional hazards model) were commonly introduced in traditional statistical textbooks, whereas modern and flexible algorithmic models (such as CART, random forests, boosting, support vector machines, neural networks, and deep learning) were introduced primarily in the fields of data mining, machine learning, and statistical learning.[56] Therefore, at that time, the two cultures were largely separated.

After nearly a quarter century, Artificial Intelligence and Machine Learning (AI/ML) have become commonly used terms. As a result, the distinction between data models and algorithmic models is no longer emphasized.

In pharmaceutical statistics, the primary objective is to construct efficient estimators for any well-defined target estimand. Within this estimand-oriented framework, all models—whether data models or algorithmic models—are regarded as tools for conducting statistical and causal inferences. Therefore, the

DOI: 10.1201/9781003635468-24

outdated divide between the two modeling paradigms should be set aside. Instead, to ensure the appropriate use of these tools, we need to pay attention to the statistical assumptions they entail and devote our efforts to identifying and, when possible, eliminating assumptions that are considered redundant. This focus forms the foundation for this book.

Gold Coin # 68. *"There are two cultures in the use of statistical modeling to reach conclusions from data. One assumes that the data are generated by a given stochastic data model. The other uses algorithmic models and treats the data mechanism as unknown."— Brieman (2001)*

20.2 Fusion of Two Cultures

In the iconic opening line of *A Tale of Two Cities*—"It was the best of times, it was the worst of times, it was the age of wisdom, it was the age of foolishness"—Charles Dickens captured the stark dualities of society during the French Revolution.

Similarly, in the today's society of statistics, we have witnessed another revolution. Years after Leo Breiman's influential paper on the two cultures of statistical modeling, we are now entering an era that embraces their fusion. This shift can also be seen as revolutionary, a revolution of causal inference.[21]

20.2.1 Target Estimand

We begin by following the data modeling culture to define the *target estimand.* Suppose we observe data $\{(X_i, A_i, Y_i)\}_{i=1}^n$, where the observations are assumed to be independent and identically distributed (i.i.d.) according to the unknown joint distribution of $(X, A, Y) \sim f_0(x, a, y)$. Here, X denotes a vector of covariates, A the binary treatment variable ($A = 1$ for treatment, $A = 0$ for control), and Y the outcome variable.

In Part II of this book, we explore the precise definition of a target estimand. Consider settings where the target estimand of interest is the *average treatment effect* (ATE), defined as

$$\theta = \mathbb{E}\left[Q(X, 1) - Q(X, 0)\right],$$

where

$$Q(X, A) = \mathbb{E}(Y \mid X, A)$$

is the conditional expectation of the outcome given covariates and treatment assignment, i.e., the outcome regression function.

20.2.2 Machine Learning

To estimate the outcome regression function $Q(X, A)$—and, when necessary, related nuisance functions such as the propensity score or non-missingness probability—we turn to the algorithmic modeling culture. We leverage flexible, state-of-the-art machine learning algorithms, including ensemble methods like Super Learner, to obtain an initial estimator $\widehat{Q}_n^0(X, A)$.

Using this estimator, we construct a plug-in estimator of the ATE:

$$\widehat{\theta}_{\mathrm{MLE}} = \frac{1}{n}\sum_{i=1}^{n}\left[\widehat{Q}_n^0(X_i, 1) - \widehat{Q}_n^0(X_i, 0)\right].$$

20.2.3 Targeted Learning

To optimally combine the strengths of both modeling cultures, we employ *targeted learning*, a methodology that refines the initial machine learning-based estimator to better target the estimand of interest.

Specifically, we update the initial estimator $\widehat{Q}_n(X, A)$ to obtain a targeted estimator $\widehat{Q}_n^*(X, A)$. This leads to the targeted maximum likelihood estimator (TMLE)[58] of the ATE:

$$\widehat{\theta}_{\mathrm{TMLE}} = \frac{1}{n}\sum_{i=1}^{n}\left[\widehat{Q}_n^*(X_i, 1) - \widehat{Q}_n^*(X_i, 0)\right].$$

As the name suggests, *targeted learning* (TL) represents the synthesis of defining a clear *target estimand* and estimating it using flexible *machine learning* methods. This fusion is underpinned by semiparametric theory in causal inference, a field significantly advanced by James Robins and collaborators.[30]

Gold Coin # 69. *We are embracing the fusion of two cultures,*

$$\textit{target estimand} + \textit{machine learning} \Rightarrow \textit{targted learning}.$$

20.3 Data Fusion

While we are fostering the fusion of the two cultures of statistical modeling, the release of the Framework for FDA's Real-World Evidence (RWE) Program in 2018 has driven a growing integration of two data types: clinical trial data and real-world data (RWD).

Two study designs that integrate clinical trial data with RWD have recently gained attention:[94] single-arm trials with external RWD controls, and randomized controlled trials (RCTs) augmented with RWD.

Gold Coin # 70. *We are embracing the fusion of two types of data: clinical trials data and real-world data.*

20.3.1 Single-Arm Trial with an External RWD Control

The framework for FDA's RWE program briefly discusses single-arm trials with external RWD controls and highlights the associated challenges:

> "Typically, the external control arm uses data from past traditional clinical trials, but in some cases, RWD have been used as the basis for external controls. Using external controls has limitations, including difficulties in reliably selecting a comparable population because of potential changes in medical practice, lack of standardized diagnostic criteria or equivalent outcome measures, and variability in follow-up procedures. Collection of RWD on patients currently receiving other treatments, together with statistical methods, such as propensity scoring, could improve the quality of the external control data that are used when randomization may not be feasible or ethical, provided there is adequate detail to capture relevant covariates."

A critical issue in single-arm trials with external RWD controls is the definition of the target estimand. Central to this process is the explicit specification of the target population.

The AT Club

Two pairs of populations may be of interest:

1. The first pair consists of:

   (1) the population underlying the single-arm trial,

   (2) the population underlying the external RWD controls.

2. The second pair consists of:

   (1) the combined population of the above two,

   (2) the overlap population between the above two.

The estimand defined on Population 1(1)—the population underlying the single-arm trial—is referred to as the *average treatment effect on the treated*

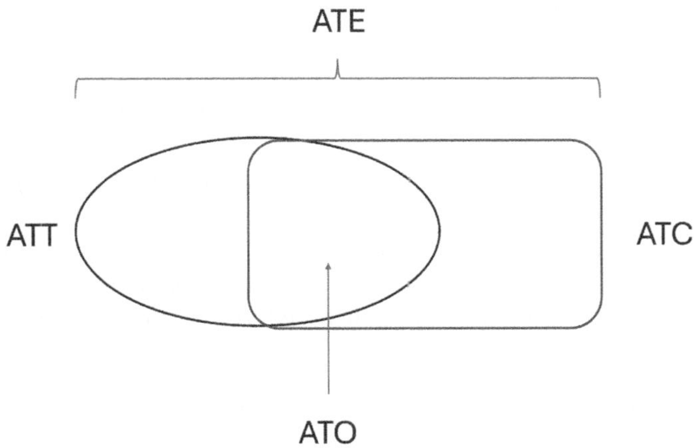

FIGURE 20.1
The AT club

(ATT). The estimand defined on Population 1(2)—the population underlying the external RWD controls—is referred to as the *average treatment effect on the controls* (ATC). The estimand defined on Population 2(1)—the combined population of the two populations—is referred to as the *average treatment effect* (ATE). The estimand defined on Population 2(2)—the overlap population between the two populations—is referred to as the *average treatment effect for the overlap population* (ATO).

We refer to this class of estimands—comprising ATT, ATC, ATE, and ATO—as the AT club, as illustrated in Figure 20.1.

These estimands in the AT club can be defined in terms of a class of weights, the so-called *balancing weights*.[95]

Consider a sample of n patients, each assigned to one of two treatment groups, for which covariate-balanced comparisons are of interest. Let $A_i = a$ denote a binary variable indicating the group membership of subject i, where $a = 1$ or $a = 0$. For each patient, an outcome Y_i and a set of cvoariates $X_i = (X_i^{(1)}, \ldots, X_i^{(d)})^\top$ are observed. The propensity score[38] is the probability of assignment to the treatment group given the covariates,

$$g(1 \mid x) = \mathbb{P}(A_i = 1 \mid X_i = x).$$

Assume that the marginal density of the covariates X exists, denoted as $p(x)$, with respect to a base measure μ, which is a product of counting measures for categorical covariates and Lebesgue measure for continuous variables. For any pre-specified function $h(x)$, the density of the form

$$p_h^*(x) = \frac{p(x)h(x)}{\int p(x)h(x)\mu(dx)}$$

defines the target population. With respect to this target population, the corresponding target estimand is defined as

$$\begin{aligned}\theta_h &= \int [\mathbb{E}(Y \mid X = x, A = 1) - \mathbb{E}(Y \mid X = x, A = 0)] p_h^*(x) \mu(dx) \\ &= \int [Q(x, 1) - Q(x, 0)] p(x) h(x) \mu(dx) \Big/ \int p(x) h(x) \mu(dx).\end{aligned}$$

Here, we omit the discussion of the identifiability assumptions—consistency, exchangeability, and positivity—in order to focus on estimation.

All estimands in the AT club can be expressed in terms of a weighting function $h(x)$. Specifically:

- When $h(x) \equiv 1$, the estimand θ_h corresponds to the ATE.
- When $h(x) = g(1 \mid x)$, the estimand θ_h corresponds to the ATT.
- When $h(x) = g(0 \mid x)$, the estimand θ_h corresponds to the ATC.
- When $h(x) = g(1 \mid x)\, g(0 \mid x)$, the estimand θ_h corresponds to the ATO.

Other estimands can also be generated by appropriately choosing the function $h(x)$.[95]

The ESTIMA Club

For a given function $h(x)$, the target estimand θ_h is well-defined. This brings us to the ESTIMA club, as illustrated in Figure 20.2.

To estimate the estimand, we attempt to construct an estimator that is aligned with it. The process of constructing an estimator aligned with the estimand is referred to as the estimand framework, while the act of applying the estimator to infer the estimand is termed the estimation process. Once data are input into the estimator, the resulting value (i.e., the realization of the estimator) is called the estimate.

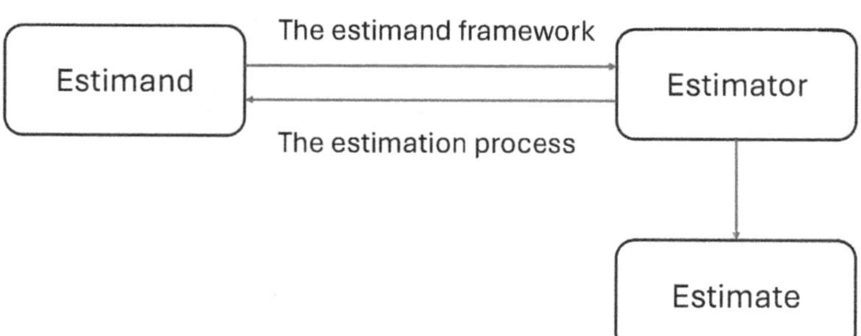

FIGURE 20.2
The ESTIMA club

In order to construct an efficient estimator for the target estimand θ_h, and following the approach proposed in Chapter 13, we can show that (see Exercise 20.1) the efficient influence function (EIF) is given by:

$$\phi_h(O) = \frac{h(x)}{\mathbb{E}[h(X)]}\left\{[Q(X,1) - Q(X,0) - \theta_h] + \frac{(2A-1)}{g(A \mid X)}[Y - Q(X,A)]\right\}.$$

Using the TL approach introduced in Chapter 17, we can outline the TMLE procedure for estimating θ_h based on $\phi_h(O)$ (see Exercise 20.2).

Although there exists a one-to-one correspondence between the class of θ_h estimands and those in the AT club, it is important to note a subtle but meaningful distinction (see Exercise 20.3).

20.3.2 RCT Augmented with RWD

We now consider another type of data fusion: RCTs augmented with RWD, motivated by the potential to improve the efficiency and precision of treatment effect estimation.

Some existing methods rely on the assumption of *mean exchangeability* across studies,[96] which posits that participation in an RCT does not affect patients' mean potential outcomes, conditional on observed baseline covariates. However, this assumption may be violated in practice for several reasons. For instance, participation in an RCT may enhance patients' adherence to their assigned treatments; the time frames of the RCT and the external RWD may differ; or outcomes may be measured inconsistently across studies.

Once again, a critical issue in analyzing such a hybrid trial is the definition of a well-specified target estimand. Central to this process is the explicit specification of the target population.

Estimands

Let $S_i = s$ denote a binary variable indicating the study membership of subject i, where $s = 1$ corresponds to the RCT and $s = 0$ to the external RWD. Let X_i denote the baseline covariates, and let $A_i = a$ be a binary treatment indicator, where $a = 1$ or $a = 0$. Let Y_i denote the outcome variable.

We may be interested in the following two estimands.[97] The first estimand is the covariate-pooled ATE:

$$\theta_1^* = \mathbb{E}_X\left[\mathbb{E}(Y^{a=1} - Y^{a=0} \mid S = 1, X)\right],$$

where the outer expectation is taken over the distribution of X in the pooled population.

The second estimand is the RCT-only ATE:

$$\theta_2^* = \mathbb{E}_{X|S=1}\left[\mathbb{E}(Y^{a=1} - Y^{a=0} \mid S = 1, X)\right],$$

where the outer expectation is taken over the covariate distribution of the RCT population.

Estimators

Following the TL roadmap, van der Laan *et al.* (2025) introduced an adaptive targeted maximum likelihood estimation (A-TMLE) framework to estimate the two estimands defined above.[97]

A full description of the A-TMLE procedure is omitted here for brevity. Readers are referred to their paper for a complete discussion. See Exercise 20.4 and Exercise 20.5 for some details on the identification step.

20.4 Final Remarks

The philosopher Hegel's dialectical method teaches us that every idea, when examined deeply, inevitably reveals internal tensions or contradictions. These tensions or contradictions are the forces that drive the growth and evolution of the idea to something more comprehensive.

In physics, the discovery of wave–particle duality fundamentally transformed our understanding of light. Today, physicists continue the attempts to unify the theories of general relativity and quantum mechanics—two pillars that describe the universe at vastly different scales.

In statistics, similar dialectical tensions arise. Consider the long-standing debate between Bayesian and frequentist approaches to inference (though not addressed in this book), the trade-off between bias and variance, the tension between exchangeability and positivity, or the evolving fusion of the two cultures of statistical modeling. Far from being obstacles, these tensions drive deeper insight and foster innovation.

This book distills seventy such pivotal ideas and concepts—symbolized as gold coins, each with two sides—essential for mastering asymptotic statistics, causal inference, semiparametric statistics, and targeted learning.

Admittedly, the book is lengthy because it carefully lays out every detail, including painstaking intermediate steps. Graduate students are encouraged to engage thoroughly with these steps, while more experienced readers may choose to focus selectively on the core ideas relevant to their work.

As we turn our attention to the rapidly advancing field of artificial intelligence (AI), it is worth noting that our founding fathers of modern statistics were, in a sense, early AI thinkers. Fisher's maximum likelihood estimation (MLE) and Neyman–Pearson's hypothesis testing procedure both resemble AI algorithms: systematic, step-by-step methods that generate estimates and inferences automatically.

Similarly, Fisher's permutation test functions like an AI system that produces p-values through automated reshuffling, while Efron's bootstrap method acts like an AI robot generating confidence intervals by repeatedly resampling data—the name "bootstrap" referring to the idea of "pulling itself up by its own bootstraps."

Building on these monumental foundations, Targeted Learning can be seen as a form of causal AI, equipped with an explicit roadmap. By following this roadmap, the estimation process becomes extraordinarily powerful when dealing with seemingly complex data-generating mechanisms.

In essence, the dialectic of tension and resolution—whether in philosophy, physics, or statistics—remains the driving force of progress, guiding us toward deeper understanding and more effective methodologies.

20.5 Exercises

Exercise 20.1

For a given $h(x)$, derive the EIF $\phi_h(O)$ for θ_h.

Exercise 20.2

For a given $h(x)$, based on the EIF $\phi_h(O)$ derived in Exercise 20.1, outline the TMLE procedure for estimating θ_h.

Exercise 20.3

Derive the EIF for estimating θ_h when $h(x) = \eta(g(x))$, where η is a known function and $g(x) = g_0(1 \mid x)$ is the unknown propensity score function.

Exercise 20.4

State the identifiability assumptions necessary to identify θ_1^* and θ_2^*.

Exercise 20.5

Under the identifiability assumptions in Exercise 20.4, identify θ_1^* and θ_2^*.

Appendix

Hints for Exercises in Chapter 1

Hints for Exercise 1.1:

Let

$$I = \int_{-\infty}^{\infty} e^{-\frac{1}{2}z^2}\, dz.$$

Then

$$\begin{aligned} I^2 &= \left(\int_{-\infty}^{\infty} e^{-\frac{1}{2}x^2}\, dx\right)\left(\int_{-\infty}^{\infty} e^{-\frac{1}{2}y^2}\, dz\right) \\ &= \int_{-\infty}^{\infty}\int_{-\infty}^{\infty} e^{-\frac{1}{2}(x^2+y^2)}\, dx\, dy. \end{aligned}$$

Then switch from Cartesian coordinates (x, y) to polar coordinates (r, θ) by the transformation $x = r\cos\theta$ and $y = r\sin\theta$.

Hints for Exercise 1.2:

Write the variance as $\mathbb{V}(Y) = \mathbb{E}(Y - \mathbb{E}Y)^2$ and expand it.

Hints for Exercise 1.3:

Note that

$$\int_{\infty}^{-\infty} ze^{-\frac{1}{2}z^2}\, dz = 0,$$

because the integrand is an odd function. Also note that

$$\frac{1}{\sqrt{2\pi}}\int_{-\infty}^{\infty} z^2 e^{-\frac{1}{2}z^2}\, dz = -\frac{1}{\sqrt{2\pi}}\int_{-\infty}^{\infty} z\, d\left(e^{-\frac{1}{2}z^2}\right) = \frac{1}{\sqrt{2\pi}}\int_{-\infty}^{\infty} e^{-\frac{1}{2}z^2}\, dz = 1,$$

where the second equality follows from integration by parts, and the final equality uses the result of Exercise 1.1.

Hints for Exercise 1.4:

Taking the derivative of the Lagrangian equation is equivalent to differentiating the following expression, which represents the part of the integrand of $\mathcal{L}$ that depends on $p(x)$,

$$-p(x)\log p(x) + \lambda_1 p(x) + \lambda_2 x p(x) + \lambda_3 (x-\mu)^2 p(x).$$

Differentiating this function with respect to $p(x)$ yields:

$$-\log p(x) - 1 + \lambda_1 + \lambda_2 x + \lambda_3 (x-\mu)^2 = 0.$$

The solution of this equation is as follows:

$$p(x) = \exp\left\{-1 + \lambda_1 + \lambda_2 x + \lambda_3 (x-\mu)^2\right\}.$$

The parameters λ_1, λ_2 and λ_3 are then determined by imposing the following constraints: $\int p(x)\,dx = 1$, $\int xp(x)\,dx = \mu$, and $\int (x-\mu)^2 p(x)\,dx = \sigma^2$.

Remark: These hints are motivated by a proof described in the Wikipedia entry on "Differential entropy."

Hints for Exercise 1.5:

The logarithm of its density is

$$\log\left[\frac{\alpha}{2\beta\Gamma(1/\alpha)}\right] - \left|\frac{x-\mu}{\beta}\right|^\alpha.$$

The entropy is then computed as

$$\begin{aligned}
H &= -\int_{-\infty}^{\infty} p(x)\log p(x)\,dx \\
&= -\log\left[\frac{\alpha}{2\beta\Gamma(1/\alpha)}\right] + \int_{-\infty}^{\infty}\left|\frac{x-\mu}{\beta}\right|^\alpha p(x)\,dx,
\end{aligned}$$

where the second term simplifies as follows:

$$\begin{aligned}
&\frac{\alpha}{2\beta\Gamma(1/\alpha)}\int_{-\infty}^{\infty}\left|\frac{x-\mu}{\beta}\right|^\alpha \exp\left\{-\left|\frac{x-\mu}{\beta}\right|^\alpha\right\}dx \\
=&\frac{\alpha}{2\Gamma(1/\alpha)}\int_{-\infty}^{\infty}|x|^\alpha e^{-|x|^\alpha}\,dx = \frac{\alpha}{\Gamma(1/\alpha)}\int_0^\infty x^\alpha e^{-x^\alpha}\,dx \\
=&-\frac{1}{\Gamma(1/\alpha)}\int_0^\infty x\,de^{-x^\alpha} = \frac{1}{\Gamma(1/\alpha)}\int_0^\infty e^{-x^\alpha}\,dx \\
=&\frac{1}{\alpha}\frac{\alpha}{\Gamma(1/\alpha)}\int_0^\infty e^{-x^\alpha}\,dx = \frac{1}{\alpha}.
\end{aligned}$$

Hints for Exercises in Chapter 2

Hints for Exercise 2.1:

Markov's inequality states that for random variable $Y > 0$ and constant $a > 0$, $\mathbb{P}(Y > a) \le \mathbb{E}Y/a$. Apply Markov's inequality to bound

$$\mathbb{P}(|X - \mu| \ge \epsilon) = \mathbb{P}((X - \mu)^2 \ge \epsilon^2).$$

Hints for Exercise 2.2:

If $S_{2n} = 2k$, then there are $n + k$ trials showing the head (1) and $n - k$ trials showing the tail (-1) among the $2n$ trials. Therefore,

$$\mathbb{P}(S_{2n} = 2k) = \binom{2n}{n+k}\left(\frac{1}{2}\right)^{2n} = \frac{(2n)!}{2^{2n}(n+k)!(n-k)!}.$$

Consider the Stirling's approximation,

$$n! \doteq \sqrt{2\pi n}\left(\frac{n}{e}\right)^n, \quad \text{as } n \to \infty,$$

and the definition of e,

$$e = \lim_{n\to\infty}\left(1 - \frac{1}{n}\right)^n.$$

Using these two tools, we can obtain the desirable approximation.

Hints for Exercise 2.3:

```
set.seed(123)
p <- 0.1  # Probability of success (or 0.5)
reps <- 10000  # Number of repetitions
n_values <- c(100, 500) # Number of trials (or c(20, 100))

for (n in n_values) {
  binom_samples <- rbinom(reps, size = n, prob = p)
  z_values <- (binom_samples - n * p) / sqrt(n * p * (1 - p))

  # Plot the histogram of standardized values
  hist(
    z_values,
    breaks = 40,
    probability = TRUE,
    main = paste("n =", n, ", p =", p, ", Reps =", reps),
    xlab = "Standardized Values",
    col = "lightblue",
    ylim = c(0, 1),
    xlim = c(-4, 4)
  )
}
```

Hints for Exercise 2.4:

Note that

$$\mathbb{E}\left[\frac{1}{n}\sum_{i=1}^{n}(X_i-\overline{X}_n)^2\right]=\mathbb{E}\left[\frac{1}{n}\sum_{i=1}^{n}X_i^2-\overline{X}_n^2\right]=\mathbb{E}(X^2)-\mathbb{E}\left(\overline{X}_n^2\right)$$
$$=\mathbb{E}(X^2)-\left[\mathbb{V}(\overline{X}_n)+\mathbb{E}^2(\overline{X}_n)\right]=\mathbb{E}(X^2)-\frac{1}{n}\mathbb{V}(X)-\mathbb{E}^2(X).$$

Hints for Exercise 2.5:

```
set.seed(123)
n <- 100
mu <- 0
num_sim <- 1000

mu1_hat <- numeric(num_sim)
mu2_hat <- numeric(num_sim)

for (i in 1:num_sim) {
  x <- rnorm(n, mean = mu, sd = 1)
  mu1_hat[i] <- mean(x)
  mu2_hat[i] <- median(x)
}

var_mu1 <- var(mu1_hat)
var_mu2 <- var(mu2_hat)
```

Hints for Exercises in Chapter 3

Hints for Exercise 3.1:

```
set.seed(1)
# Generate Bernoulli sequence for coins
coin_sequence <- rbinom(20, size = 1, prob = 0.5)
print(coin_sequence)
#[1] 0 0 1 1 0 1 1 1 1 0 0 0 1 0 1 0 1 1 0 1

# Generate Bernoulli sequence for outcomes
outcome_sequence <- rbinom(20, size = 1, prob = 0.6)
print(outcome_sequence)
#[1] 0 1 0 1 1 1 1 1 0 1 1 1 1 1 0 0 0 1 0 1
```

Hints for Exercise 3.2:

Continue the hints of Exercise 3.1 with the following codes.

```
# Generate 1000 random permutations
permutations <- replicate(1000, sample(coin_sequence, size = 20,
    replace = FALSE))
# View the first permutation
print(permutations[, 1])
#[1] 1 1 1 0 1 0 0 1 0 1 0 0 1 0 0 1 0 1 1 1
```

Hints for Exercise 3.3:

Use the following R function.

```
pbinom(k, n, p_values)
```

Hints for Exercise 3.4:

Method 1: Use some non-parametric method (say kernel density estimation) to estimate $p_0(x)$, with the estimator denoted by $\widehat{p}(x)$. Then $p_0(\theta_0)$ can be estimated by $\widehat{p}(\widehat{\theta}_n)$. Method 2: The bootstrap method.

Hints for Exercise 3.5:

Continue the hints of Exercise 3.1 with the following codes.

```
# Generate 1000 bootstrap samples
bootstraps <- replicate(1000, sample(1:20, size = 20, replace =
    TRUE))
# View the first first bootstrap sample
print(bootstraps[, 1])
#[1]  5 12 10  9 16 13  3 14 18  6  5 16 17 18  5 15  2 10  6 20
```

Hints for Exercises in Chapter 4

Hints for Exercise 4.1:

The likelihood is

$$L(\mu, b; X_1, \dots, X_n) = \frac{1}{(2b)^n} \exp\left(-\frac{\sum_{i=1}^n |X_i - \mu|}{b}\right).$$

The log-likelihood is

$$\ell(\mu, b; X_1, \dots, X_n) = -n\log(2b) - \sum_{i=1}^n |X_i - \mu|/b.$$

The first derivatives of the log-likelihood with respect to μ and b are

$$\frac{\partial \ell(\mu, b)}{\partial \mu} = \frac{1}{b} \sum_{i=1}^{n} \operatorname{sign}(X_i - \mu),$$

where $\operatorname{sign}(\cdot)$ is the sign function, and

$$\frac{\partial \ell(\mu, b)}{\partial b} = -\frac{n}{b} + \frac{\sum_{i=1}^{n} |X_i - \mu|}{b^2}.$$

Note that, as μ moves from $-\infty$ to ∞ and b moves from 0 to ∞, the above two derivatives are positive first, implying that the log-likelihood is increasing first, then reaches zero, and later stays negative—implying that the log-likelihood is decreasing after reaching its maximum. Therefore, the MLEs of μ and b are the solutions to the following two equations,

$$\frac{1}{b} \sum_{i=1}^{n} \operatorname{sign}(X_i - \mu) = 0 \quad \text{and} \quad -\frac{n}{b} + \frac{\sum_{i=1}^{n} |X_i - \mu|}{b^2} = 0.$$

Hints for Exercise 4.2:

The likelihood is

$$L(p; X_1, \ldots, X_n) = \prod_{i=1}^{n} p^{X_i} (1-p)^{1-X_i}.$$

The log-likelihood is

$$\ell(p; X_1, \ldots, X_n) = \sum_{i=1}^{n} X_i \log p + (1 - X_i) \log(1-p).$$

The first derivative of the log-likelihood is

$$\frac{\partial \ell(p)}{\partial p} = \frac{1}{p} \sum_{i=1}^{n} X_i - \frac{1}{1-p} \sum_{i=1}^{n} (1 - X_i).$$

The second derivative of the log-likelihood is

$$\frac{\partial^2 \ell(p)}{\partial p^2} = -\frac{\sum_{i=1} X_i}{p^2} - \frac{\sum_{i=1}^{n} (1 - X_i)}{(1-p)^2} < 0.$$

This implies that the log-likelihood is a convex function and the MLE of p is the solution to the following equation,

$$\frac{1}{p} \sum_{i=1}^{n} X_i - \frac{1}{1-p} \sum_{i=1}^{n} (1 - X_i) = 0.$$

Hints for Exercise 4.3:

The MLE of p_j is

$$\widehat{p}_{jn} = \sum_{i=1}^{n} I(X_i = p^j)/n, \quad j = 1, \ldots, J.$$

Verify that

$$\mathbb{E}(\widehat{p}_{jn}) = p_j, \; \mathbb{V}(\widehat{p}_{jn}) = p_j(1 - p_j), \text{ and } \operatorname{Cov}(\widehat{p}_{j_1 n}, \widehat{p}_{j_2 n}) = -p_{j_1} p_{j_2}/n.$$

Hints for Exercise 4.4:

Use the following facts of exp(1):

$$\int_0^\infty e^{-x} dx = 1, \quad \int_0^\infty x e^{-x} dx = 1, \quad \text{and} \quad \int_0^\infty (x-1)^2 e^{-x} dx = 1.$$

Hints for Exercise 4.5:

If $X \sim \exp(\lambda_0)$, then

$$\begin{aligned}
\mathbb{E}_0(X) &= \lambda_0^{-1}, \\
\mathbb{V}_0(X) &= \lambda_0^{-2}, \\
\mathbb{E}_0\left[(X - \lambda_0^{-1})^3\right] &= 2\lambda_0^{-3}, \\
\mathbb{E}_0\left[(X - \lambda_0^{-1})^4\right] &= 9\lambda_0^{-4}.
\end{aligned}$$

Hints for Exercises in Chapter 5

Hints for Exercise 5.1:

Note that

$$\begin{aligned}
R^2 &= \frac{\left[\Sigma_x^{-1} \sum_{i=1}^n (X_i - \overline{X})(Y_i - \overline{Y})\right]^\top \sum_{i=1}^n (X_i - \overline{X})(Y_i - \overline{Y})}{\sum_{i=1}^n (Y_i - \overline{Y})^2} \\
&= \frac{\sum_{i=1}^n \widehat{\beta}^\top (X_i - \overline{X})(Y_i - \overline{Y})}{\sum_{i=1}^n (Y_i - \overline{Y})^2} \\
&= \frac{\sum_{i=1}^n \left\{(Y_i - \overline{Y}) - \left[Y_i - \overline{Y} - \widehat{\beta}^\top (X_i - \overline{X})\right]\right\} (Y_i - \overline{Y})}{\sum_{i=1}^n (Y_i - \overline{Y})^2} \\
&= 1 - \frac{\sum_{i=1}^n \left[Y_i - \overline{Y} - \widehat{\beta}^\top (X_i - \overline{X})\right] (Y_i - \overline{Y})}{\sum_{i=1}^n (Y_i - \overline{Y})^2} \\
&= 1 - \frac{\sum_{i=1}^n \left[Y_i - \overline{Y} - \widehat{\beta}^\top (X_i - \overline{X})\right]^2}{\sum_{i=1}^n (Y_i - \overline{Y})^2},
\end{aligned}$$

where the last equality follows from the following equation:

$$\begin{aligned}&\sum_{i=1}^{n}[Y_i - \overline{Y} - \widehat{\beta}^\top (X_i - \overline{X})][\widehat{\beta}^\top (X_i - \overline{X})]\\ =&\widehat{\beta}^\top \sum_{i=1}^{n}[Y_i - \overline{Y} - \widehat{\beta}^\top (X_i - \overline{X})](X_i - \overline{X}) = 0.\end{aligned}$$

Hints for Exercise 5.2:

The joint likelihood of X and Y is equal to the product of the marginal likelihood of X and the conditional likelihood of Y conditional on X. Since the marginal distribution of X is independent of β and σ^2, we consider the MLE of β and σ^2 based on the following conditional likelihood:

$$\prod_{i=1}^{n} \frac{1}{\sqrt{2\pi\sigma^2}} \exp\left\{-\frac{[Y_i - Q(X_i;\beta)]^2}{2}\right\}.$$

The log-likelihood is

$$-\frac{n}{2}\log(2\pi\sigma^2) - \frac{1}{2}\sum_{i=1}^{n}[Y_i - Q(X_i;\beta)]^2.$$

Hints for Exercise 5.3:

Note that $Q(X;\beta) = \mathbb{P}(Y = 1 \mid X) = \mathbb{E}(Y \mid X)$. The $\mathcal{L}_2$ loss is

$$\sum_{i=1}^{n}[Y_i - Q(X_i;\beta)]^2.$$

Hints for Exercise 5.4:

The joint likelihood of X and Y is equal to the product of the marginal likelihood of X and the conditional likelihood of Y conditional on X. Since the marginal distribution of X is independent of β, we consider the MLE of β based on the following conditional likelihood,

$$\prod_{i=1}^{n}[Q(X_i;\beta)]^{Y_i}[1 - Q(X_i;\beta)]^{1-Y_i}.$$

The log-likelihood is

$$\sum_{i=1}^{n}\left\{Y_i \log[Q(X_i;\beta)] + (1 - Y_i)\log[1 - Q(X_i;\beta)]\right\}.$$

Hints for Exercise 5.5:

Consider the linear transformation $\widetilde{Y} = (Y - a)/(b - a)$ such that $\widetilde{Y} \in [0, 1]$. Consequently, the regression function of $\widetilde{Y}$ becomes $\widetilde{Q}(X;\beta) = [Q(X;\beta) - a]/(b - a)$. Then,

$$\widehat{\beta}_n = \arg\min_{\beta} \left\{ -\sum_{i=1}^{n} [Y_i \log(Q(X_i;\beta)) + (1 - Y_i)\log(1 - Q(X_i;\beta))] \right\}.$$

Hints for Exercises in Chapter 6

Hints for Exercise 6.1:

If M2T is the intervention of interest, then T2M is the comparator. Or vice versa.

Hints for Exercise 6.2:

```
# Set seed for reproducibility
set.seed(123)

# Sex variable
sex <- rbinom(20, 1, 0.6)
# Baseline value
y.b <- abs(rnorm(20, 0, 1))
# Values for T2M
y.0 <- y.b + 0.5*rnorm(20, 1, 1) + 0.4*sex
# Values for M2T
y.1 <- y.b + 0.5*rnorm(20, 2, 1) - 0.1*sex

# Tuning values to ordinal
y <- c(y.b, y.0, y.1)
ov <- as.numeric(cut(y, breaks = 7, labels = 1:7)) + 3
ov.b <- ov[1:20]
ov.0 <- ov[21:40]
ov.1 <- ov[41:60]

# Final dataset
tea.df<-data.frame(sex, ov.b, ov.0, ov.1)
summary(tea.df)
```

Hints for Exercise 6.3:

Revise the above R code.

Hints for Exercise 6.4:

An estimator is said to be asymptotically consistent if it converges to the true value of the estimand as $n \to \infty$.

Hints for Exercise 6.5:

For the 4th subject, compare Y with Y^A.

Hints for Exercises in Chapter 7

Hints for Exercise 7.1:

Use the full dataset generated in Exercise 6.2.

```
set.seed(5)
# Generate randomized assignments
A.ran <- rbinom(20, 1, 0.5)

# Generate non-randomized assignments
ps <- 1-1/(1+exp(1 - sex - 0.8*(ov.b>5)))
A.ps <- rbinom(20, 1, ps)
```

Hints for Exercise 7.2:

```
sum(A.ran*ov.1)/sum(A.ran) - sum((1-A.ran)*ov.0)/sum(1-A.ran)
sum(A.ps*ov.1)/sum(A.ps) - sum((1-A.ps)*ov.0)/sum(1-A.ps)
```

Hints for Exercise 7.3:

See Exercise 7.1.

Hints for Exercise 7.4:

See Exercise 7.2.

Hints for Exercise 7.5:

Proof of $A \perp\!\!\!\perp Y \mid M$ in the chain: Let $p_{A,M,Y}(a,m,y)$ be the joint distribution of (A,M,Y). By the chain rule, $p_{A,M,Y}(a,m,y) = p_A(a)p_{M|a}(m \mid a)p_{Y|m}(y \mid m)$. Thus,

$$\begin{aligned}p_{A,Y|m}(a,y \mid m) &= p_{A,M,Y}(a,m,y)/p_M(m)\\ &= p_A(a)p_{M|a}(m \mid a)p_{Y|m}(y \mid m)/p_M(m)\\ &= 1/p_M(m) \cdot p_A(a)p_{M|a}(m \mid a) \cdot p_{Y|m}(y \mid m).\end{aligned}$$

Since $p_{A,Y|m}(a,y \mid m)$ is the product of one function depending on a only and the other function on y only, we show that $A \perp\!\!\!\perp Y \mid M$.

Proof of $A \perp\!\!\!\perp Y \mid X$ in the fork: Let $p_{C,A,Y}(c,a,y)$ be the joint distribution of (C,A,Y). By the chain rule, $p_{C,A,Y}(c,a,y) = p_C(c)p_{A|c}(a \mid c)p_{Y|c}(y \mid c)$.

Thus,

$$p_{A,Y|c}(a, y \mid c) = p_{C,A,Y}(c, a, y)/p_C(c)$$
$$= p_C(c)p_{A|c}(a \mid c)p_{Y|c}(y \mid c)/p_C(c)$$
$$= p_{A|c}(a \mid c) \cdot p_{Y|c}(y \mid c).$$

Since $p_{A,Y|c}(a, y| \mid c)$ is the product of one function depending on a only and the other function on y only, we show that $A \perp\!\!\!\perp Y \mid C$.

Proof of $A \not\perp\!\!\!\perp Y \mid E$ *in the collider*: Let $p_{A,Y,E}(a, y, e)$ be the joint distribution of (A, Y, E). By the chain rule, $p_{A,Y,E}(a, y, e) = p_A(a)p_Y(y)p_{E|a,y}(e \mid a, y)$. Thus,

$$p_{A,Y|e}(a, y \mid e) = p_{A,Y,E}(a, y, e)/p_E(e)$$
$$= p_A(a)p_Y(y)p_{E|a,y}(e \mid a, y)/p_E(e).$$

Since $p_{E|a,y}(e \mid a, y)$ depends on both a and y, which cannot be written as a product of a function that depends on a only and the other function on y only, we show that $A \not\perp\!\!\!\perp Y \mid E$.

Hints for Exercises in Chapter 8

Hints for Exercise 8.1:

```
y.1 <- c(6, 6, 10, 5, 10, 7, 6, 6, 9, 10)
y.0 <- c(8, 5, 6, 6, 6, 5, 5, 8, 7, 6)
var.1 <- var(y.1)
var.0 <- var(y.2)
```

Hints for Exercise 8.2:

Consider the mean of the last column of Table 6.1.

Hints for Exercise 8.3:

```
#ATT
mean(c(0, 1, 2, 0, 0, 2, 2, 2, 1, 1, 3))
#[1] 1.272727
```

Hints for Exercise 8.4:

```
#ATC
mean(c(-1, 0, -2, 2, 1, 0, -3, -3, 1))
[#1] -0.5555556
```

Hints for Exercise 8.5:

$$
\begin{aligned}
&\mathbb{E}\left\{\frac{\mathbb{P}(A=1 \mid X)\mathbb{E}[Y^{a=0} \mid X]}{\mathbb{P}(A=1)}\right\} \\
=&\mathbb{E}\left\{\frac{\mathbb{P}(A=1 \mid X)\mathbb{E}[Y^{a=0} \mid X, A=1]}{\mathbb{P}(A=1)}\right\} \\
=&\int \frac{g(1|x)\mathbb{E}[Y^{a=0} \mid x, A=1]}{g(1)} f_X(x)dx \\
=&\int f_{X|A=1}(x|1)\mathbb{E}[Y^{a=0} \mid x, A=1]dx,
\end{aligned}
$$

where the last equality uses Bayes' theorem,

$$\frac{g(1|x)f_X(x)}{g(1)} = f_{X|A=1}(x|1).$$

Hints for Exercises in Chapter 9

Hints for Exercise 9.1:

By the law of iterative expectations,

$$\mathbb{E}[M_1(O;\theta_0,\beta,\gamma_0)] = \mathbb{E}\left\{\mathbb{E}[M_1(O;\theta_0,\beta,\gamma_0)] \mid X\right\} = 0.$$

Hints for Exercise 9.2:

By the law of iterative expectations,

$$\mathbb{E}[M_1(O;\theta_0,\beta_0,\gamma)] = \mathbb{E}\left\{\mathbb{E}[M_1(O;\theta_0,\beta_0,\gamma)] \mid X, A\right\} = 0.$$

Hints for Exercise 9.3:

```
lm0 <- lm(Y ~ X1*X2 + A*X1 + A*X2)
new_data <- data.frame(X1 = c(0, 0, 1, 0, 0, 1, 1, 1 ),
                       X2 = c(0, 0, 1, 0, 0, 1, 1, 1),
                       A = c(1, 0, 1, 0, 1, 0, 1, 0))
# Predict y for the new x values
predict(lm0, newdata = new_data)
```

Hints for Exercise 9.4:

```
new_data1 <- data.frame(X1 = X1, X2 = X2, A = rep(1, 20))
new_data0 <- data.frame(X1 = X1, X2 = X2, A = rep(0, 20))
Q1 <- predict(lm0, newdata = new_data1)
Q0 <- predict(lm0, newdata = new_data0)
mean(Q1 - Q0)
```

Hints for Exercise 9.5:

```
logit0 <- glm(A ~ X1 + X2, family = binomial)
new_data0 <- data.frame(X1 = c(0, 1, 0, 1 ), X2 = c(0, 0, 1, 1))
g1 <- predict(logit0, newdata = new_data0, type = "response")
g0 <- 1-g1
```

Hints for Exercise 9.6:

```
new_data <- data.frame(X1 = X1, X2 = X2)
g1 <- predict(logit0, newdata = new_data, type = "response")
g0 <- 1-g0
mean(A * Y / g1) - mean((1 - A)* Y / g0)
```

Hints for Exercise 9.7:

```
IPW <- mean(A * Y / g1) - mean((1 - A)* Y / g0)
D <- mean((A - g1) * (Q1 / g1 + Q0 / g0))
AIPW <- IPW + D
```

Hints for Exercise 9.8:

```
set.seed(123)
nn<-100
sex <- rbinom(nn, 1, 0.6)

y0 <- abs(rnorm(nn, 0, 1))
y1 <- y0 + 0.5*rnorm(nn, 1, 1) + 0.4 * sex
y2 <- y0 + 0.5*rnorm(nn, 2, 1) - 0.1 * sex
ov.y <- cut(c(y0, y1, y2), breaks = 7, labels = 1:7)
ov <- as.numeric(ov.y) + 3
ov0 <- ov[1 : nn]
ov1 <- ov[(nn+1) : (2*nn)]
ov2 <- ov[(2*nn+1) : (3*nn)]

baseline <- 1 * (ov0 > 5)

# True value of estimand
mean(ov2-ov1)

set.seed(5)
# Generate non-randomized assignments
ps <- 1 - 1 / (1 + exp(1 - sex - 0.8 * baseline))
A <- rbinom(nn, 1, ps)

# Unadjsuted estimate
sum(A * ov2) / sum(A) - sum((1 - A) * ov1) / sum(1 - A)

# The dataset to be used for three adjusted methods
X1 <- sex
X2 <- baseline
Y <- A * ov2 + (1 - A) * ov1
data.frame(X1 = X1, X2 = X2, A = A, Y = Y)
```

Hints for Exercises in Chapter 10

Hints for Exercise 10.1:

$1.68 + \sqrt{1.68(1.68-1)} = 2.75$ and $1.55 + \sqrt{1.55(1.55-1)} = 2.47$.

Hints for Exercise 10.2:

$1.25 + \sqrt{1.25(1.25-1)} = 1.81$.

Hints for Exercise 10.3:

$B = 1.87/1.68 = 1.11$ and $\Delta = 1.11 + \sqrt{1.11(1.11-1)} = 1.46$.

Hints for Exercise 10.4:

$B = 1.68/1.43 = 1.17$ and $\Delta = 1.17 + \sqrt{1.17(1.17-1)} = 1.62$.

Hints for Exercise 10.5:

See the following table:

An augmented dataset for sensitivity analysis of positivity violation

ID	X_1 (Sex)	X_2 (Baseline)	A (Treatment)	Y (Outcome)
...				
21*	0	0	0	$6-\delta$
22*	0	0	0	$7-\delta$
23*	0	0	0	$6-\delta$
24*	0	0	0	$7-\delta$

Hints for Exercises in Chapter 11

Hints for Exercise 11.1:

```
set.seed(123)
n <- 50
N_sim <- 10000
theta <- seq(-5, 5, length.out = 100)
mse_hodges <- rep(0, 100)
for (j in 1:100) {
  mse <- rep(0, N_sim)
  for (i in 1:N_sim) {
    data <- rnorm(n, mean = theta[j], sd = 1)
    mle <- mean(data)
```

```
    delta <- n^(-1/4)
    hodges <- ifelse(abs(mle <= delta), 0, mle)
    mse[i] <- (hodges - theta[j])^2
  }
  mse_hodges[j] <- mean(mse)
}
```

Hints for Exercise 11.2:

The inverse of the above matrix is

$$\begin{pmatrix} *, & ** \\ **^\top, & S^{-1} \end{pmatrix} > 0,$$

where $*$ and $**$ are certain matrices that can be expressed explicitly. This is known as the Schur complement inequality in matrix theory.

Hints for Exercise 11.3:

The CRLB is

$$\left(-\frac{\sigma}{\mu^2}\right)^2 \cdot \frac{\sigma^2}{n} + \left(\frac{1}{\mu}\right)^2 \cdot \frac{\sigma^2}{2n}.$$

Hints for Exercise 11.4:

Take the derivative of the log-likelihood, $\log\{p_0(x)[1 + \nu s(X)]\}$, and evaluate it at $\nu = 0$.

Hints for Exercise 11.4:

Take the derivative of the log-likelihood, $\log\{p_0(x) \exp[\nu s(x)]\} - \log\{\int p_0(t) \exp[\nu s(t)]dt\}$, and evaluate it at $\nu = 0$.

Hints for Exercises in Chapter 12

Hints for Exercise 12.1:

Rewrite $\theta_a(\nu)$ as

$$\theta_a(\nu) = \sum_{k=1}^{K} (1)_k \times (2)_k,$$

where

$$(1)_k = \frac{\sum_{m=1}^{M} y_m \nu_{k,a,m}}{\sum_{m=1}^{M} \nu_{k,a,m}},$$
$$(2)_k = \sum_{m=1}^{M} (\nu_{k,a,m} + \nu_{k,a',m}).$$

Then,

$$\frac{\partial \theta_a(\nu_0)}{\partial \nu_{k,a,m}} = \frac{\partial (1)_k}{\partial \nu_{k,a,m}} \times (2)_k + (1)_k \times \frac{\partial (2)_k}{\partial \nu_{k,a,m}},$$

where

$$\frac{\partial (1)_k}{\partial \nu_{k,a,m}} = \frac{y_m}{\mathbb{P}(A = a, X = x_k)} - \frac{\sum_{m=1}^{M} y_m \nu_{k,a,m}}{\mathbb{P}^2(A = a, X = x_k)},$$
$$\frac{\partial (2)_k}{\partial \nu_{k,a,m}} = 1.$$

Thus,

$$\frac{\partial \theta_a(\nu_0)}{\partial \nu_{k,a,m}} = \left[\frac{y_m}{\mathbb{P}(A = a, X = x_k)} - \frac{\mathbb{E}(Y \mid A = a, X = x_k)}{\mathbb{P}(A = a, X = x_k)}\right] \mathbb{P}(X = x_k) + \mathbb{E}(Y|A = a, X = x_k).$$

Hints for Exercise 12.2:

Rewrite $\theta_a(\nu)$ as

$$\theta_a(\nu) = \sum_{k=1}^{K} [\mathbb{E}(Y \mid A = a, X = x_k)] \sum_{m=1}^{M} (\nu_{k,a,m} + \nu_{k,a',m}).$$

Hints for Exercise 12.3:

Compare them conditional on $A = 1$ and $A = 0$, respectively.

Hints for Exercise 12.4:

$$\begin{aligned}\theta_{\text{ATE}}(\nu_1, \nu_2, \nu_3) =& \mathbb{E}_\nu[\mathbb{E}_\nu(Y \mid A = 1, X)] - \mathbb{E}_\nu[\mathbb{E}_\nu(Y \mid A = 0, X)] \\ =& \int \left[\int y f(y|1, x; \nu_3) dy\right] p(x; \nu_1) dx \\ &- \int \left[\int y f(y|0, x; \nu_3) dy\right] p(x; \nu_1) dx \\ =& \theta_{\text{ATE}}(\nu_1, \nu_3)\end{aligned}$$

Hints for Exercise 12.5:

The equation is the same as

$$
\begin{aligned}
&\mathbb{E}\left(\frac{(2A-1)Y}{\mathbb{P}(A\mid X)}[A-\mathbb{P}(A=1\mid X)]\,\Big|\, X\right)\\
=&-T(X)\mathbb{E}\left([A-\mathbb{P}(A=1\mid X)][A-\mathbb{P}(A=1\mid X)]\mid X\right)\\
=&-T(X)\mathbb{P}(A=1\mid X)\mathbb{P}(A=0\mid X).
\end{aligned}
$$

Note that

$$
\begin{aligned}
&\mathbb{E}\left(\frac{(2A-1)Y}{\mathbb{P}(A\mid X)}[A-\mathbb{P}(A=1\mid X)]\,\Big|\, X\right)\\
=&\mathbb{E}\left[\mathbb{E}\left(\frac{(2A-1)Y}{\mathbb{P}(A\mid X)}[A-\mathbb{P}(A=1\mid X)]\,\Big|\, A,X\right)\,\Big|\, X\right]\\
=&\mathbb{E}\left(\frac{(2A-1)\mathbb{E}[Y\mid A,X]}{\mathbb{P}(A\mid X)}[A-\mathbb{P}(A=1\mid X)]\,\Big|\, X\right)\\
=&\frac{\mathbb{E}[Y\mid 1,X]}{\mathbb{P}(A=1\mid X)}[1-\mathbb{P}(A=1\mid X)]\mathbb{P}(A=1\mid X)\\
&+\frac{-\mathbb{E}[Y\mid 0,X]}{\mathbb{P}(A=0\mid X)}[0-\mathbb{P}(A=1\mid X)]\mathbb{P}(A=0\mid X).
\end{aligned}
$$

Hints for Exercises in Chapter 13

Hints for Exercise 13.1:

When $A=1$, verify that both are equal to

$$
\begin{aligned}
&[\mathbb{E}(Y\mid A=1,X)-\mathbb{E}(Y\mid A=0,X)]-\theta_{\text{ATE},0}+\frac{Y-\mathbb{E}(Y\mid A=1,X)}{\mathbb{P}(A=1\mid X)}\\
=&\frac{Y}{\mathbb{P}(A=1\mid X)}-\theta_{\text{ATE},0}-\frac{\mathbb{P}(A=0\mid X)}{\mathbb{P}(A=1\mid X)}\mathbb{E}(Y\mid A=1,X)-\mathbb{E}(Y\mid A=0,X).
\end{aligned}
$$

When $A=0$, verify that both are equal to

$$
\begin{aligned}
&[\mathbb{E}(Y\mid A=1,X)-\mathbb{E}(Y\mid A=0,X)]-\theta_{\text{ATE},0}-\frac{Y-\mathbb{E}(Y\mid A=0,X)}{\mathbb{P}(A=0\mid X)}\\
=&\frac{-Y}{\mathbb{P}(A=0\mid X)}-\theta_{\text{ATE},0}+\mathbb{E}(Y\mid A=1,X)+\frac{\mathbb{P}(A=1\mid X)}{\mathbb{P}(A=0\mid X)}\mathbb{E}(Y\mid A=1,X).
\end{aligned}
$$

Hints for Exercise 13.2:

To show the first equation, note that

$$\begin{aligned}\mathbb{E}\{\phi_2(X,A,Y)S_{0;\nu_1}(X)\} &= \mathbb{E}\{\mathbb{E}[\phi_2(X,A,Y)S_{0;\nu_1}(X)\mid A,X]\}\\ &= \mathbb{E}\{S_{0;\nu_1}(X)\mathbb{E}[\phi_2(X,A,Y)\mid A,X]\}=0.\end{aligned}$$

To show the second equation, note that

$$\begin{aligned}\mathbb{E}\{\phi_1(X)S_{0;\nu_2}(X,A)\} &= \mathbb{E}\{\mathbb{E}[\phi_1(X)S_{0;\nu_2}(X,A)\mid X]\}\\ &= \mathbb{E}\{\phi_1(X)(X)\mathbb{E}[S_{0;\nu_2}(X,A)\mid X]\}=0.\end{aligned}$$

To show the third equation, note that

$$\begin{aligned}\mathbb{E}\{\phi_2(X,A,Y)S_{0;\nu_2}(X,A)\} &= \mathbb{E}\{\mathbb{E}[\phi_2(X,A,Y)S_{0;\nu_2}(X,A)\mid A,X]\}\\ &= \mathbb{E}\{S_{0;\nu_2}(X,A)\mathbb{E}[\phi_2(X,A,Y)\mid A,X]\}=0.\end{aligned}$$

To show the fourth equation, note that

$$\begin{aligned}\mathbb{E}\{\phi_1(X)S_{0;\nu_3}(X,A,Y)\} &= \mathbb{E}\{\mathbb{E}[\phi_1(X)S_{0;\nu_3}(X,A,Y)\mid A,X]\}\\ &= \mathbb{E}\{\phi_1(X)(X)\mathbb{E}[S_{0;\nu_3}(X,A,Y)\mid A,X]\}=0.\end{aligned}$$

Hints for Exercise 13.3:

Consider the following log-likelihood,

$$\begin{aligned}\ell(\nu_1,\nu_2,\nu_3) &= A\log[g(1;\nu_1)]+(1-A)\log[1-g(1;\nu_1)]\\ &\quad+\log p(X\mid A;\nu_2)+\log f(Y\mid A,X;\nu_3).\end{aligned}$$

Thus,

$$\begin{aligned}S_{0;\nu_1}(A) &= \left.\frac{\partial\ell(\nu_1,\nu_2,\nu_3)}{\partial\nu_1}\right|_0 = \frac{A\partial g(1;0)/\partial\nu_1}{g_0(1)} - \frac{(1-A)\partial g(1;0)/\partial\nu_1}{1-g_0(1)}\\ &= c\left[\frac{A}{g_0(1)}-\frac{1-A}{1-g_0(1)}\right],\end{aligned}$$

where $c=\partial g(1;0)/\partial\nu_1$ is a constant.

Hints for Exercise 13.4:

On the one hand,

$$\begin{aligned}&\partial\int[Q_0(x,1)-Q_0(x,0)]\,p(x\mid A=1;\nu_2)dx/\partial\nu_2\big|_{\nu_2=0}\\ &=\int[Q_0(x,1)-Q_0(x,0)]\,\partial p(x\mid A=1;0)dx/\partial\nu_2\\ &=\int[Q_0(x,1)-Q_0(x,0)]\,S_{0;\nu_2}(x,1)p_0(x\mid A=1)dx.\end{aligned}$$

On the other hand,

$$\begin{aligned}&\mathbb{E}\left\{\phi_1(X,A)S_{0;\nu_2}(X,A)\right\}\\=&\mathbb{E}\left\{\phi_1(X,1)S_{0;\nu_2}(X,1)\mid A=1\right\}\mathbb{P}(A=1)\\&+\mathbb{E}\left\{0\times S_{0;\nu_2}(X,0)\mid A=0\right\}\mathbb{P}(A=0)\\=&\mathbb{E}\{[Q_0(X,1)-Q_0(X,0)]S_{0;\nu_2}(X,1)\}.\end{aligned}$$

Hints for Exercise 13.5:

Note that

$$\begin{aligned}&\partial\mathbb{E}_{\nu_3}\left\{\left[\frac{I(A=1)}{\mathbb{P}(A=1)}-\frac{I(A=0)\mathbb{P}(A=1\mid X)}{\mathbb{P}(A=1)\mathbb{P}(A=0\mid X)}\right]Y\right\}/\partial\nu_3\Big|_{\nu_3=0}\\=&\partial\mathbb{E}\left\{\left[\frac{I(A=1)}{\mathbb{P}(A=1)}-\frac{I(A=0)\mathbb{P}(A=1\mid X)}{\mathbb{P}(A=1)\mathbb{P}(A=0\mid X)}\right]\mathbb{E}_{\nu_3}(Y|X,A)\right\}/\partial\nu_3\Big|_{\nu_3=0}\\=&\mathbb{E}\left\{\left[\frac{I(A=1)}{\mathbb{P}(A=1)}-\frac{I(A=0)\mathbb{P}(A=1\mid X)}{\mathbb{P}(A=1)\mathbb{P}(A=0\mid X)}\right]\partial\mathbb{E}_{\nu_3}(Y|X,A)/\partial\nu_3\Big|_{\nu_3=0}\right\}\end{aligned}$$

and

$$\partial\mathbb{E}_{\nu_3}(Y|X,A)/\partial\nu_3\Big|_{\nu_3=0}=\mathbb{E}\{[Y-Q_0(X,A)]S_{0;\nu_3}(X,A,Y)\mid X,A\}.$$

Hints for Exercises in Chapter 14

Hints for Exercise 14.1:

To show (1) $\Rightarrow$ (2), note that (1) implies that

$$\begin{aligned}\mathbb{P}(X=x\mid Y=y)&=\mathbb{P}(X=x),\\\mathbb{P}(X=x,Z=z\mid Y=y)&=\mathbb{P}(X=x\mid Y=y)\mathbb{P}(Z=z\mid Y=y).\end{aligned}$$

Thus,

$$\begin{aligned}\mathbb{P}(X=x,Y=y,Z=z)=&\mathbb{P}(X=x,Z=z\mid Y=y)\mathbb{P}(Y=y)\\=&\mathbb{P}(X=x\mid Y=y)\mathbb{P}(Z=z\mid Y=y)\mathbb{P}(Y=y)\\=&\mathbb{P}(X=x)\mathbb{P}(Z=z\mid Y=y)\mathbb{P}(Y=y)\\=&\mathbb{P}(X=x)\mathbb{P}(Y=y,Z=z).\end{aligned}$$

To show (2) $\Rightarrow$ (1), note that (2) implies that

$$\mathbb{P}(X=x,Y=y,Z=z)=\mathbb{P}(X=x)\mathbb{P}(Y=y,Z=z).$$

Thus,

$$\sum_z\mathbb{P}(X=x,Y=y,Z=z)=\sum_z\mathbb{P}(X=x)\mathbb{P}(Y=y,Z=z),$$

which becomes

$$\mathbb{P}(X = x, Y = y) = \mathbb{P}(X = x)\mathbb{P}(Y = y).$$

Therefore, $X \perp\!\!\!\perp Y$. In addition,

$$\mathbb{P}(X = x, Y = y, Z = z) = \mathbb{P}(X = x)\mathbb{P}(Y = y, Z = z).$$

implies that

$$\mathbb{P}(X = x, Y = y, Z = z)/\mathbb{P}(Y = y) = \mathbb{P}(X = x)\mathbb{P}(Y = y, Z = z)/\mathbb{P}(Y = y),$$

which is the same as

$$\begin{aligned}\mathbb{P}(X = x, Z = z \mid Y = y) =& \mathbb{P}(X = x)\mathbb{P}(Z = z \mid Y = y)\\ =& \mathbb{P}(X = x \mid Y = y)\mathbb{P}(Z = z \mid Y = y).\end{aligned}$$

Hints for Exercise 14.2:

The log-likelihood is

$$\begin{aligned}\ell(\boldsymbol{\nu}) = & \log[p(x;\nu_1)]\\ & + A\log[g(1 \mid X;\nu_2)] + (1-A)[1-g(1 \mid X;\nu_2)]\\ & + \Delta\log[h(X,A;\nu_3)] + (1-\Delta)\log[1-h(X,A;\nu_3)]\\ & + \Delta\log[f(y \mid X,A,\Delta=1;\nu_4)].\end{aligned}$$

First,

$$\left.\frac{\partial \ell}{\partial \nu_1}\right|_{\boldsymbol{\nu}=0} = \left.\frac{\partial p(x;\nu_1)/\partial\nu_1}{p(x;\nu_1)}\right|_{\nu_1=0} \triangleq S(X).$$

Second,

$$\begin{aligned}\left.\frac{\partial \ell}{\partial \nu_2}\right|_{\boldsymbol{\nu}=0} &= \left.\left[\frac{A\partial g(1 \mid X;\nu_2)/\partial\nu_2}{g(1 \mid X;\nu_2)} - \frac{(1-A)\partial g(1 \mid X;\nu_2)/\partial\nu_2}{1-g(1 \mid X;\nu_2)}\right]\right|_{\nu_2=0}\\ &= \left.\frac{\partial g/\partial\nu_2}{g(1-g)}[A - g(1 \mid X;\nu_2)]\right|_{\nu_2=0},\end{aligned}$$

$$S'(X) \triangleq \left.\frac{\partial g(1 \mid X;\nu_2)/\partial\nu_2}{g(1 \mid X;\nu_2)g(0 \mid X;\nu_2)}\right|_{\nu_2=0}.$$

Third,

$$\begin{aligned}\left.\frac{\partial \ell}{\partial \nu_3}\right|_{\boldsymbol{\nu}=0} &= \left.\left[\frac{\Delta\partial h(X,A;\nu_3)/\partial\nu_3}{h(X,A;\nu_3)} - \frac{(1-\Delta)\partial h(X,A;\nu_3)/\partial\nu_3}{1-h(X,A;\nu_3)}\right]\right|_{\nu_3=0}\\ &= \left.\frac{\partial h/\partial\nu_3}{h(1-h)}[\Delta - h(X,A;\nu_3)]\right|_{\nu_3=0},\end{aligned}$$

$$S'(X) \triangleq \left.\frac{\partial h(X,A;\nu_3)/\partial\nu_3}{h(X,A;\nu_3)[h(X,A;\nu_3)]}\right|_{\nu_3=0}.$$

Fourth,

$$\left.\frac{\partial \ell}{\partial \nu_4}\right|_{\boldsymbol{\nu}=0} = \Delta \left.\frac{\partial f(y \mid X, A, \Delta = 1; \nu_4)/\partial \nu_4}{f(y \mid X, A, \Delta = 1; \nu_4)}\right|_{\nu_4=0},$$

$$S'''(X, A, \Delta, Y) = \left.\frac{\partial f(y \mid X, A, \Delta = 1; \nu_4)/\partial \nu_4}{f(y \mid X, A, \Delta = 1; \nu_4)}\right|_{\nu_4=0}.$$

Hints for Exercise 14.3:

By the LIE,

$$\begin{aligned}
&\mathbb{E}[\phi_2(X, A, \Delta, Y) S_{0;\nu_1}(X)] \\
=&\mathbb{E}\{\mathbb{E}[\phi_2(X, A, \Delta, Y) S_{0;\nu_1}(X) \mid X, A, \Delta = 1]\} \\
=&\mathbb{E}\{S_{0;\nu_1}(X)\mathbb{E}[\phi_2(X, A, \Delta, Y) \mid X, A, \Delta = 1]\} \\
=&\mathbb{E}\{S_{0;\nu_1}(X) \times 0\} = 0,
\end{aligned}$$

$$\begin{aligned}
&\mathbb{E}[\phi_1(X) S_{0;\nu_2}(X, A)] \\
=&\mathbb{E}\{\mathbb{E}[\phi_1(X) S_{0;\nu_2}(X, A) \mid X]\} \\
=&\mathbb{E}\{\phi_1(X)\mathbb{E}[S_{0;\nu_2}(X, A) \mid X]\} \\
=&\mathbb{E}\{\phi_1(X) \times 0\} = 0,
\end{aligned}$$

$$\begin{aligned}
&\mathbb{E}[\phi_2(X, A, \Delta, Y) S_{0;\nu_2}(X, A)] \\
=&\mathbb{E}\{\mathbb{E}[\phi_2(X, A, \Delta, Y) S_{0;\nu_2}(X, A) \mid X, A, \Delta = 1]\} \\
=&\mathbb{E}\{S_{0;\nu_2}(X, A)\mathbb{E}[\phi_2(X, A, \Delta, Y) \mid X, A, \Delta = 1]\} \\
=&\mathbb{E}\{S_{0;\nu_2}(X, A) \times 0\} = 0,
\end{aligned}$$

$$\begin{aligned}
&\mathbb{E}[\phi_1(X) S_{0;\nu_3}(X, A, \Delta)] \\
=&\mathbb{E}\{\mathbb{E}[\phi_1(X) S_{0;\nu_3}(X, A, \Delta) \mid X, A]\} \\
=&\mathbb{E}\{\phi_1(X)\mathbb{E}[S_{0;\nu_3}(X, A, \Delta) \mid X, A]\} \\
=&\mathbb{E}\{\phi_1(X) \times 0\} = 0,
\end{aligned}$$

$$\begin{aligned}
&\mathbb{E}[\phi_2(X, A, \Delta, Y) S_{0;\nu_3}(X, A, \Delta)] \\
=&\mathbb{E}\{\mathbb{E}[\phi_2(X, A, \Delta, Y) S_{0;\nu_3}(X, A, \Delta) \mid X, A, \Delta = 1]\} \\
=&\mathbb{E}\{S_{0;\nu_3}(X, A, \Delta)\mathbb{E}[\phi_2(X, A, \Delta, Y) \mid X, A, \Delta = 1]\} \\
=&\mathbb{E}\{S_{0;\nu_3}(X, A, \Delta) \times 0\} = 0,
\end{aligned}$$

$$\begin{aligned}
&\mathbb{E}[\phi_1(X) S_{0;\nu_4}(X, A, \Delta, Y)] \\
=&\mathbb{E}\{\mathbb{E}[\phi_1(X) S_{0;\nu_4}(X, A, \Delta, Y) \mid X, A, \Delta = 1]\} \\
=&\mathbb{E}\{\phi_1(X)\mathbb{E}[S_{0;\nu_4}(X, A, \Delta, Y) \mid X, A, \Delta = 1]\} \\
=&\mathbb{E}\{\phi_1(X) \times 0\} = 0,
\end{aligned}$$

Hints for Exercise 14.4:

Note that

$$
\begin{aligned}
&\left.\frac{\partial \theta_{\text{ATE}}(\boldsymbol{\nu})}{\partial \nu_1}\right|_{\boldsymbol{\nu}=0} \\
&= \left.\frac{\partial \mathbb{E}_{\nu_1}\left[\mathbb{E}_{\nu_4}(Y \mid \Delta=1, A=1, X) - \mathbb{E}_{\nu_4}(Y \mid \Delta=1, A=0, X)\right]}{\partial \nu_1}\right|_{\nu_1=0},
\end{aligned}
$$

where

$$
\begin{aligned}
&\mathbb{E}_{\nu_1}\left[\mathbb{E}_{\nu_4}(Y \mid \Delta=1, A=1, X) - \mathbb{E}_{\nu_4}(Y \mid \Delta=1, A=0, X)\right] \\
&= \int \left[\int y f(y \mid x, 1, \Delta=1; \nu_4)\, dy - \int y f(y \mid x, 0, \Delta=1; \nu_4)\, dy\right] p(x; \nu_1)\, dx.
\end{aligned}
$$

Hints for Exercise 14.5:

Note that

$$
\begin{aligned}
\left.\frac{\partial \theta_{\text{ATE}}(\boldsymbol{\nu})}{\partial \nu_4}\right|_{\boldsymbol{\nu}=0} &= \partial\left\{\mathbb{E}_{\boldsymbol{\nu}} \frac{(2A-1)\Delta Y}{\mathbb{P}(A, \Delta=1 \mid X)}\right\} \Big/ \partial \nu_4 \Big|_{\boldsymbol{\nu}=0} \\
&= \partial\left\{\mathbb{E}_{\boldsymbol{\nu}} \mathbb{E}_{\boldsymbol{\nu}}\left[\frac{(2A-1)\Delta Y}{\mathbb{P}(A, \Delta=1 \mid X)} \;\Big|\; X, A, \Delta=1\right]\right\} \Big/ \partial \nu_4 \Big|_{\boldsymbol{\nu}=0} \\
&= \partial\left\{\mathbb{E}_{\nu_1, \nu_2, \nu_3} \frac{(2A-1)\Delta \mathbb{E}_{\nu_4}(Y \mid \Delta=1, A, X)}{\mathbb{P}(A, \Delta=1 \mid X)}\right\} \Big/ \partial \nu_4 \Big|_{\nu_4=0} \\
&= \partial\left\{\mathbb{E} \frac{(2A-1)\Delta \mathbb{E}_{\nu_4}(Y \mid \Delta=1, A, X)}{\mathbb{P}(A, \Delta=1 \mid X)}\right\} \Big/ \partial \nu_4 \Big|_{\nu_4=0}.
\end{aligned}
$$

Hints for Exercises in Chapter 15

Hints for Exercise 15.1:

By the chain rule, the likelihood can be written as

$$
\begin{aligned}
L =& p^{(0)}(X(0)) \\
&\times [g^{(0)}(1 \mid X(0))]^{A(0)}[1 - g^{(0)}(1 \mid X(0))]^{1-A(0)} \\
&\times p^{(1)}(X(1) \mid X(0), A(0)) \\
&\times [g^{(1)}(1 \mid X(0), A(0), X(1))]^{A(1)}[1 - g^{(1)}(1 \mid X(0), A(0), X(1))]^{1-A(1)} \\
&\cdots \\
&\times p^{(T-1)}(X(T-1) \mid \overline{X}(T-2), \overline{A}(T-2)) \\
&\times [g^{(T-1)}(1 \mid \overline{X}(T-1), \overline{A}(T-2))]^{A(T-1)}[1 - g^{(1)}(1 \mid \cdots)]^{1-A(T-1)} \\
&\times f(Y \mid \overline{X}(T-1), \overline{A}(T-1)).
\end{aligned}
$$

Hints for Exercise 15.2:

Denote the following true models for $p^{(t)}, g^{(t)}, g^{\delta(t)}$, and f:

$$
\begin{aligned}
&p_0^{(0)}(X(0)),\\
&g_0^{(0)}(1 \mid X(0)),\\
&p_0^{(1)}(X(1) \mid (X(0), A(0)),\\
&g_0^{(1)}(1 \mid X(0), A(0), X(1)),\\
&p_0^{(t)}(X(t) \mid \overline{X}(t-1), \overline{A}(t-1)), t = 2, \ldots, T-1,\\
&g_0^{(t)}(1 \mid \overline{X}(t), \overline{A}(t-1)), t = 2, \ldots, T-1,\\
&f_0(Y \mid \overline{X}(t), \overline{A}(t)).
\end{aligned}
$$

Specify any family of models of the following parametric forms:

$$
\begin{aligned}
&p^{(0)}(X(0); \nu^p(0)),\\
&g^{(0)}(1 \mid X(0); \nu^g(0)),\\
&p^{(1)}(X(1) \mid (X(0), A(0); \nu^p(1)),\\
&g^{(1)}(1 \mid X(0), A(0), X(1); \nu^g(1)),\\
&p^{(t)}(X(t) \mid \overline{X}(t-1), \overline{A}(t-1); \nu^p(t)), t = 2, \ldots, T-1,\\
&g^{(t)}(1 \mid \overline{X}(t), \overline{A}(t-1), \nu^g(t)), t = 2, \ldots, T-1,\\
&f(Y \mid \overline{X}(t), \overline{A}(t); \nu^f),
\end{aligned}
$$

which are equal to the corresponding true models when the parameters are equal to zero, respectively.

Hints for Exercise 15.3:

The log-likelihood is

$$
\begin{aligned}
\ell = {} & \log p^{(0)}(X(0); \nu^p(0))\\
& + A(0) \log g^{(0)}(1 \mid X(0); \nu^g(0)) + (1 - A(0)) \log[1 - g^{(0)}(1 \mid X(0); \nu^g(0))]\\
& + \ldots\\
& + \log f(Y \mid \overline{X}(t), \overline{A}(t); \nu^f).
\end{aligned}
$$

The score functions evaluated at $\boldsymbol{\nu} = \mathbf{0}$ are

$$
\begin{aligned}
\left.\frac{\partial \ell}{\partial \nu^p(0)}\right|_{\boldsymbol{\nu}=\mathbf{0}} &= \left.\frac{\partial p^{(0)}(X(0); \nu^p(0))}{p^{(0)}(X(0); \nu^p(0))}\right|_{\nu^p(0)=0},\\
\left.\frac{\partial \ell}{\partial \nu^g(0)}\right|_{\boldsymbol{\nu}=\mathbf{0}} &= \left.\frac{\partial g^{(0)}(1 \mid X(0); \nu^g(0))[A(0) - g^{(0)}(1 \mid X(0); \nu^g(0))]}{g^{(0)}(1 \mid X(0); \nu^g(0))[1 - g^{(0)}(1 \mid X(0); \nu^g(0))]}\right|_{\nu^p(0)=0},\\
&\ldots\\
\left.\frac{\partial \ell}{\partial \nu^f}\right|_{\boldsymbol{\nu}=\mathbf{0}} &= \left.\frac{\partial f(Y \mid \overline{X}(t), \overline{A}(t); \nu^f)}{f(Y \mid \overline{X}(t), \overline{A}(t); \nu^f)}\right|_{\nu^f=0}.
\end{aligned}
$$

Hints for Exercise 15.4:

Obtain the following derivatives:

$$\begin{aligned}\left.\frac{\partial\theta(\boldsymbol{\nu})}{\partial\nu^p(0)}\right|_{\boldsymbol{\nu}=0} &= \left.\frac{\partial\mathbb{E}_{\boldsymbol{\nu}}\left\{Q_0\left(X(0),a(0)\right)\right\}}{\partial\nu^p(0)}\right|_{\boldsymbol{\nu}=0}\\ &= \left.\frac{\partial\mathbb{E}_{\nu^p(0)}\left\{Q_0\left(X(0),a(0)\right)\right\}}{\partial\nu^p(0)}\right|_{\boldsymbol{\nu}=0}\\ &= \mathbb{E}\left[\phi_0(O)\times\left.\frac{\partial\ell}{\partial\nu^p(0)}\right|_{\boldsymbol{\nu}=\mathbf{0}}\right],\end{aligned}$$

$$\begin{aligned}\left.\frac{\partial\theta(\boldsymbol{\nu})}{\partial\nu^p(t)}\right|_{\boldsymbol{\nu}=0} &= \left.\frac{\partial\mathbb{E}_{\boldsymbol{\nu}}\left\{\frac{I\{\overline{A}(t-1)=\overline{a}(t-1)\}}{\prod_{s=0}^{t-1}g(1|\overline{X}(s),\overline{A}(s-1);\nu^g(s))}Q_t(\overline{X}(t),\overline{A}(t-1),a(t))\right\}}{\partial\nu^p(t)}\right|_{\boldsymbol{\nu}=0}\\ &= \left.\frac{\partial\mathbb{E}_{\nu^p(t)}\left\{\frac{I\{\overline{A}(t-1)=\overline{a}(t-1)\}}{\prod_{s=0}^{t-1}g_0(1|\overline{X}(s),\overline{A}(s-1))}Q_t(\overline{X}(t),\overline{A}(t-1),a(t))\right\}}{\partial\nu^p(t)}\right|_{\boldsymbol{\nu}=0}\\ &= \mathbb{E}\left[\phi_t(O)\times\left.\frac{\partial\ell}{\partial\nu^p(t)}\right|_{\boldsymbol{\nu}=\mathbf{0}}\right],\quad t=1,\ldots,T-1,\end{aligned}$$

$$\begin{aligned}\left.\frac{\partial\theta(\boldsymbol{\nu})}{\partial\nu^g(t)}\right|_{\boldsymbol{\nu}=0} &= \left.\frac{\partial\mathbb{E}_{\boldsymbol{\nu}}\left\{Q_0\left(X(0),a(0)\right)\right\}}{\partial\nu^g(t)}\right|_{\boldsymbol{\nu}=0}\\ &= \left.\frac{\partial\mathbb{E}_{\nu^g(t)}\left\{Q_0\left(X(0),a(0)\right)\right\}}{\partial\nu^g(t)}\right|_{\boldsymbol{\nu}=0}\\ &= 0,\quad t=0,\ldots,T-1,\end{aligned}$$

$$\begin{aligned}\left.\frac{\partial\theta(\boldsymbol{\nu})}{\partial\nu^f}\right|_{\boldsymbol{\nu}=0} &= \left.\frac{\partial\mathbb{E}_{\boldsymbol{\nu}}\left\{\frac{I\{\overline{A}(T-1)=\overline{a}(T-1)\}}{\prod_{t=0}^{T-1}g(1|\overline{X}(t),\overline{A}(t-1);\nu^g(t))}Y\right\}}{\partial\nu^f}\right|_{\boldsymbol{\nu}=0}\\ &= \left.\frac{\partial\mathbb{E}_{\nu^f}\left\{\frac{I\{\overline{A}(T-1)=\overline{a}(T-1)\}}{\prod_{t=0}^{T-1}g_0(1|\overline{X}(t),\overline{A}(t-1))}Y\right\}}{\partial\nu^f}\right|_{\boldsymbol{\nu}=0}\\ &= \mathbb{E}\left[\phi_T(O)\times\left.\frac{\partial\ell}{\partial\nu^f}\right|_{\boldsymbol{\nu}=\mathbf{0}}\right].\end{aligned}$$

Hints for Exercise 15.5:

First, to verify that

$$\left.\frac{\partial\theta(\boldsymbol{\nu})}{\partial\nu^p(t)}\right|_{\boldsymbol{\nu}=0} = \mathbb{E}\left[\sum_{s=0}^{T}\phi_s(O)\times\left.\frac{\partial\ell}{\partial\nu^p(t)}\right|_{\boldsymbol{\nu}=\mathbf{0}}\right], \quad t=0,\ldots,T-1,$$

it suffices to verify that

$$\mathbb{E}\left[\phi_s(O)\times\left.\frac{\partial\ell}{\partial\nu^p(t)}\right|_{\boldsymbol{\nu}=\mathbf{0}}\right] = 0, \quad s\neq t.$$

In fact, if $s<t$,

$$\begin{aligned}
\mathbb{E}\left[\phi_s(O)\times\left.\frac{\partial\ell}{\partial\nu^p(t)}\right|_{\boldsymbol{\nu}=\mathbf{0}}\right] &= \mathbb{E}\left\{\mathbb{E}\left[\phi_s(O)\left.\frac{\partial\ell}{\partial\nu^p(t)}\right|_{\boldsymbol{\nu}=\mathbf{0}}\;\middle|\;\overline{X}(t-1),\overline{A}(t-1)\right]\right\}\\
&= \mathbb{E}\left\{\phi_s(O)\,\mathbb{E}\left[\left.\frac{\partial\ell}{\partial\nu^p(t)}\right|_{\boldsymbol{\nu}=\mathbf{0}}\;\middle|\;\overline{X}(t-1),\overline{A}(t-1)\right]\right\}\\
&= \mathbb{E}\left\{\phi_s(O)\times 0\right\}=0.
\end{aligned}$$

If $s>t$,

$$\begin{aligned}
\mathbb{E}\left[\phi_s(O)\times\left.\frac{\partial\ell}{\partial\nu^p(t)}\right|_{\boldsymbol{\nu}=\mathbf{0}}\right] &= \mathbb{E}\left\{\mathbb{E}\left[\phi_s(O)\left.\frac{\partial\ell}{\partial\nu^p(t)}\right|_{\boldsymbol{\nu}=\mathbf{0}}\;\middle|\;\overline{X}(t),\overline{A}(t)\right]\right\}\\
&= \mathbb{E}\left\{\left.\frac{\partial\ell}{\partial\nu^p(t)}\right|_{\boldsymbol{\nu}=\mathbf{0}}\mathbb{E}\left[\phi_s(O)\;\middle|\;\overline{X}(t),\overline{A}(t)\right]\right\}\\
&= \mathbb{E}\left\{\left.\frac{\partial\ell}{\partial\nu^p(t)}\right|_{\boldsymbol{\nu}=\mathbf{0}}\times 0\right\}=0.
\end{aligned}$$

Second, to verify that

$$\left.\frac{\partial\theta(\boldsymbol{\nu})}{\partial\nu^g(t)}\right|_{\boldsymbol{\nu}=0} = \mathbb{E}\left[\sum_{s=0}^{T}\phi_s(O)\times\left.\frac{\partial\ell}{\partial\nu^g(t)}\right|_{\boldsymbol{\nu}=\mathbf{0}}\right], \quad t=0,\ldots,T-1,$$

it suffices to verify that

$$\mathbb{E}\left[\phi_s(O)\times\left.\frac{\partial\ell}{\partial\nu^g(t)}\right|_{\boldsymbol{\nu}=\mathbf{0}}\right] = 0, \quad s=0,\ldots,T.$$

In fact, if $s\leq t$,

$$\begin{aligned}
\mathbb{E}\left[\phi_s(O)\times\left.\frac{\partial\ell}{\partial\nu^g(t)}\right|_{\boldsymbol{\nu}=\mathbf{0}}\right] &= \mathbb{E}\left\{\mathbb{E}\left[\phi_s(O)\left.\frac{\partial\ell}{\partial\nu^g(t)}\right|_{\boldsymbol{\nu}=\mathbf{0}}\;\middle|\;\overline{X}(t),\overline{A}(t-1)\right]\right\}\\
&= \mathbb{E}\left\{\phi_s(O)\,\mathbb{E}\left[\left.\frac{\partial\ell}{\partial\nu^g(t)}\right|_{\boldsymbol{\nu}=\mathbf{0}}\;\middle|\;\overline{X}(t),\overline{A}(t-1)\right]\right\}\\
&= \mathbb{E}\left\{\phi_s(O)\times 0\right\}=0.
\end{aligned}$$

If $s > t$,

$$\begin{aligned}\mathbb{E}\left[\phi_s(O) \times \left.\frac{\partial \ell}{\partial \nu^g(t)}\right|_{\boldsymbol{\nu}=\mathbf{0}}\right] &= \mathbb{E}\left\{\mathbb{E}\left[\phi_s(O) \left.\frac{\partial \ell}{\partial \nu^g(t)}\right|_{\boldsymbol{\nu}=\mathbf{0}} \,\middle|\, \overline{X}(t), \overline{A}(t)\right]\right\} \\ &= \mathbb{E}\left\{\left.\frac{\partial \ell}{\partial \nu^g(t)}\right|_{\boldsymbol{\nu}=\mathbf{0}} \mathbb{E}\left[\phi_s(O) \,\middle|\, \overline{X}(t), \overline{A}(t)\right]\right\} \\ &= \mathbb{E}\left\{\left.\frac{\partial \ell}{\partial \nu^g(t)}\right|_{\boldsymbol{\nu}=\mathbf{0}} \times 0\right\} = 0.\end{aligned}$$

Finally, to verify

$$\left.\frac{\partial \theta(\boldsymbol{\nu})}{\partial \nu^f}\right|_{\boldsymbol{\nu}=0} = \mathbb{E}\left[\sum_{s=0}^{T} \phi_s(O) \times \left.\frac{\partial \ell}{\partial \nu^f(t)}\right|_{\boldsymbol{\nu}=\mathbf{0}}\right],$$

it suffices to verify that

$$\mathbb{E}\left[\phi_s(O) \times \left.\frac{\partial \ell}{\partial \nu^f}\right|_{\boldsymbol{\nu}=\mathbf{0}}\right] = 0, \quad s = 0, \ldots, T-1.$$

In fact, for $s = 0, \ldots, T-1$,

$$\begin{aligned}\mathbb{E}\left[\phi_s(O) \times \left.\frac{\partial \ell}{\partial \nu^f}\right|_{\boldsymbol{\nu}=\mathbf{0}}\right] &= \mathbb{E}\left\{\mathbb{E}\left[\phi_s(O) \left.\frac{\partial \ell}{\partial \nu^f}\right|_{\boldsymbol{\nu}=\mathbf{0}} \,\middle|\, \overline{X}(T-1), \overline{A}(T-1)\right]\right\} \\ &= \mathbb{E}\left\{\phi_s(O)\, \mathbb{E}\left[\left.\frac{\partial \ell}{\partial \nu^f}\right|_{\boldsymbol{\nu}=\mathbf{0}} \,\middle|\, \overline{X}(T-1), \overline{A}(T-1)\right]\right\} \\ &= \mathbb{E}\left\{\phi_s(O) \times 0\right\} = 0.\end{aligned}$$

Hints for Exercises in Chapter 16

Hints for Exercise 16.1:

Consider the following two Taylor expansions:

$$\begin{aligned}&-\sum_{i=1}^{n} 2 \log f(O_i; \widehat{\theta}_n(k)) \\ &\overset{(1)}{=} -\sum_{i=1}^{n} 2 \log f(O_i; \theta_0) - \sum_{i=1}^{n} 2(\widehat{\theta}_n(k) - \theta_0)^\top \frac{\partial \log f(O_i; \theta_0)}{\partial \theta} \\ &\quad -\sum_{i=1}^{n} (\widehat{\theta}_n(k) - \theta_0)^\top \frac{\partial^2 \log f(O_i; \theta_0)}{\partial \theta \partial \theta^\top} (\widehat{\theta}_n(k) - \theta_0),\end{aligned}$$

and

$$
\begin{aligned}
&-\sum_{i=1}^{n} 2\log f(O_i;\theta_0) \\
&\overset{(2)}{=} -\sum_{i=1}^{n} 2\log f(O_i;\widehat{\theta}_n(k)) - \sum_{i=1}^{n} 2(\theta_0 - \widehat{\theta}_n(k))^\top \frac{\partial \log f(O_i;\widehat{\theta}_n(k))}{\partial\theta} \\
&\quad -\sum_{i=1}^{n} (\widehat{\theta}_n(k) - \theta_0)^\top \frac{\partial^2 \log f(O_i;\widehat{\theta}_n(k))}{\partial\theta\partial\theta^\top} (\widehat{\theta}_n(k) - \theta_0),
\end{aligned}
$$

In the first Taylor expansion, the conditional expectation of the second term given $\widehat{\theta}_n(k)$ is zero because

$$
\mathbb{E}\frac{\partial \log f(O_i;\theta_0)}{\partial\theta} = 0.
$$

In the second Taylor expansion, the second term is zero because, according to the definition of the maximum likelihood estimator,

$$
\sum_{i=1}^{n} \frac{\partial \log f(O_i;\widehat{\theta}_n(k))}{\partial\theta} = 0.
$$

Therefore, we have

$$
\begin{aligned}
D(\widehat{\theta}_n(k)) \doteq & \mathbb{E}\left[-\sum_{i=1}^{n} 2\log f(O_i;\widehat{\theta}_n(k))\right] \\
&+ \mathbb{E}\left[-\sum_{i=1}^{n} (\widehat{\theta}_n(k) - \theta_0)^\top \frac{\partial^2 \log f(O_i;\theta_0)}{\partial\theta\partial\theta^\top} (\widehat{\theta}_n(k) - \theta_0)\right] \\
&+ \mathbb{E}\left[-\sum_{i=1}^{n} (\widehat{\theta}_n(k) - \theta_0)^\top \frac{\partial^2 \log f(O_i;\widehat{\theta}_n(k))}{\partial\theta\partial\theta^\top} (\widehat{\theta}_n(k) - \theta_0)\right].
\end{aligned}
$$

Next, by the asymptotic normality of the maximum likelihood estimator, we see that

$$
-\sum_{i=1}^{n} (\widehat{\theta}_n(k) - \theta_0)^\top \frac{\partial^2 \log f(O_i;\theta_0)}{\partial\theta\partial\theta^\top} (\widehat{\theta}_n(k) - \theta_0) \xrightarrow{d} \chi_k^2,
$$

$$
-\sum_{i=1}^{n} (\widehat{\theta}_n(k) - \theta_0)^\top \frac{\partial^2 \log f(O_i;\widehat{\theta}_n(k))}{\partial\theta\partial\theta^\top} (\widehat{\theta}_n(k) - \theta_0) \xrightarrow{d} \chi_k^2.
$$

Here, χ_k^2 denotes the chi-square distribution with k degrees of freedom, and the expectation of χ_k^2 is k.

Hints for Exercise 16.2:

The likelihood is

$$L = \prod_{i=1}^{n} \frac{1}{\sqrt{2\pi\sigma^2}} \exp\left\{-\frac{(Y_i - X_i^\top \beta)^2}{2\sigma^2}\right\},$$

and the log-likelihood is

$$\ell = -\frac{n}{2}\log(2\pi) - \frac{n}{2}\log(\sigma^2) - \frac{1}{2\sigma^2}\sum_{i=1}^{n}(Y_i - X_i^\top \beta)^2.$$

Thus,

$$-2\ell = n\log(2\pi) + n\log(\sigma^2) + \frac{1}{\sigma^2}\sum_{i=1}^{n}(Y_i - X_i^\top \beta)^2.$$

Plugging in the MLEs of β and σ^2, it becomes

$$-2\ell = n\log(2\pi) + n\log\left(\frac{\text{RSS}}{n}\right) + n.$$

Therefore, ignoring the constants $n\log(2\pi)$ and n in the expression of -2ℓ, the AIC can be written as

$$\text{AIC} = n\log\left(\frac{\text{RSS}}{n}\right) + 2k,$$

where $k = (d+1) + 1 = d + 2$.

Hints for Exercise 16.3:

Define a smart vector $\mathbf{Y}^*$,

$$\mathbf{Y}^* = \begin{pmatrix} Y_1 \\ \cdots \\ Y_{i-1} \\ X_i^\top \widehat{\beta}_n^{-i} \\ Y_{i+1} \\ \cdots \\ Y_n \end{pmatrix}.$$

That is, we replace the ith element of $\mathbf{Y}$ by its prediction $X_i^\top \widehat{\beta}_n^{-i}$. Since

$$\|\mathbf{Y}^* - \mathbf{X}\beta\|^2 \geq \|\mathbf{Y}_{-i} - \mathbf{X}_{-i}\beta\|^2 \geq \|\mathbf{Y}_{-i} - \mathbf{X}_{-i}\widehat{\beta}_n^{-i}\|^2 = \|\mathbf{Y}^* - \mathbf{X}\widehat{\beta}_n^{-i}\|^2,$$

$$\widehat{\beta}_n^{-i} = (\mathbf{X}^\top \mathbf{X})^{-1}\mathbf{X}^\top \mathbf{Y}^*.$$

Thus, we have

$$\mathbf{X}\widehat{\beta}_n^{-i} = H\mathbf{Y}^*.$$

Looking at the ith row of the above equation, we have

$$\begin{aligned}
X_i^\top \widehat{\beta}_n^{-i} &= \sum_{i' \neq i} h_{i'i} Y_{i'} + h_{ii} X_i^\top \widehat{\beta}_n^{-i} \\
&= \sum_{i'=1}^{n} h_{i'i} Y_i - h_{ii} Y_i + h_{ii} X_i^\top \widehat{\beta}_n^{-i} \\
&= X_i^\top \widehat{\beta}_n - h_{ii} Y_i + h_{ii} X_i^\top \widehat{\beta}_n^{-i} \\
&= X_i^\top \widehat{\beta}_n - Y_i + (1 - h_{ii}) Y_i + h_{ii} X_i^\top \widehat{\beta}_n^{-i}
\end{aligned}$$

leading to

$$Y_i - X_i^\top \widehat{\beta}_n = (1 - h_{ii})(Y_i - X_i^\top \widehat{\beta}_n^{-i}).$$

Hints for Exercise 16.4:

Let $Z_i^{(j)} = \widehat{Q}^{j,k}(X_i)$, for $(X_i, Y_i) \in \mathcal{D}_k$. Thus,

$$CV(w_1, \ldots, w_J) = \frac{1}{n} \sum_{i=1}^{n} \left(Y_i - \sum_{j=1}^{J} w_j Z_i^{(j)} \right)^2.$$

Hints for Exercise 16.5:

Let $Z_i^{(j)} = \text{logit}[\widehat{Q}^{j,k}(X_i)]$, for $(X_i, Y_i) \in \mathcal{D}_k$. Let $w = (w_1, \ldots, w_J)^\top$ and $Z_i = (Z_i^{(1)}, \ldots, Z_i^{(J)})^\top$. Thus,

$$\begin{aligned}
CV(w) &= \frac{1}{n} \sum_{i=1}^{n} L\left(Y_i, \text{expit}(w^\top Z_i)\right) \\
&= -\frac{1}{n} \sum_{i=1}^{n} \left\{ Y_i \log\left[\text{expit}(w^\top Z_i)\right] + (1 - Y_i) \log\left[1 - \text{expit}(w^\top Z_i)\right] \right\}.
\end{aligned}$$

Hints for Exercises in Chapter 17

Hints for Exercise 17.1:

By the LIE,

$$\begin{aligned}
\mathbb{E}\{D_{Y|X,A}\, S_{0;\nu_1}(X)\} &= \mathbb{E}\{S_{0;\nu_1}(X)\, \mathbb{E}[D_{Y|X,A} \mid X]\} = 0, \\
\mathbb{E}\{D_X\, S_{0;\nu_3}(X, A, Y)\} &= \mathbb{E}\{D_X\, \mathbb{E}[S_{0;\nu_3}(X, A, Y) \mid X, A]\} = 0.
\end{aligned}$$

Hints for Exercise 17.2:

For a given submodel $p(X;\nu_1)$, the C-R lower bound is

$$\mathbb{E}[D_X\, S_{0;\nu_1}(X)]\,\mathbb{V}^{-1}[S_{0;\nu_1}(X)]\,\mathbb{E}D_X\, S_{0;\nu_1}(X).$$

By the Cauchy-Schwarz inequality,

$$\mathbb{E}[D_X\, S_{0;\nu_1}(X)]\,\mathbb{V}^{-1}[S_{0;\nu_1}(X)]\,\mathbb{E}[D_X\, S_{0;\nu_1}(X)] \le \mathbb{V}(D_X),$$

where the equality holds if and only if $S_{0;\nu_1}(X) = c\cdot D_X$ for a constant c. This shows that a least favorable parametric submodel is any parametric submodel with $S_{0;\nu_1}(X) = c\cdot D_X$.

Similarly, for a given submodel $f(Y \mid X, A; \nu_3)$, the C-R lower bound is

$$\mathbb{E}[D_{Y|X,A}\, S_{0;\nu_3}(X,A,Y)]\,\mathbb{V}^{-1}[S_{0;\nu_3}(X,A,Y)]\,\mathbb{E}[D_X\, S_{0;\nu_3}(X,A,Y)],$$

achieving maximum if and only if $S_{0;\nu_3}(X,A,Y) = c\cdot D_{Y|X,A}$.

Hints for Exercise 17.3:

The score function $S_{0;\nu_1}(X)$ is

$$\left.\frac{\partial \log p(X;\nu_1)}{\partial \nu_1}\right|_{\nu_1=0} = \left.\frac{\partial p(X;\nu_1)/\partial\nu_1}{p(X;\nu_1)}\right|_{\nu_1=0} = D_X.$$

The score function $S_{0;\nu_2}(X)$ is zero. The score function $S_{0;\nu_3}(X,A,Y)$ is

$$\begin{aligned}
&\left.\frac{\partial\left\{Y\log[Q(X,A;\nu_3)] + (1-Y)\log[1-Q(X,A;\nu_3)]\right\}}{\partial\nu_3}\right|_{\nu_3=0}\\
=&\left.\frac{\partial\left\{\log[1-Q(X,A;\nu_3)] + Y\mathrm{logit}[Q(X,A;\nu_3)]\right\}}{\partial\nu_3}\right|_{\nu_3=0}\\
=&\partial\log\left[\frac{1}{1+\exp(q_0+\nu_3)}\right]/\partial\nu_3\Big|_{\nu_3=0} + \partial\{Y(q_0+\nu_3 H)\}\partial\nu_3\Big|_{\nu_3=0}\\
=&-\left.\frac{\exp(q_0+\nu_3 H)H}{1+\exp(q_0+\nu_3 H)}\right|_{\nu_3=0} + YH\\
=&-Q_0(X,A)H(X,A) + YH(X,A) = H(X,A)[Y - Q_0(X,A)] = D_{Y|X,A}.
\end{aligned}$$

Hints for Exercise 17.4:

The log-likelihood is

$$\begin{aligned}
\ell(\nu_3) &= \sum_{i=1}^{n} Y_i\log[Q(X_i,A_i;\nu_3)] + (1-Y_i)\log[1-Q(X_i,A_i;\nu_3)]\\
&= \sum_{i=1}^{n}\log[1-Q(X_i,A_i;\nu_3)] + Y_i\mathrm{logit}[Q(X_i,A_i;\nu_3)].
\end{aligned}$$

Thus,

$$\frac{\partial \ell(\nu_3)}{\partial \nu_3} = \sum_{i=1}^{n} \left[-\widehat{H}_n^0(X_i, A_i) Q(X_i, A_i; \nu_3) + Y_i \widehat{H}_n^0(X_i, A_i) \right]$$
$$= \sum_{i=1}^{n} \widehat{H}_n^0(X_i, A_i) \left[Y_i - Q(X_i, A_i; \nu_3) \right].$$

Hints for Exercise 17.5:

The TMLE procedure described in Subsection 17.4.1 is referred to as the *point-treatment TMLE*. The LTMLE can then be interpreted as a backward algorithm consisting of T sequential applications of the point-treatment TMLE, with one application performed at each time point. Moreover, as shown in Subsection 17.4.1, no further updates are required for a point-treatment TMLE after the initial targeting step.

Hints for Exercises in Chapter 18

Hints for Exercise 18.1:

$W_1 = X_2/X_1^2$, $W_2 = X_3$, $W_3 = I(X_3 = \texttt{NA})$, and $\Delta = I(Y \neq \texttt{NA})$:

ID	W_1	W_2	W_3	A	Δ	Y
1	22.5	40	1	1	1	18
2	33.2	35	0	0	0	NA
...						
200	17.9	42	0	1	1	14

Hints for Exercise 18.2:

The R code implies the following SCM:

$$\begin{aligned}
L0 &= f_{L0}(U_{L0}), \\
A0 &= f_{A0}(L0, U_{A0}), \\
C0 &= f_{C0}(A0, U_{C0}), \\
L1 &= f_{L1}(L0, A0, U_{L1}), \\
Y1 &= f_{Y1}(L0, A0, U_{Y1}), \\
A1 &= f_{A1}(L0, A0, U_{A1}), \\
C1 &= f_{C1}(A1, U_{C1}), \\
L2 &= f_{L2}(L1, A1, U_{L2}), \\
Y2 &= f_{Y2}(L1, Y1, A1, U_{Y2}), \\
A2 &= f_{A2}(L1, A1, U_{A2}), \\
C2 &= f_{C2}(A2, U_{C2}), \\
Y3 &= f_{Y3}(L2, Y2, A2, U_{Y3}).
\end{aligned}$$

Hints for Exercise 18.3:

Use the R function `ltmle()`.

Hints for Exercise 18.4:

Apply the R function `ltmle()` to each DTR rule and compare the resulting estimates.

Hints for Exercise 18.5:

In a four-arm study, the dummy variable pairs $(A.a, A.b) = (0,0), (1,0), (0,1), (1,1)$ can represent treatments 0, 1, 2, and 3, respectively. In a five-arm study, one possible encoding using three dummy variables is: $(A.a, A.b, A.c) = (0,0,0),\ (1,0,0),\ (0,1,0),\ (0,0,1),\ (1,1,1)$.

Hints for Exercises in Chapter 19

Hints for Exercise 19.1:

If the intercurrent event is ignored for the fifth subject, the data would be represented as: $C_5(0) = \cdots = C_5(11) = \texttt{uncensored}$, $Y_5(1) = \cdots = Y_5(6) = 0$, $Y_5(7) = \cdots = Y_5(12) = 1$. Similarly, if the intercurrent event is ignored for the sixth subject, the data would be: $C_6(0) = \cdots = C_6(6) = \texttt{uncensored}$, $C_6(7) = \cdots = C_6(11) = \texttt{censored}$, $Y_6(1) = \cdots = Y_6(7) = 0$, $Y_6(8) = \cdots = Y_6(12) = \texttt{NA}$.

Hints for Exercise 19.2:

Summarize the values of $C_i(t)$ and $Y(_it)$, $i = 1, \ldots, 6$, as described in detail in Subsection 19.3.1, in a 6×24 table, where 12 columns for C nodes and 12 for Y nodes.

Hints for Exercise 19.3:

Revise the original values of $A(t)$ to obtain $\widetilde{A}(t)$ under the treatment policy strategy, and construct a 2×36 table accordingly.

Hints for Exercise 19.4:

Revise the original values of $C(t)$ to obtain $\widetilde{C}(t)$ under the hypothetical strategy, and construct a 2×36 table accordingly.

Hints for Exercise 19.5:

Revise the original values of $Y(t)$ to obtain $\widetilde{Y}(t)$ under the composite variable strategy, and construct a 2×36 table accordingly.

Hints for Exercises in Chapter 20

Hints for Exercise 20.1:

Two equivalent expressions of θ_h are

$$\begin{aligned}\theta_h &\overset{(1)}{=} \frac{1}{\mathbb{E}[h(X)]}\mathbb{E}\{[Q(X,1) - Q(X,0)]h(X)\} \\ &\overset{(2)}{=} \frac{1}{\mathbb{E}[h(X)]}\mathbb{E}\left[\frac{(2A-1)h(X)Y}{g(1 \mid X)}\right],\end{aligned}$$

where (1) is via the standardization strategy and (2) is via the weighting strategy. Write the estimand θ_h as the ratio of two estimands,

$$\theta_h = \alpha_h/\beta_h,$$

where

$$\alpha_h = \mathbb{E}\{[Q(X,1) - Q(X,0)]h(X)\} \quad \text{and} \quad \beta_h = \mathbb{E}[h(X)].$$

To derive the EIF for α_h, we apply the convenient approach in Chapter 13. The first step is to center the above two equivalent expressions as:

$$\begin{aligned}&h(X)[Q(X,1) - Q(X,0)] - \alpha_h, \\ &\frac{h(X)(2A-1)}{g(A \mid X)}[Y - Q(X,A)].\end{aligned}$$

The second step is to add them, leading to the following EIF for α_h:

$$\phi_1 = h(X)[Q(X,1) - Q(X,0)] - \alpha_h + \frac{h(X)(2A-1)}{g(A \mid X)}[Y - Q(X,A)].$$

Since $\beta_h = \mathbb{E}[h(X)]$, the EIF for β_h is:

$$\phi_2 = h(X_i) - \beta_h.$$

For any given parametric submodel indexed by ϵ, we have

$$\left.\frac{\partial\theta_h(\epsilon)}{\partial\epsilon}\right|_{\epsilon=0} = \frac{1}{\beta_h}\left[\left.\frac{\partial\alpha_h(\epsilon)}{\partial\epsilon}\right|_{\epsilon=0} - \theta_h\left.\frac{\partial\beta_h(\epsilon)}{\partial\epsilon}\right|_{\epsilon=0}\right].$$

Thus, by fundamental theorem of regularity, for the ratio-type estimand, the EIF for θ_h equals:

$$\phi_h(X, A, Y) = \frac{1}{\beta_h}\left[\phi_1(X, A, Y) - \theta_h \phi_2(X)\right].$$

Plugging ϕ_1 and ϕ_2 in to the above expression, we derive the EIF for θ_h.

Hints for Exercise 20.2:

In the initial step, we fit a generalized linear model or a super learner for the outcome regression $Q(x, a)$ and propensity score $g(1 \mid x)$, yielding $\widehat{Q}^0(x, a)$ and $\widehat{g}^0(1 \mid x)$. The initial estimator is

$$\widehat{\theta}_h^0 = \frac{1}{\sum_{j=1}^N h(X_j)} \sum_{i=1}^N \left[\widehat{Q}^0(X_i, 1) - \widehat{Q}^0(X_i, 0)\right].$$

In the targeting step, using the efficient influence function $\phi_h(X, A, Y)$, consider the submodel

$$\begin{aligned} p(X; \epsilon_1) &= \widehat{p}^0(X)\{1 + \epsilon_1 D(X)\}, \\ g(1 \mid X; \epsilon_2) &= \operatorname{expit}\big(\operatorname{logit}[\widehat{g}^0(1 \mid X)]\big), \\ Q(X, A; \epsilon_3) &= \operatorname{expit}\big(\operatorname{logit}[\widehat{Q}^0(X, A)] + \epsilon_3 H(X, A)\big), \end{aligned}$$

where $\widehat{p}^0$ is the empirical distribution,

$$D(X) = h(X)\big[\widehat{Q}^0(X, 1) - \widehat{Q}^0(X, 0) - \widehat{\theta}_h^0\big], \quad H(X, A) = \frac{h(X)(2A - 1)}{\widehat{g}^0(A \mid X)}.$$

Since $\widehat{p}^0$ is the NPMLE and the score for g at $\epsilon_2 = 0$ is zero, only Q is updated via

$$\operatorname{logit} Q(X, A; \epsilon_3) = \operatorname{logit} \widehat{Q}^0(X, A) + \epsilon_3 H(X, A).$$

Let $\widehat{\epsilon}_3$ be the estimate, giving

$$\widehat{Q}^*(X, A) = \operatorname{expit}\big(\operatorname{logit}[\widehat{Q}^0(X, A)] + \widehat{\epsilon}_3 H(X, A)\big).$$

A single update suffices, yielding the TMLE

$$\widehat{\theta}_h = \frac{1}{\sum_{j=1}^N h(X_j)} \sum_{i=1}^N h(X_i)\big[\widehat{Q}^*(X_i, 1) - \widehat{Q}^*(X_i, 0)\big].$$

Hints for Exercise 20.3:

The estimand θ_h can be expressed as the ratio of two population parameters,

$$\theta_h = \frac{\theta_1}{\theta_2},$$

where

$$\theta_1 = \mathbb{E}\{\eta(g(X))[Q(X,1) - Q(X,0)]\},$$
$$\theta_2 = \mathbb{E}\{\eta(g(X))\}.$$

Define

$$\dot{\eta}(g(X)) = \frac{d}{dg}\eta(g)\Big|_{g=g(X)}.$$

The EIF of θ_1 is

$$\phi_1(O) = \eta(g(X))\left[\frac{A}{g(X)}(Y - Q(X,1)) - \frac{1-A}{1-g(X)}(Y - Q(X,0))\right]$$
$$+ \eta(g(X))[Q(X,1) - Q(X,0)] - \theta_1$$
$$+ \dot{\eta}(g(X))[Q(X,1) - Q(X,0)](A - g(X)).$$

The EIF of θ_2 is

$$\phi_2(O) = \eta(g(X)) - \theta_2 + \dot{\eta}(g(X))(A - g(X)).$$

Thus, the EIF of τ_η is

$$\phi_\eta(O) = \frac{\phi_1(O) - \theta_h\phi_2(O)}{\theta_2}.$$

Plugging in ϕ_1 and ϕ_2, the EIF of θ_h is

$$\phi_h(O) = \frac{\eta(g(X))}{\mathbb{E}\{\eta(g(X))\}}\left[\frac{A}{g(X)}(Y - Q(X,1)) - \frac{1-A}{1-g(X)}(Y - Q(X,0))\right]$$
$$+ \frac{\eta(g(X))}{\mathbb{E}\{\eta(g(X))\}}[Q(X,1) - Q(X,0) - \theta_h]$$
$$+ \frac{\dot{\eta}(g(X))}{\mathbb{E}\{\eta(g(X))\}}[Q(X,1) - Q(X,0) - \theta_h](A - g(X)).$$

Remark: The hints for Exercises 20.1–20.3 are motivated by Fang and He (2026).[98]

Hints for Exercise 20.4:

First, SUTVA and consistency. Second, exchangeability and positivity within the RCT. Third, positivity of RCT enrollment in the pooled population:

$$\mathbb{P}(S = 1 \mid X) > 0.$$

Hints for Exercise 20.5:

The corresponding statistical estimands are

$$\theta_1 = \mathbb{E}\left[\mathbb{E}(Y \mid S = 1, X, A = 1) - \mathbb{E}(Y \mid S = 1, X, A = 0)\right],$$
$$\theta_2 = \mathbb{E}\left\{[\mathbb{E}(Y \mid S = 1, X, A = 1) - \mathbb{E}(Y \mid S = 1, X, A = 0)] \mid S = 1\right\}.$$

Bibliography

[1] D. Stipp, *A Most Elegant Equation: Euler's Formula and the Beauty of Mathematics.* Hachette UK, 2017.

[2] K.-L. Chung, *A Course in Probability Theory.* Academic Press, 2001.

[3] B. Efron, "Bootstrap methods: Another look at the jackknife," *The Annals of Statistics*, vol. 7, no. 2, pp. 1–26, 1979.

[4] B. Efron and R. Tibshirani, *An Introduction to the Bootstrap.* CRC Press, 1994.

[5] S. Hawking, *A Brief History of Time: From the Big Bang to Black Holes.* Bantam Books, 1988.

[6] S. Kullback, *Information Theory and Statistics.* Courier Corporation, 1997.

[7] C. E. Shannon, "A mathematical theory of communication," *The Bell System Technical Journal*, vol. 27, no. 3, pp. 379–423, 1948.

[8] W. Feller, *An Introduction to Probability Theory and Its Applications.* John Wiley & Sons, 1971.

[9] S. Nadarajah, "A generalized normal distribution," *Journal of Applied Statistics*, vol. 32, no. 7, pp. 685–694, 2005.

[10] W. G. Cochran, *Sampling Techniques.* John Wiley & Sons, 1977.

[11] S. M. Ross, *A First Course in Probability.* Macmillan, 1976.

[12] A. van der Vaart, *Asymptotic Statistics.* Cambridge University Press, 2000.

[13] E. L. Lehmann and G. Casella, *Theory of Point Estimation.* Springer Science & Business Media, 2006.

[14] E. L. Lehmann, *Fisher, Neyman, and the Creation of Classical Statistics.* Springer Science & Business Media, 2011.

[15] D. Salsburg, *The Lady Tasting Tea: How Statistics Revolutionized Science in the Twentieth Century.* Macmillan, 2002.

[16] D. R. Cox and D. V. Hinkley, *Theoretical Statistics.* CRC Press, 1979.

[17] E. L. Lehmann and J. P. Romano, *Testing Statistical Hypotheses.* Springer, 2005.

[18] R. L. Wasserstein and N. A. Lazar, "The ASA statement on p-values: Context, process, and purpose," *The American Statistician*, vol. 70, no. 2, pp. 129–133, 2016.

[19] J. Aldrich, "RA Fisher and the making of maximum likelihood 1912-1922," *Statistical Science*, vol. 12, no. 3, pp. 162–176, 1997.

[20] P. McCullagh and J. Nelder, *Generalized Linear Models.* CRC Press, 1989.

[21] J. Pearl and D. Mackenzie, *The Book of Why: The New Science of Cause and Effect.* Basic Books, 2018.

[22] Y. Fang, *Causal Inference in Pharmaceutical Statistics.* Chapman and Hall/CRC Press, 2024.

[23] C. Schardt, M. B. Adams, T. Owens, S. Keitz, and P. Fontelo, "Utilization of the PICO framework to improve searching PubMed for clinical questions," *BMC Medical Informatics and Decision Making*, vol. 7, pp. 1–6, 2007.

[24] A. Dmitrienko, A. C. Tamhane, and F. Bretz, *Multiple Testing Problems in Pharmaceutical Statistics.* CRC Press, 2009.

[25] J. J. Riva, K. M. Malik, S. J. Burnie, A. R. Endicott, and J. W. Busse, "What is your research question? An introduction to the PICOT format for clinicians," *The Journal of the Canadian Chiropractic Association*, vol. 56, no. 3, p. 167, 2012.

[26] D. B. Rubin, "Estimating causal effects of treatments in randomized and nonrandomized studies," *Journal of Educational Psychology*, vol. 66, no. 5, p. 688, 1974.

[27] A. Gelman and A. Vehtari, "What are the most important statistical ideas of the past 50 years?," *Journal of the American Statistical Association*, vol. 116, no. 536, pp. 2087–2097, 2021.

[28] J. Pearl, *Causality.* Cambridge University Press, 2009.

[29] D. Rubin, "Discussion on 'randomization analysis of experimental data: The Fisher randomization test' by Basu," *Journal of the American Statistical Association*, vol. 75, no. 371, pp. 591–593, 1980.

[30] M. Hernán and J. Robins, *Causal Inference: What If.* Chapman & Hall/CRC Press, 2020.

[31] E. H. Simpson, "The interpretation of interaction in contingency tables," *Journal of the Royal Statistical Society: Series B (Statistical Methodology)*, vol. 13, no. 2, pp. 238–241, 1951.

[32] J. Berkson, "Limitations of the application of fourfold table analysis to hospital data," *Biometrics Bulletin*, vol. 2, no. 3, pp. 47–53, 1946.

[33] J. Pearl, M. Glymour, and N. Jewell, *Causal Inference in Statistics: A Primer.* John Wiley & Sons, 2016.

[34] S. Greenland, J. Pearl, and J. Robins, "Causal diagrams for epidemiologic research," *Epidemiology*, vol. 10, no. 1, pp. 37–48, 1999.

[35] G. Imbens and D. Rubin, *Causal Inference in Statistics, Social, and Biomedical Sciences.* Cambridge University Press, 2015.

[36] Y. Fang, "Two basic statistical strategies of conducting causal inference in real-world studies," *Contemporary Clinical Trials*, vol. 99, p. 106193, 2020.

[37] D. Phillippo, T. Ades, S. Dias, S. Palmer, K. R. Abrams, and N. Welton, "NICE DSU technical support document 18: Methods for population-adjusted indirect comparisons in submissions to NICE," *NICE Decision Support Unit*, 2016.

[38] P. Rosenbaum and D. Rubin, "The central role of the propensity score in observational studies for causal effects," *Biometrika*, vol. 70, no. 1, pp. 41–55, 1983.

[39] J. Robins, "A new approach to causal inference in mortality studies with a sustained exposure period—application to control of the healthy worker survivor effect," *Mathematical Modelling*, vol. 7, no. 9–12, pp. 1393–1512, 1986.

[40] J. M. Robins, "Correcting for non-compliance in randomized trials using structural nested mean models," *Communications in Statistics—Theory and Methods*, vol. 23, no. 8, pp. 2379–2412, 1994.

[41] P. Bickel, C. Klaassen, Y. Ritov, and J. Wellner, *Efficient and Adaptive Estimation for Semiparametric Models.* Springer, 1993.

[42] A. Tsiatis, *Semiparametric Theory and Missing Data.* Springer, 2006.

[43] J. M. Robins, A. Rotnitzky, and L. P. Zhao, "Estimation of regression coefficients when some regressors are not always observed," *Journal of the American statistical Association*, vol. 89, no. 427, pp. 846–866, 1994.

[44] SAS Institute Inc., "SAS/STAT 14.2 user's guide," SAS Institute Inc, 2016.

[45] P. Diaconis, S. Holmes, and R. Montgomery, "Dynamical bias in the coin toss," *SIAM Review*, vol. 49, no. 2, pp. 211–235, 2007.

[46] A. Brookhart, S. Schneeweiss, K. Rothman, R. Glynn, J. Avorn, and T. Stürmer, "Variable selection for propensity score models," *American Journal of Epidemiology*, vol. 163, no. 12, pp. 1149–1156, 2006.

[47] S. Schneeweiss, "Sensitivity analysis and external adjustment for unmeasured confounders in epidemiologic database studies of therapeutics," *Pharmacoepidemiology and Drug Safety*, vol. 15, no. 5, pp. 291–303, 2006.

[48] X. Zhang, D. Faries, H. Li, J. Stamey, and G. Imbens, "Addressing unmeasured confounding in comparative observational research," *Pharmacoepidemiology and Drug Safety*, vol. 27, no. 4, pp. 373–382, 2018.

[49] Y. Fang and W. He, "Sensitivity analysis in the analysis of real-world data," in *Real-World Evidence in Medical Product Development*, pp. 271–287, Springer, 2023.

[50] Y. Fang, P. Mishra-Kalyani, X. Zhang, S. Gruber, S. Yang, P. Ding, M. Shang, J.-Y. Lee, M. van der Laan, D. Faries, and H. Lee, "Sensitivity analysis for unmeasured confounding in medical product development and evaluation using real-world evidence," *Statistics in Biopharmaceutical Research*, 2025.

[51] T. VanderWeele and P. Ding, "Sensitivity analysis in observational research: Introducing the E-value," *Annals of Internal Medicine*, vol. 167, no. 4, pp. 268–274, 2017.

[52] P. Ding and T. VanderWeele, "Sensitivity analysis without assumptions," *Epidemiology*, vol. 27, no. 3, p. 368, 2016.

[53] A. Linden, M. Mathur, and T. VanderWeele, "Conducting sensitivity analysis for unmeasured confounding in observational studies using E-values: The evalue package," *The Stata Journal*, vol. 20, no. 1, pp. 162–175, 2020.

[54] M. van der Laan and S. Rose, *Targeted Learning: Causal Inference for Observational and Experimental Data.* Springer Science & Business Media, 2011.

[55] C. Stein, "Efficient nonparametric testing and estimation," in *Proceedings of the Third Berkeley Symposium on Mathematical Statistics and Probability, Volume 1: Contributions to the Theory of Statistics*, vol. 3, pp. 187–196, University of California Press, 1956.

[56] T. Hastie, R. Tibshirani, and J. Friedman, *The Elements of Statistical Learning: Data Mining, Inference, and Prediction.* Springer, 2009.

[57] W. Newey, "Semiparametric efficiency bounds," *Journal of Applied Econometrics*, vol. 5, no. 2, pp. 99–135, 1990.

[58] M. van der Laan and D. Rubin, "Targeted maximum likelihood learning," *The International Journal of Biostatistics*, vol. 2, no. 1, 2006.

[59] D. Rubin, "Inference and missing data," *Biometrika*, vol. 63, no. 3, pp. 581–592, 1976.

[60] Y. Fang and M. Jin, "Sequential modeling for a class of reference-based imputation methods in clinical trials with quantitative or binary outcomes," *Statistics in Medicine*, vol. 41, no. 8, pp. 1525–1540, 2022.

[61] A. Tsiatis, M. Davidian, S. Holloway, and E. Laber, *Dynamic Treatment Regimes: Statistical Methods for Precision Medicine.* Chapman & Hall/CRC Press, 2020.

[62] L. Breiman, J. Friedman, R. A. Olshen, and C. J. Stone, *Classification and Regression Trees.* Chapman and Hall/CRC Press, 1984.

[63] L. Breiman, "Random forests," *Machine Learning*, vol. 45, no. 1, pp. 5–32, 2001.

[64] M. Stone, "Cross-validatory choice and assessment of statistical predictions," *Journal of the Royal Statistical Society: Series B (Statistical Methodology)*, vol. 36, no. 2, pp. 111–133, 1974.

[65] H. Akaike, "A new look at the statistical model identification," *IEEE Transactions on Automatic Control*, vol. 19, no. 6, pp. 716–723, 1974.

[66] G. Wahba, *Spline Models for Observational Data.* SIAM, 1990.

[67] Y. Fang, "Asymptotic equivalence between cross-validations and akaike information criteria in mixed-effects models," *Journal of Data Science*, vol. 9, no. 1, pp. 15–21, 2011.

[68] G. H. Golub, M. Heath, and G. Wahba, "Generalized cross-validation as a method for choosing a good ridge parameter," *Technometrics*, vol. 21, no. 2, pp. 215–223, 1979.

[69] R. Tibshirani, "Regression shrinkage and selection via the lasso," *Journal of the Royal Statistical Society Series B (Statistical Methodology)*, vol. 58, no. 1, pp. 267–288, 1996.

[70] A. E. Hoerl and R. W. Kennard, "Ridge regression: Biased estimation for nonorthogonal problems," *Technometrics*, vol. 12, no. 1, pp. 55–67, 1970.

[71] H. Zou and T. Hastie, "Regularization and variable selection via the elastic net," *Journal of the Royal Statistical Society Series B (Statistical Methodology)*, vol. 67, no. 2, pp. 301–320, 2005.

[72] M. van der Laan, E. Polley, and A. Hubbard, "Super learner," *Statistical Applications in Genetics and Molecular Biology*, vol. 6, no. 1, 2007.

[73] Y. Fang and S. Zhong, "The targeted virtual control approach for single-arm clinical trials with external controls," *Statistics in Biopharmaceutical Research*, vol. 15, no. 4, pp. 802–811, 2023.

[74] SAS Institute Inc., "SAS Viya workbench: Statistical procedures," SAS Institute Inc, 2024.

[75] R. V. Phillips, M. J. van der Laan, H. Lee, and S. Gruber, "Practical considerations for specifying a super learner," *International Journal of Epidemiology*, vol. 52, no. 4, pp. 1276–1285, 2023.

[76] E. Polley, E. LeDell, C. Kennedy, S. Lendle, and M. van der Laan, "Package 'SuperLearner'," *CRAN*, 2019.

[77] S. Gruber and M. van der Laan, "TMLE: An R package for targeted maximum likelihood estimation," *Journal of Statistical Software*, vol. 51, pp. 1–35, 2012.

[78] S. Lendle, J. Schwab, M. Petersen, and M. van der Laan, "ltmle: An R package implementing targeted minimum loss-based estimation for longitudinal data," *Journal of Statistical Software*, vol. 81, pp. 1–21, 2017.

[79] C. Mallinckrodt, G. Molenberghs, I. Lipkovich, and B. Ratitch, *Estimands, Estimators and Sensitivity Analysis in Clinical Trials.* CRC Press, 2020.

[80] C. Frangakis and D. Rubin, "Principal stratification in causal inference," *Biometrics*, vol. 58, no. 1, pp. 21–29, 2002.

[81] J. Luo, S. J. Ruberg, and Y. Qu, "Estimating the treatment effect for adherers using multiple imputation," *Pharmaceutical Statistics*, vol. 21, no. 3, pp. 525–534, 2022.

[82] D. Cox, "Partial likelihood," *Biometrika*, vol. 62, no. 2, pp. 269–276, 1975.

[83] M. Hernán, "The hazards of hazard ratios," *Epidemiology*, vol. 21, no. 1, p. 13, 2010.

[84] J. Irwin, "The standard error of an estimate of expectation of life, with special reference to expectation of tumourless life in experiments with mice," *Epidemiology & Infection*, vol. 47, no. 2, pp. 188–189, 1949.

[85] T. Karrison, "Restricted mean life with adjustment for covariates," *Journal of the American Statistical Association*, vol. 82, no. 400, pp. 1169–1176, 1987.

[86] P. K. Andersen, M. G. Hansen, and J. P. Klein, "Regression analysis of restricted mean survival time based on pseudo-observations," *Lifetime Data Analysis*, vol. 10, pp. 335–350, 2004.

[87] M. Jin and Y. Fang, "A targeted learning framework for estimating restricted mean survival time difference using pseudo-observations," *arXiv preprint arXiv:2601.06296*, 2026.

[88] O. Stitelman, V. De Gruttola, and M. van der Laan, "A general implementation of TMLE for longitudinal data applied to causal inference in survival analysis," *The International Journal of Biostatistics*, vol. 8, no. 1, 2012.

[89] D. Benkeser, M. Carone, and P. Gilbert, "Improved estimation of the cumulative incidence of rare outcomes," *Statistics in Medicine*, vol. 37, no. 2, pp. 280–293, 2018.

[90] Y. Fang and M. Jin, "Estimand framework and intercurrent events handling for clinical trials with time-to-event outcomes," *arXiv preprint arXiv:2510.15000*, 2025.

[91] P. Austin, D. Lee, and J. Fine, "Introduction to the analysis of survival data in the presence of competing risks," *Circulation*, vol. 133, no. 6, pp. 601–609, 2016.

[92] J. Zhang and D. Rubin, "Estimation of causal effects via principal stratification when some outcomes are truncated by 'death'," *Journal of Educational and Behavioral Statistics*, vol. 28, no. 4, pp. 353–368, 2003.

[93] L. Breiman, "Statistical modeling: The two cultures (with comments and a rejoinder by the author)," *Statistical Science*, vol. 16, no. 3, pp. 199–231, 2001.

[94] W. He, Y. Fang, and H. Wang, *Real-World Evidence in Medical Product Development.* Springer, 2023.

[95] F. Li, K. L. Morgan, and A. M. Zaslavsky, "Balancing covariates via propensity score weighting," *Journal of the American Statistical Association*, vol. 113, no. 521, pp. 390–400, 2018.

[96] X. Li, W. Miao, F. Lu, and X.-H. Zhou, "Improving efficiency of inference in clinical trials with external control data," *Biometrics*, vol. 79, no. 1, pp. 394–403, 2023.

[97] M. van der Laan, S. Qiu, J. M. Tarp, and L. van der Laan, "Adaptive-TMLE for the average treatment effect based on randomized controlled trial augmented with real-world data," *arXiv preprint arXiv:2405.07186*, 2025.

[98] Y. Fang and W. He, "Targeted learning for externally controlled trials and beyond," *Submitted*, 2026.

Index

For Product Safety Concerns and Information please contact our EU representative GPSR@taylorandfrancis.com Taylor & Francis Verlag GmbH, Kaufingerstraße 24, 80331 München, Germany